Tutorials, Schools, and Workshops in the Mathematical Sciences

 Birkhäuser

This series will serve as a resource for the publication of results and developments presented at summer or winter schools, workshops, tutorials, and seminars. Written in an informal and accessible style, they present important and emerging topics in scientific research for PhD students and researchers. Filling a gap between traditional lecture notes, proceedings, and standard textbooks, the titles included in TSWMS present material from the forefront of research.

More information about this series at http://www.springer.com/series/15641

Dorothea Bahns • Anke Pohl • Ingo Witt
Editors

Open Quantum Systems

A Mathematical Perspective

 Birkhäuser

Editors

Dorothea Bahns
Mathematisches Institut
Universität Göttingen
Göttingen, Germany

Anke Pohl
Department 3 – Mathematik
Universität Bremen
Bremen, Germany

Ingo Witt
Mathematisches Institut
Universität Göttingen
Göttingen, Germany

ISSN 2522-0969 ISSN 2522-0977 (electronic)
Tutorials, Schools, and Workshops in the Mathematical Sciences
ISBN 978-3-030-13048-0 ISBN 978-3-030-13046-6 (eBook)
https://doi.org/10.1007/978-3-030-13046-6

Mathematics Subject Classification (2010): 81S22, 82C10, 81P16, 81P45, 81P40, 60J35, 46L55, 81Q93, 47D03, 94A17

This book is published under the imprint Birkhäuser, www.birkhauser-science.com by the registered company Springer Nature Switzerland AG.
The registered company address is: Gewerbestrasse 11, 6330 Cham, Switzerland

Preface

The theory of open quantum systems, that is, quantum systems interacting with an environment, is a fascinating and multifaceted field. It is guided by the demands of models for the actual quantum physical experiments and their interpretational questions, and it draws heavily upon many different mathematical areas in order to merge and combine tools and methods to form a new theory. Of particular importance are functional analysis, evolution equations, semigroups, C^*-algebras, and probability theory.

The mathematical theory of open quantum systems is rather young and still in a formative stage, with a few first streamlines emerging. An intensive and rich development in future is to be expected. This book is a compilation of four articles by internationally leading experts which provide an introduction to some fundamental mathematical aspects relevant to understanding open quantum systems.

The article by Alexander Belton provides a self-contained introduction to Markovian semigroups on both a classical and quantum-mechanical level, eventually characterizing the generators of certain quantum Feller semigroups.

Dariusz Chruściński discusses non-Markovian quantum dynamics, with both time-local and memory kernel master equations. It is mostly based on the example of n-level quantum systems. A recurrent theme in both the articles of Alexander Belton and Dariusz Chruściński is the concept of completely positive maps, which is central to the theory of open quantum systems.

Niels Jacob and Elian Rhind study the generators of Feller semigroups by taking advantage of techniques from microlocal analysis. Their article constitutes a first attempt to include geometric aspects into the framework. It focuses on the classical theory, and it is an interesting open problem to generalize this formalism to the quantum case.

Also Vojkan Jaksic's article focuses on the classical theory. It is the first part of a series of articles devoted to the notion of entropy. He introduces and carefully analyzes notions of classical entropy that allow for counterparts in quantum information theory or quantum statistical mechanics. The forthcoming parts in this series will be published elsewhere.

The original stimulus for this volume is a winter school on dynamical methods in open quantum systems which was held at the Mathematical Institute of the University of Göttingen in November 2016 and where Alexander Belton, Dariusz Chruściński, Niels Jacob, and Vojkan Jaksic were lecturers.

Göttingen, Germany   Dorothea Bahns
Bremen, Germany   Anke Pohl
Göttingen, Germany   Ingo Witt
December 2018

Contents

Contributors

Alexander C. R. Belton Department of Mathematics and Statistics, Lancaster University, Lancaster, UK

Dariusz Chruściński Institute of Physics, Faculty of Physics, Astronomy and Informatics Nicolaus Copernicus University, Torun, Poland

Niels Jacob Swansea University, Swansea, Wales, UK

Vojkan Jakšić Department of Mathematics and Statistics, McGill University, Montreal, QC, Canada

Elian O. T. Rhind Swansea University, Swansea, Wales, UK

Introduction to Classical and Quantum Markov Semigroups

Alexander C. R. Belton

Abstract We provide a self-contained and fast-paced introduction to the theories of operator semigroups, Markov semigroups and quantum dynamical semigroups. The level is appropriate for well-motivated graduate students who have a background in analysis or probability theory, with the focus on the characterisation of infinitesimal generators for various classes of semigroups. The theorems of Hille–Yosida, Hille–Yosida–Ray, Lumer–Phillips and Gorini–Kossakowski–Sudarshan–Lindblad are all proved, with the necessary technical prerequisites explained in full. Exercises are provided throughout.

1 Introduction

These notes are an extension of a series of lectures given at the Winter School on Dynamical Methods in Open Quantum Systems held at Georg-August-Universität Göttingen during November 2016. These lectures were aimed at graduate students with a background in analysis or probability theory. The aim has been to make the notes self-contained but brief, so that they are widely accessible. Exercises are provided throughout.

We begin with the basics of the theory of operator semigroups on Banach spaces, and develop this up to the Hille–Yosida and Lumer–Phillips theorems; these provide characterisations for the generators of strongly continuous semigroups and strongly continuous contraction semigroups, respectively. As those with a background in probability theory may not be comfortable with all of the necessary material from functional analysis, this is covered rapidly at the start. The reader can find much more on these topics in Davies's book [9].

After these fundamentals, we recall some key ideas from probability theory. The correspondence between time-homogeneous Markov processes and Markov

A. C. R. Belton (✉)
Department of Mathematics and Statistics, Lancaster University, Lancaster, UK
e-mail: a.belton@lancaster.ac.uk

© Springer Nature Switzerland AG 2019
D. Bahns et al. (eds.), *Open Quantum Systems*, Tutorials, Schools, and Workshops
in the Mathematical Sciences, https://doi.org/10.1007/978-3-030-13046-6_1

1

semigroups is explained, and we explore the concepts of Feller semigroups and Lévy processes. We conclude with the Hille–Yosida–Ray theorem, which characterises generators of Feller semigroups via the positive maximum principle. Applebaum [3, Chapter 3] provides another view of much of this material, as do Liggett [20, Chapter 3] and Rogers and Williams [26, Chapter III].

The final part of these notes addresses the theory of quantum Markov semigroups, and builds to the characterisation of the generators of uniformly continuous conservative semigroups, and the Gorini–Kossakowski–Sudarshan–Lindblad form. En route, we establish Stinespring dilation and Kraus decomposition for linear maps defined on unital C^* algebras and von Neumann algebras, respectively, which are important results in the theories of open quantum systems and quantum information. The lecture notes of Alicki and Lendi [2] provide a useful complement, and those of Fagnola [14] study quantum Markov semigroups from the fruitful perspective of quantum probability. There is much scope, and demand, for further developments in this subject.

1.1 Acknowledgements

The author is grateful to the organisers of the winter school, Prof. Dr. Dorothea Bahns (Göttingen), Prof. Dr. Anke Pohl (Jena) and Prof. Dr. Ingo Witt (Göttingen), for the opportunity to give these lectures, and for their hospitality during his time in Göttingen. He is also grateful to Mr. Jason Hancox, for his comments on a previous version of these notes.

1.2 Conventions

The notation "$P := Q$" means that the quantity P is defined to equal Q.

The sets of natural numbers, non-negative integers, non-negative real numbers, real numbers and complex numbers are denoted $\mathbb{N} := \{1, 2, 3, \ldots\}$, $\mathbb{Z}_+ := \{0\} \cup \mathbb{N}$, $\mathbb{R}_+ := [0, \infty)$, $\mathbb{R}$ and $\mathbb{C}$, respectively; the square root of -1 is denoted i. Note that we follow the Anglophone rather than Francophone convention, in that 0 is both non-negative and non-positive but is neither positive nor negative.

The indicator function of the set A is denoted 1_A, with the domain determined by context. If $f : A \to B$ and $C \subseteq A$, then $f|_C : C \to B$, the restriction of f to C, takes the same value at any point in C as f does.

Inner products on complex vector spaces are taken to be linear on the right and conjugate linear on the left. Given our final destination, we work with complex vector spaces and complex-valued functions by default.

2 Operator Semigroups

2.1 Functional-Analytic Preliminaries

Throughout their development, there has been a fruitful interplay between abstract functional analysis and the theory of operator semigroups. Here we give a rapid introduction to some of the basic ideas of the former. We cover a little more material that will be used in the sequel, but the reader will find it useful for their further studies in semigroup theory.

Definition 2.1 In these notes, a *normed vector space* V is a vector space with complex scalar field, equipped with a *norm* $\| \cdot \| : V \to \mathbb{R}_+$ which is

(i) subadditive: $\|u + v\| \leqslant \|u\| + \|v\|$ for all $u, v \in V$;

(ii) homogeneous: $\|\lambda v\| = |\lambda|\, \|v\|$ for all $v \in V$ and $\lambda \in \mathbb{C}$; and

(iii) faithful: $\|v\| = 0$ if and only if $v = 0$, for all $v \in V$.

The normed vector space V is *complete* if, whenever $(v_n)_{n \in \mathbb{N}} \subseteq V$ is a Cauchy sequence, there exists $v_\infty \in V$ such that $v_n \to v_\infty$ as $n \to \infty$. A complete normed vector space is called a *Banach space*. Thus Banach spaces are those normed vector spaces in which every Cauchy sequence is convergent.

[Recall that a sequence $(v_n)_{n \in \mathbb{N}} \subseteq V$ is *Cauchy* if, for all $\varepsilon > 0$, there exists $N \in \mathbb{N}$ such that $\|v_m - v_n\| < \varepsilon$ for all $m, n \geqslant N$.]

Exercise 2.2 (Banach's Criterion) Let $\| \cdot \|$ be a norm on the complex vector space V. Prove that V is complete for this norm if and only if every absolutely convergent series in V is convergent.

[Given $(v_n)_{n \in \mathbb{N}} \subseteq V$, the series $\sum_{n=1}^{\infty} v_n$ is said to be *convergent* precisely when the sequence of partial sums $(\sum_{j=1}^{n} v_j)_{n \in \mathbb{N}}$ is convergent, and *absolutely convergent* when $(\sum_{j=1}^{n} \|v_j\|)_{n \in \mathbb{N}}$ is convergent.]

Example 2.3 If $n \in \mathbb{N}$, then the finite-dimensional vector space $\mathbb{C}^n$ is a Banach space for any of the ℓ^p norms, where $p \in [1, \infty]$ and

$$\|(v_1, \ldots, v_n)\|_p := \begin{cases} \left(\sum_{j=1}^{n} |v_j|^p\right)^{1/p} & \text{if } p < \infty, \\ \max\{|v_j| : j = 1, \ldots, n\} & \text{if } p = \infty. \end{cases}$$

These norms are all *equivalent*: for all $p, q \in [1, \infty]$ there exists $C_{p,q} > 1$ such that

$$C_{p,q}^{-1}\|v\|_q \leqslant \|v\|_p \leqslant C_{p,q}\|v\|_q \qquad \text{for all } v \in \mathbb{C}^n.$$

Example 2.4 For all $p \in [1, \infty]$, let the *sequence space*

$$\ell^p := \{v = (v_n)_{n \in \mathbb{Z}_+} \subseteq \mathbb{C} : \|v\|_p < \infty\},$$

where

$$\|v\|_p := \begin{cases} \left(\sum_{n=0}^{\infty} |v_n|^p\right)^{1/p} & \text{if } p \in [1, \infty), \\ \sup\{|v_n| : n \in \mathbb{Z}_+\} & \text{if } p = \infty, \end{cases}$$

and the vector-space operations are defined coordinate-wise: if $u, v \in \ell^p$ and $\lambda \in \mathbb{C}$, then

$$(u + v)_n := u_n + v_n \quad \text{and} \quad (\lambda v)_n := \lambda v_n \qquad \text{for all } n \in \mathbb{Z}_+.$$

These are Banach spaces, with $\ell^p \subseteq \ell^q$ if $p, q \in [1, \infty]$ are such that $p \leqslant q$. If $p \in [1, \infty)$, then $\ell^p \subseteq c_0 \subseteq \ell^\infty$, where

$$c_0 := \{v = (v_n)_{n \in \mathbb{Z}_+} \subseteq \mathbb{C} : \lim_{n \to \infty} v_n = 0\}$$

is itself a Banach space for the norm $\|\cdot\|_\infty$.

Example 2.5 An *inner product* on the complex vector space V is a form

$$\langle \cdot, \cdot \rangle : V \times V \to \mathbb{C}; \ (u, v) \mapsto \langle u, v \rangle$$

which is

    (i) linear in the second argument: the map $V \to \mathbb{C}; v \mapsto \langle u, v \rangle$ is linear for all $u \in V$;

    (ii) Hermitian: $\overline{\langle u, v \rangle} = \langle v, u \rangle$ for all $u, v \in V$; and

    (iii) positive definite: $\langle v, v \rangle \geqslant 0$ for all $v \in V$, with equality if and only if $v = 0$.

Any inner product determines a norm on V, by setting $\|v\| := \langle v, v \rangle^{1/2}$ for all $v \in V$. Furthermore, the inner product can be recovered from the norm by *polarisation*: if $q : V \times V \to \mathbb{C}$ is a sesquilinear form on V, so is conjugate linear in the first argument and linear in the second, then

$$q(u, v) = \sum_{j=0}^{3} i^{-j} q(u + iv, u + iv) \qquad \text{for all } u, v \in V.$$

A Banach space with norm which comes from an inner product is a *Hilbert space*. For example, the sequence space ℓ^2 is a sequence space, since setting

$$\langle u, v \rangle := \sum_{n=0}^{\infty} \overline{u_n} v_n \qquad \text{for all } u, v \in \ell^2$$

defines an inner product on ℓ^2 such that $\langle v, v \rangle = \|v\|^2$ for all $v \in \ell^2$. In any Hilbert space H, the *Cauchy–Schwarz inequality* holds:

$$|\langle u, v \rangle| \leqslant \|u\| \, \|v\| \qquad \text{for all } u, v \in \mathsf{H}.$$

It may be shown that a Banach space V is a Hilbert space if and only if the norm satisfies the *parallelogram law*:

$$\|u + v\|^2 + \|u - v\|^2 = 2\|u\|^2 + 2\|v\|^2 \qquad \text{for all } u, v \in V.$$

Exercise 2.6 Let H be a Hilbert space. Given any set $S \subseteq H$, prove that its *orthogonal complement*

$$S^\perp := \{x \in H : \langle x, y \rangle = 0 \text{ for all } y \in S\}$$

is a closed linear subspace of H. Prove further that $L \subseteq H$ is a closed linear subspace of H if and only if $L = (L^\perp)^\perp$.

Example 2.7 Let $C(K)$ denote the complex vector space of complex-valued functions on the compact Hausdorff space K, with vector-space operations defined pointwise: if $x \in K$ then

$$(f + g)(x) := f(x) + g(x) \quad \text{and} \quad (\lambda f)(x) := \lambda f(x)$$

for all $f, g \in C(K)$ and $\lambda \in \mathbb{C}$. The *supremum norm*

$$\| \cdot \| : f \mapsto \|f\|_\infty := \sup\{f(x) : |x| \in K\}$$

makes $C(K)$ a Banach space. [Completeness is the undergraduate-level fact that uniform convergence preserves continuity.]

Example 2.8 Let $(\Omega, \mathcal{F}, \mu)$ be a σ-finite measure space, so that $\mu : \mathcal{F} \to [0, \infty]$ is a measure and there exists a countable cover of Ω with elements in $\mathcal{F}$ of finite measure.

For all $p \in [1, \infty]$, the L^p *space*

$$L^p(\Omega, \mathcal{F}, \mu) := \{f : \Omega \to \mathbb{C} \mid \|f\|_p < \infty\}$$

is a Banach space when equipped with the L^p *norm*

$$\|f\|_p := \begin{cases} \left(\int_\Omega |f(x)|^p \, \mu(\mathrm{d}x) \right)^{1/p} & \text{if } p \in [1, \infty), \\ \inf\{\sup\{\, |f(x)| : x \in \Omega \setminus V\} : V \subseteq \Omega \text{ is a null set}\} & \text{if } p = \infty, \end{cases}$$

and where functions are identified if they differ on a null set. [Note that if $f \in L^p(\Omega, \mathcal{F}, \mu)$ then $\|f\|_p = 0$ if and only if $f = 0$ on a null set.]

The space $L^2(\Omega, \mathcal{F}, \mu)$ is a Hilbert space, with inner product such that

$$\langle f, g \rangle := \int_\Omega \overline{f(x)} g(x) \, \mu(dx) \qquad \text{for all } f, g \in L^2(\Omega, \mathcal{F}, \mu).$$

If $p, q, r \in [1, \infty]$ are such that $p^{-1} + q^{-1} = r^{-1}$, where $\infty^{-1} := 0$, then

$$\|fg\|_r \leqslant \|f\|_p \|g\|_q \qquad \text{for all } f \in L^p(\Omega, \mathcal{F}, \mu) \text{ and } g \in L^q(\Omega, \mathcal{F}, \mu); \tag{2.1}$$

this is *Hölder's inequality*. The subadditivity of the L^p norm, known as *Minkowski's inequality*, may be deduced from Hölder's inequality. When $r = 1$ and $p = q = 2$, Hölder's inequality is known as the *Cauchy–Bunyakovsky–Schwarz inequality*.

Example 2.9 Let $d \geqslant 1$. The space $C_c^\infty(\mathbb{R}^d)$ of continuous functions on $\mathbb{R}^d$ with compact support is a subspace of $L^p(\mathbb{R}^d)$ for all $p \in [1, \infty]$, and is dense for $p \in [1, \infty)$, when $\mathbb{R}^d$ is equipped with Lebesgue measure.

Given a *multi-index* $\alpha = (\alpha_1, \ldots, \alpha_d) \in \mathbb{Z}_+^d$, let $|\alpha| := \alpha_1 + \cdots + \alpha_d$ and

$$D^\alpha f := \frac{\partial^{\alpha_1}}{\partial x_1} \cdots \frac{\partial^{\alpha_d}}{\partial x_d} f \qquad \text{for all } f \in C_c^\infty(\mathbb{R}^d).$$

Note that $D^\alpha f \in C_c^\infty(\mathbb{R}^d)$ for all $f \in C_c^\infty(\mathbb{R}^d)$ and $\alpha \in \mathbb{Z}_+^d$.

Let $f \in L^p(\mathbb{R}^d)$, where $p \in [1, \infty]$, and note that $fg \in L^1(\mathbb{R}^d)$ for all $g \in C_c^\infty(\mathbb{R}^d)$, by Hölder's inequality. If there exists $F \in L^p(\mathbb{R}^d)$ such that

$$\int_{\mathbb{R}^d} f(x) D^\alpha g(x) \, dx = (-1)^{|\alpha|} \int_{\mathbb{R}^d} F(x) g(x) \, dx \qquad \text{for all } g \in C_c^\infty(\mathbb{R}^d)$$

then F is the *weak derivative* of f, and we write $F = D^\alpha f$. [It is a straightforward exercise to verify that the weak derivative is unique, and that this agrees with the previous definition if $f \in C_c^\infty(\mathbb{R}^d)$.]

Given $p \in [1, \infty)$ and $k \in \mathbb{Z}_+$, the *Sobolev space*

$$W^{k,p}(\mathbb{R}^d) := \{ f \in L^p(\mathbb{R}^d) : D^\alpha f \in L^p(\mathbb{R}^d) \text{ whenever } |\alpha| \leqslant k \}$$

is a Banach space when equipped with the norm

$$f \mapsto \|f\| := \left(\sum_{|\alpha| \leqslant k} \|D^\alpha f\|_p^p \right)^{1/p}$$

and contains $C_c^\infty(\mathbb{R}^d)$ as a dense subspace.

The Sobolev space $W^{k,2}(\mathbb{R}^d)$ is usually abbreviated to $H^k(\mathbb{R}^d)$ and is a Hilbert space, with inner product such that

$$\langle f, g \rangle := \sum_{|\alpha| \leqslant k} \langle D^\alpha f, D^\alpha g \rangle \qquad \text{for all } f, g \in H^k(\mathbb{R}^d).$$

Exercise 2.10 Prove that the normed vector space $W^{k,p}(\mathbb{R}^d)$, as defined in Example 2.9, is complete.

Example 2.11 Let U and V be normed vector spaces. A linear operator $T : U \to V$ is *bounded* if

$$\|T\| := \{\|Tu\| : u \in U\} < \infty.$$

If T is bounded, then $\|Tu\| \leqslant \|T\| \|u\|$ for all $u \in U$, and $\|T\|$ is the smallest constant with this property.

The set of all such linear operators is denoted by $B(U; V)$, or $B(U)$ if U and V are equal.

This set is a normed vector space, with *operator norm* $T \mapsto \|T\|$ and algebraic operations defined pointwise, so that

$$(S + T)u = Su + Tu \quad \text{and} \quad (\lambda T)u := \lambda Tu$$

for all $S, T \in B(U; V)$, $\lambda \in \mathbb{C}$ and $U \in U$. Furthermore, the space $B(U; V)$ is a Banach space whenever V is.

Exercise 2.12 Prove the claims in Example 2.11.

Exercise 2.13 Let V be a normed vector space. Prove that the norm on $B(V)$ is *submultiplicative*: if $S, T \in B(V)$, then $ST : v \mapsto S(Tv) \in B(V)$, with $\|ST\| \leqslant \|S\| \|T\|$.

Exercise 2.14 Let U and V be normed vector spaces and let $T : U \to V$ be a linear operator. Prove that T is bounded if and only if it is continuous when U and V are equipped with their norm topologies.

Example 2.15 Given any normed space V, its *topological dual* or *dual space* is the Banach space $V^* := B(V; \mathbb{C})$. An element of V^* is called a *linear functional* or simply a *functional*.

If $p, q \in (1, \infty)$ are *conjugate indices*, so that such that $p^{-1} + q^{-1} = 1$, then $(\ell^p)^*$ is naturally isomorphic to ℓ^q via the *dual pairing*

$$[u, v] := \sum_{n=0}^{\infty} u_n v_n \qquad \text{for all } u \in \ell^p \text{ and } v \in \ell^q.$$

Hölder's inequality shows that $u \mapsto [u, v]$ is an element of $(\ell^p)^*$ for any $v \in \ell^q$; proving that every functional arises this way is an exercise. Furthermore, the same pairing gives an isomorphism between $(\ell^1)^*$ and ℓ^∞. [The dual of ℓ^∞ is much larger than ℓ^1; it is isomorphic to the space $M(\beta\mathbb{N})$ of regular complex Borel measures on the Stone–Čech compactification of the natural numbers.]

Similarly, for conjugate indices $p, q \in (1, \infty)$, the dual of $L^p(\Omega, \mathcal{F}, \mu)$ is identified with $L^q(\Omega, \mathcal{F}, \mu)$, and the dual of $L^1(\Omega, \mathcal{F}, \mu)$ with $L^\infty(\Omega, \mathcal{F}, \mu)$, via the pairing

$$[f, g] := \int_\Omega f(x)g(x)\,\mu(\mathrm{d}x).$$

In particular, ℓ^2 and $L^2(\Omega, \mathcal{F}, \mu)$ are conjugate-linearly isomorphic to their dual spaces. This is a general fact about Hilbert spaces, known as the *Riesz–Fréchet theorem*: if H is a Hilbert space, then

$$H^* = \{\langle u| : u \in H\}, \qquad \text{where } \langle u|v := \langle u, v \rangle \quad \text{for all } v \in H.$$

If K is a compact Hausdorff space, then the dual of $C(K)$ is naturally isomorphic to the space $M(K)$ of regular complex Borel measures on K, with dual pairing

$$[f, \mu] := \int_K f(x)\,\mu(\mathrm{d}x) \qquad \text{for all } f \in C(K) \text{ and } \mu \in M(K).$$

The Hahn–Banach theorem [25, Corollary 2 to Theorem III.6] implies that the dual space separates points: if $v \in V$, then there exists $\phi \in V^*$ such that $\|\phi\| = 1$ and $\phi(v) = \|v\|$.

Example 2.16 Duality makes an appearance at the level of operators. If U and V are normed spaces and $T \in B(U; V)$, then there exists a unique *dual operator* $T' \in B(V^*; U^*)$ such that

$$(T'\psi)(v) = \psi(Tu) \qquad \text{for all } u \in U \text{ and } \psi \in V^*.$$

The map $T \mapsto T'$ from $B(U; V)$ to $B(V^*; U^*)$ is linear and reverses the order of products: if $S \in B(U; V)$ and $T \in B(V; W)$, then $(TS)' = S'T'$.

If H and K are Hilbert spaces, and we identify each of these with its dual via the Riesz–Fréchet theorem, then the operator dual to $T \in B(\mathsf{H}; \mathsf{K})$ is identified with the *adjoint operator* $T^* \in B(\mathsf{K}; \mathsf{H})$, since

$$\bigl(T'\langle v|\bigr)u = \langle v, Tu \rangle_\mathsf{K} = \langle T^*v, u \rangle_\mathsf{H} = \langle T^*v|u \qquad \text{for all } u \in \mathsf{H} \text{ and } v \in \mathsf{K}.$$

2.2 Semigroups on Banach Spaces

Definition 2.17 A family of operators $T = (T_t)_{t \in \mathbb{R}_+} \subseteq B(V)$ is a *one-parameter semigroup* on V, or a *semigroup* for short, if

(i) $T_0 = I$ the identity operator   and   (ii) $T_s T_t = T_{s+t}$ for all $s, t \in \mathbb{R}_+$.

The semigroup T is *strongly continuous* if

$$\lim_{t \to 0+} \|T_t v - v\| = 0 \qquad \text{for all } v \in V,$$

and is *uniformly continuous* if

$$\lim_{t \to 0+} \|T_t - I\| = 0.$$

Exercise 2.18 Prove that a uniformly continuous semigroup is strongly continuous. [The converse is false: see Exercise 2.29.]

Theorem 2.19 *Let T be a strongly continuous semigroup on the Banach space V. There exist constants $M \geqslant 1$ and $a \in \mathbb{R}$ such that $\|T_t\| \leqslant Me^{at}$ for all $t \in \mathbb{R}_+$.*

Proof See [9, Theorem 6.2.1]. □

Remark 2.20 The semigroup T of Theorem 2.19 is said to be of *type (M, a)*. A semigroup of type $(1, 0)$ is also called a *contraction semigroup*.

By replacing T_t with $e^{-at} T_t$, one can often reduce to the case of semigroups with uniformly bounded norm. However, it is not always possible to go further and reduce to contraction semigroups; see [9, Example 6.2.3 and Theorem 6.3.8].

Exercise 2.21 Prove that a strongly continuous semigroup is strongly continuous at every point: if $t \geqslant 0$, then $\lim_{h \to 0} \|T_{t+h} x - T_t x\| = 0$. Prove further that the same is true if "strongly" is replaced by "uniformly".

Exercise 2.22 Given any $A \in B(V)$, let $\exp(A) := \sum_{n=0}^{\infty} \frac{1}{n!} A^n$.

(i) Prove that this series is convergent, so that $\exp(A) \in B(V)$. Prove further that $\| \exp(A) \| \leqslant \exp \|A\|$.

(ii) Prove that if $B \in B(V)$ commutes with A, so that that $AB = BA$, then $\exp(A)$ and $\exp(B)$ also commute, with $\exp(A) \exp(B) = \exp(A + B)$. [Hint: consider the derivatives of

$$t \mapsto \exp(tA) \exp(-tA) \quad \text{and} \quad t \mapsto \exp(tA) \exp(tB) \exp(-t(A+B)).]$$

(iii) Prove that setting $T_t := \exp(tA)$ for all $t \in \mathbb{R}_+$ produces a uniformly continuous one-parameter semigroup T.

The converse of Exercise 2.22(iii) is true, and we state it as a theorem.

Theorem 2.23 *If T is a uniformly continuous one-parameter semigroup, then there exists an operator $A \in B(V)$ such that $T_t = \exp(tA)$ for all $t \in \mathbb{R}_+$.*

Proof By continuity at the origin, there exists $t_0 > 0$ such that

$$\|T_s - I\| < 1/2 \qquad \text{for all } s \in [0, t_0].$$

Then

$$\left\| t_0^{-1} \int_0^{t_0} T_s \, ds - I \right\| = t_0^{-1} \left\| \int_0^{t_0} T_s - I \, ds \right\| \leqslant 1/2 < 1.$$

Thus $X := t_0^{-1} \int_0^{t_0} T_s \, ds \in B(V)$ is invertible, because the Neumann series

$$\sum_{n=0}^{\infty} (I - X)^n = I + (I - X) + (I - X)^2 + \dots$$

is absolutely convergent, so convergent, by Banach's criterion. Furthermore,

$$h^{-1}(T_h - I) \int_0^{t_0} T_s \, ds = h^{-1} \int_0^{t_0} T_{s+h} - T_s \, ds$$

$$= h^{-1} \int_h^{t_0+h} T_s \, ds - h^{-1} \int_0^{t_0} T_s \, ds$$

$$= h^{-1} \int_{t_0}^{t_0+h} T_s \, ds - h^{-1} \int_0^h T_s \, ds$$

$$\to T_{t_0} - I$$

as $h \to 0+$. Hence

$$A := \lim_{h \to 0+} h^{-1}(T_h - I) = (T_{t_0} - I)(t_0 X)^{-1}.$$

Moreover, for any $t \in [0, t_0]$,

$$T_{t_0} = I + A \int_0^t T_{t_1} \, dt_1 = I + A\left(tI + \int_0^t \int_0^{t_1} T_{t_2} \, dt_2 \, dt_1 \right)$$

$$= I + tA + \frac{t^2}{2} A^2 + \dots$$

$$+ A^n \int_0^t \cdots \int_0^{t_n} T_{t_{n+1}} \, dt_{n+1} \cdots dt_1$$

$$\to \sum_{n \geqslant 0} \frac{1}{n!} (tA)^n = \exp(tA)$$

as $n \to \infty$, since

$$\left\| A^n \int_0^t \cdots \int_0^{t_n} T_{t_{n+1}} \, dt_{n+1} \cdots dt_1 \right\| \leqslant \frac{3 t^{n+1} \|A\|^n}{2(n+1)!}.$$

This working shows that $T_t = \exp(tA)$ for any $t \in [0, t_0]$, so for all $t \in \mathbb{R}_+$, by the semigroup property: there exists $n \in \mathbb{Z}_+$ and $s \in [0, t_0)$ such that $t = n t_0 + s$, and

$$T_t = T_{t_0}^n T_s = \exp(n t_0 A + sA) = \exp(tA).$$

$\square$

Remark 2.24 The integrals in the previous proof are *Bochner integrals*; they are an extension of the Lebesgue integral to functions which take values in a Banach space. We will only be concerned with continuous functions, so do not need to concern ourselves with notions of measurability. All the standard theorems carry over from the Lebesgue to the Bochner setting, such as the inequality $\| \int f(t) \, dt \| \leqslant \int \|f(t)\| \, dt$, and if T is a bounded operator then $T \int f(t) \, dt = \int T f(t) \, dt$.

Definition 2.25 If T is a uniformly continuous semigroup, then the operator $A \in B(V)$ such that $T_t = \exp(tA)$ for all $t \in \mathbb{R}_+$ is the *generator* of the semigroup.

Exercise 2.26 Prove that the generator of a uniformly continuous one-parameter semigroup T is unique. [Hint: consider the limit of $t^{-1}(T_t - I)$ as $t \to 0+$.]

Example 2.27 Given $t \in \mathbb{R}_+$ and $f \in V := L^p(\mathbb{R}_+)$, where $p \in [1, \infty)$, let

$$(T_t f)(x) := f(x + t) \qquad \text{for all } x \in \mathbb{R}_+.$$

Then $T_t \in B(V)$, with $\|T_t\| = 1$, and $T = (T_t)_{t \in \mathbb{R}_+}$ is a one-parameter semigroup. If f is continuous and has compact support, then an application of the Dominated Convergence Theorem gives that $T_t f \to f$ as $t \to 0+$; since such functions are dense in V, it follows that T is strongly continuous.

Exercise 2.28 Prove the assertions in Example 2.27. Prove also that if $f \in V = L^p(\mathbb{R}_+)$ is absolutely continuous, with $f' \in V$ such that

$$f(x) = f(0) + \int_0^x f'(y) \, dy \qquad \text{for all } x \in \mathbb{R}_+,$$

then

$$\lim_{t \to 0+} t^{-1}(T_t f - f) = f',$$

where the limit exists in V. [Hint: show that

$$\|t^{-1}(T_t f - f) - f'\|_p^p = t^{-1} \int_0^t \|T_y f' - f'\|_p^p \, \mathrm{d}y$$

and then use the strong continuity of T at the origin.]

Exercise 2.29 Prove that the semigroup of Example 2.27 is not uniformly continuous. [Hint: let $f_n = \lambda_n 1_{[n^{-1}, 2n^{-1}]}$, where the positive constant λ_n is chosen to make f_n a unit vector in V, and consider $\|T_t f_n - f_n\|$ for $n > t^{-1}$.]

2.3   Beyond Uniform Continuity

As shown above, uniformly continuous one-parameter semigroups are in one-to-one correspondence with bounded linear operators. To move beyond this situation, we need to introduce linear operators which are only partially defined on the ambient Banach space V.

Definition 2.30 An *unbounded operator* in V is a linear transformation A defined on a linear subspace $V_0 \subseteq V$, its *domain*; we write $\operatorname{dom} A = V_0$.

An *extension* of A is an unbounded operator B in V such that $\operatorname{dom} A \subseteq \operatorname{dom} B$ and the restriction $B|_{\operatorname{dom} A} = A$. In this case, we write $A \subseteq B$.

An unbounded operator A in V is *densely defined* if $\operatorname{dom} A$ is dense in V for the norm topology.

Definition 2.31 Given operators A and B, let $A + B$ and AB be defined by setting

$$\operatorname{dom}(A + B) := \operatorname{dom} A \cap \operatorname{dom} B, \quad (A + B)v := Av + Bv$$

and

$$\operatorname{dom} AB := \{v \in \operatorname{dom} A : Av \in \operatorname{dom} B\}, \quad (AB)v := A(Bv).$$

Note that neither $A + B$ nor AB need be densely defined, even if both A and B are.

Definition 2.32 Let T be a strongly continuous one-parameter semigroup on V. Its *generator* A is an unbounded operator with domain

$$\operatorname{dom} A := \left\{ v \in V : \lim_{t \to 0+} t^{-1}(T_t v - v) \text{ exists in } V \right\}$$

and action

$$Av := \frac{\mathrm{d}}{\mathrm{d}t} T_t v \Big|_{t=0} := \lim_{t \to 0+} t^{-1}(T_t v - v) \qquad \text{for all } v \in \mathrm{dom}\, A.$$

It is readily verified that A is an unbounded operator.

Exercise 2.33 Prove that if $v \in V$ and $t \in \mathbb{R}_+$ then

$$\int_0^t T_s v \, \mathrm{d}s \in \mathrm{dom}\, A \qquad \text{and} \qquad (T_t - I)v = A \int_0^t T_s v \, \mathrm{d}s.$$

Deduce that $\mathrm{dom}\, A$ is dense in V. [Hint: begin by imitating the proof of Theorem 2.23.]

Lemma 2.34 *Let the strongly continuous semigroup T have generator A. If $v \in \mathrm{dom}\, A$ and $t \in \mathbb{R}_+$, then $T_t v \in \mathrm{dom}\, A$ and $T_t A v = A T_t v$; thus, $T_t(\mathrm{dom}\, A) \subseteq \mathrm{dom}\, A$. Furthermore,*

$$(T_t - I)v = \int_0^t T_s A v \, \mathrm{d}s = \int_0^t A T_s v \, \mathrm{d}s.$$

Proof First, note that

$$h^{-1}(T_h - I)T_t v = T_t h^{-1}(T_h - I)v \to T_t A v \quad \text{as } h \to 0+,$$

by the boundedness of T_t, so $T_t v \in \mathrm{dom}\, A$ and $A T_t v = T_t A v$, as claimed. For the second part, let

$$F : \mathbb{R}_+ \to V; \ t \mapsto (T_t - I)v - \int_0^t T_s A v \, \mathrm{d}s.$$

Note that F is continuous and $F(0) = 0$; furthermore, if $t > 0$, then

$$h^{-1}(F(t+h) - F(t)) = T_t h^{-1}(T_h - I)v - h^{-1}\int_0^h T_{s+t} A v \, \mathrm{d}s \to T_t A v - T_t A v = 0$$

as $h \to 0+$, whence $F \equiv 0$. $\qquad\qquad\qquad\qquad\qquad\qquad\qquad\qquad\qquad\qquad$ $\square$

Definition 2.35 An operator A in V is *closed* if, whenever $(v_n)_{n \in \mathbb{N}} \subseteq \mathrm{dom}\, A$ is such that $v_n \to v \in V$ and $A v_n \to u \in V$, it follows that $v \in \mathrm{dom}\, A$ and $A v = u$. Note that a bounded operator is automatically closed.

The operator A is *closable* if it has a closed extension, in which case the *closure* $\overline{A}$ is the smallest closed extension of A, where the ordering of operators is given in Definition 2.30.

Exercise 2.36 Prove that the *graph*

$$\mathcal{G}(A) := \{(v, Av) : v \in \text{dom } A\}$$

of an unbounded operator A in V is a normed vector space for the product norm

$$\| \cdot \| : (v, Av)\| \mapsto \|v\| + \|Av\|.$$

Prove further that A is closed if and only if $\mathcal{G}(A)$ is a Banach space, and that A is closable if and only if the closure of its graph in $V \oplus V$ is the graph of some operator. Finally, prove that if A is closable then $\mathcal{G}(\overline{A})$ is the intersection of the graphs of all closed extensions of A.

Exercise 2.37 Let A be the generator of the strongly continuous one-parameter semigroup T. Use Lemma 2.34 and Theorem 2.19 to show that A is closed.

Proof Suppose $(v_n)_{n \in \mathbb{N}} \subseteq \text{dom } A$ is such that $v_n \to v$ and $Av_n \to u$. Let $t > 0$ and note that

$$T_t v_n - v_n = \int_0^t T_s A v_n \, ds \qquad \text{for all } n \geqslant 1.$$

Furthermore,

$$\left\| \int_0^t T_s A v_n \, ds - \int_0^t T_s u \, ds \right\| \leqslant \int_0^t M e^{as} \|Av_n - u\| \, ds \leqslant M t e^{\max\{a,0\}t} \|Av_n - u\| \to 0$$

as $n \to \infty$, so

$$T_t v - v = \int_0^t T_s u \, ds.$$

Dividing both sides by t and letting $t \to 0+$ gives that $v \in \text{dom } A$ and $Av = u$, as required. $\qquad\qquad\square$

Definition 2.38 Let H be Hilbert space. If A is a densely defined operator in H, then the *adjoint* A^* is defined by setting

$$\text{dom } A^* := \{u \in \mathsf{H} : \text{there exists } v \in \mathsf{H} \text{ such that } \langle u, Aw \rangle$$
$$= \langle v, w \rangle \text{ for all } w \in \text{dom } A\}$$

and

$$A^* u = v, \qquad \text{where } v \text{ is as in the definition of dom } A^*.$$

When A is bounded, this agrees with the earlier definition. If A is not densely defined, then there may be no unique choice for v, so this definition cannot immediately be extended further.

It is readily verified that the adjoint A^* is always closed: if $(u_n)_{n \in \mathbb{N}} \subseteq \operatorname{dom} A^*$ is such that $u_n \to u \in \mathsf{H}$ and $A^* u_n \to v \in \mathsf{H}$ then

$$\langle u, Aw \rangle = \lim_{n \to \infty} \langle u_n, Aw \rangle = \lim_{n \to \infty} \langle A^* u_n, w \rangle = \lim_{n \to \infty} \langle v, w \rangle \qquad \text{for all } w \in \operatorname{dom} A,$$

so $x \in \operatorname{dom} A^*$ and $A^* u = v$.

Exercise 2.39 Prove that a densely defined operator A is closable if and only if its adjoint A^* is densely defined, in which case $\overline{A} = (A^*)^*$ and $\overline{A}^* = A^*$.

Definition 2.40 A densely defined operator A in a Hilbert space is *self-adjoint* if and only if $A^* = A$. This is stronger than the condition that

$$\langle u, Av \rangle = \langle Au, v \rangle \qquad \text{for all } u, v \in \operatorname{dom} A,$$

which is merely the condition that $A \subseteq A^*$. An operator satisfying this inclusion is called *symmetric*.

Exercise 2.41 Let A be a densely defined operator in the Hilbert space H. Prove that A is self-adjoint if and only if A is symmetric and such that both $A + iI$ and $A - iI$ are surjective, so that

$$\{ Av + iv : v \in \operatorname{dom} A \} = \{ Av - iv : v \in \operatorname{dom} A \} = \mathsf{H}.$$

Proof Suppose first that A is symmetric and the range conditions hold. Let $u, v \in \mathsf{H}$ be such that

$$\langle u, Aw \rangle = \langle v, w \rangle \qquad \text{for all } w \in \operatorname{dom} A,$$

so that $u \in \operatorname{dom} A^*$ and $A^* u = v$. We wish to prove that $u \in \operatorname{dom} A$ and $Au = v$.

Let $x, y \in \operatorname{dom} A$ be such that $(A - iI)x = v - iu$ and $(A + iI)y = u - x$. Then

$$\langle u, u - x \rangle = \langle u, (A + iI)y \rangle = \langle v - iu, y \rangle$$
$$= \langle (A - iI)x, y \rangle = \langle x, (A + iI)y \rangle = \langle x, u - x \rangle,$$

where the penultimate equality holds because A is symmetric and $x, y \in \operatorname{dom} A$. It follows that $\| u - x \|^2 = 0$, so $u = x \in \operatorname{dom} A$ and $Au = Ax = v - iu + ix = v$.

Now suppose that A is self-adjoint, and note that it suffices to prove that $A + iI$ is surjective, since $-A$ is self-adjoint whenever A is.

Note first that

$$\| (A + iI)v \|^2 = \| Av \|^2 + \| v \|^2 \qquad \text{for all } v \in \operatorname{dom} A, \tag{2.2}$$

which implies that ran$(A + iI)$ is closed: if the sequence $(v_n)_{n \in \mathbb{N}} \subseteq \operatorname{dom} A$ is such that $\big((A + iI)v_n\big)_{n \in \mathbb{N}}$ is convergent, then both $(v_n)_{n \in \mathbb{N}}$ and $(Av_n)_{n \in \mathbb{N}}$ are Cauchy, so convergent, with $v_n \to v \in \mathsf{H}$ and $Av_n \to u \in \mathsf{H}$. Since A is closed, it follows that $v \in \operatorname{dom} A$ and $Av = u$, from which we see that $(A + iI)v_n \to u + iv = (A + iI)v$.

It is also follows from (2.2), with A replaced by $-A$, that $A - iI$ is injective. As

$$u \in \ker(A - iI) \iff \langle (A - iI)u, v \rangle = 0 \qquad \text{for all } v \in \operatorname{dom} A$$

$$\iff \langle u, (A + iI)v \rangle = 0 \qquad \text{for all } v \in \operatorname{dom} A = \operatorname{dom} A^*$$

$$\iff u \in \operatorname{ran}(A + iI)^\perp,$$

so

$$\operatorname{ran}(A + iI) = (\operatorname{ran}(A + iI)^\perp)^\perp = \ker(A - iI)^\perp = \{0\}^\perp = \mathsf{H}.$$

$\square$

Definition 2.42 Let A be an unbounded operator in V. Its *spectrum* is the set

$$\sigma(A) := \{\lambda \in \mathbb{C} : \lambda I - A \text{ has no inverse in } B(V)\}$$

and its *resolvent* is the map

$$\mathbb{C} \setminus \sigma(A) \to B(V); \ \lambda \mapsto (\lambda I - A)^{-1}.$$

In other words, $\lambda \in \mathbb{C}$ is not in the spectrum of A if and only if there exists a bounded operator $B \in B(V)$ such that $B(\lambda I - A) = I_{\operatorname{dom} A}$ and $(\lambda I - A)B = I_V$; in particular, the operator $\lambda I - A$ is a bijection from $\operatorname{dom} A$ onto V.

Remark 2.43 If the operator $T : V \to V$ is bounded, then its spectrum $\sigma(T)$ is contained in the closed disc $\{\lambda \in \mathbb{C} : |\lambda| \leqslant \|T\|\}$ [22, Lemma 1.2.4].

Exercise 2.44 Let A be an unbounded operator in V and suppose $\lambda \in \mathbb{C}$ is such that $\lambda I - A$ is a bijection from $\operatorname{dom} A$ onto V. Prove that $(\lambda I - A)^{-1}$ is bounded if and only if A is closed. [Thus algebraic invertibility of $\lambda I - A$ is equivalent to its topological invertibility if and only if A is closed.]

The following theorem shows that the resolvent of a semigroup generator may be thought of as the Laplace transform of the semigroup.

Theorem 2.45 *Let A be the generator of a one-parameter semigroup T of type (M, a) on V. Then $\sigma(A) \subseteq \{\lambda \in \mathbb{C} : \operatorname{Re} \lambda \leqslant a\}$. Furthermore, if $\operatorname{Re} \lambda > a$, then*

$$(\lambda I - A)^{-1} v = \int_0^\infty e^{-\lambda t} T_t v \, dt \qquad \text{for all } v \in V \tag{2.3}$$

and $\|(\lambda I - A)^{-1}\| \leqslant M(\operatorname{Re} \lambda - a)^{-1}$.

Proof Fix $\lambda \in \mathbb{C}$ with $\operatorname{Re} \lambda > a$ and note first that

$$R : V \mapsto V; \quad v \mapsto \int_0^\infty e^{-\lambda t} T_t v \, dt$$

is a bounded linear operator, with $\|R\| \leqslant M(\operatorname{Re} \lambda - a)^{-1}$.

If $v \in V$ and $u = Rv$, then

$$T_t u = \int_0^\infty e^{-\lambda s} T_{s+t} v \, ds = \int_t^\infty e^{-\lambda(r-t)} T_u v \, dr = e^{\lambda t} \int_t^\infty e^{-\lambda u} T_r v \, dr,$$

and therefore, if $t > 0$,

$$t^{-1}(T_t - I)u = t^{-1} e^{\lambda t} \int_t^\infty e^{-\lambda s} T_s v \, ds - t^{-1} \int_0^\infty e^{-\lambda s} T_s v \, ds$$

$$= -t^{-1} e^{\lambda t} \int_0^t e^{-\lambda s} T_s v \, ds + t^{-1}(e^{\lambda t} - 1) \int_0^\infty e^{-\lambda s} T_s v \, ds$$

$$\to -v + \lambda u \qquad \text{as } t \to 0+.$$

Thus $u \in \operatorname{dom} A$ and $(\lambda I - A)u = v$. It follows that $\operatorname{ran} R \subseteq \operatorname{dom} A$ and $(\lambda I - A)R = I_V$.

However, since $(T_t - I)R = R(T_t - I)$ and R is bounded, the same working shows that

$$RAu = -u + \lambda Ru \iff R(\lambda I - A)u = u \qquad \text{for all } u \in \operatorname{dom} A.$$

Thus $R(\lambda I - A) = I_{\operatorname{dom} A}$ and $R = (\lambda I - A)^{-1}$, as claimed. $\qquad \square$

The Laplace-transform formula of Theorem 2.45 allows one to recover a semigroup from its resolvent.

Theorem 2.46 *Let A be the generator of a one-parameter semigroup T of type (M, a) on V, and let $\lambda \in \mathbb{C}$ with $\operatorname{Re} \lambda > a$. Then*

$$(\lambda I - A)^{-n} v = \int_0^\infty \frac{t^{n-1}}{(n-1)!} e^{-\lambda t} T_t v \, dt \qquad \text{for all } n \in \mathbb{N} \text{ and } v \in V,$$

and

$$T_t v = \lim_{n \to \infty} (I - n^{-1} t A)^{-n} v$$

$$= \lim_{n \to \infty} (n/t)^n \big((n/t)I - A\big)^{-n} v \qquad \text{for all } t > 0 \text{ and } v \in V.$$

Proof The first claim follows by induction, with Theorem 2.45 giving the case $n = 1$.

As noted by Hille and Phillips [16, Theorem 11.6.6], the second follows from the Post–Widder inversion formula for the Laplace transform. For all $n \in \mathbb{N}$, let

$$f_n : \mathbb{R}_+ \to \mathbb{R}_+; \ t \mapsto \frac{n^n}{(n-1)!} t^n e^{-nt},$$

and note that f_n is strictly increasing on $[0, 1]$ and strictly decreasing on $[1, \infty)$, and its integral $\int_0^\infty f_n(t)\,dt = 1$; this last fact may be proved by induction. If n is sufficiently large, then a short calculation shows that

$$(n/t)^n\big((n/t)I - A\big)^{-n} v = (1 - n^{-1})^{-n} \int_0^\infty f_{n-1}(r) e^{-r} T_{tr} v \, dr.$$

The result follows by splitting the integral into three parts. Fix $\varepsilon \in (0, 1)$ and note first that $f_n(r) \leqslant n e^n r^n e^{-nr}$ for all $r \in \mathbb{R}_+$, with the latter function strictly increasing on $[0, 1]$, so

$$\left\| \int_0^{1-\varepsilon} f_n(r) e^{-r} T_{tr} v \, dr \right\| \leqslant n(1 - \varepsilon)^{n+1} e^{n\varepsilon} M \max\{1, e^{at(1-\varepsilon)}\} \|v\| \to 0 \quad \text{as } n \to \infty.$$

Similarly, if $b := \varepsilon/(1 + \varepsilon)$, then $f_n(r) e^{bnr} \leqslant n e^n (1 + \varepsilon)^n e^{(b-1)n(1+\varepsilon)} \leqslant n(1+\varepsilon)^n$ for all $r \geqslant 1 + \varepsilon$, and so

$$\left\| \int_{1+\varepsilon}^\infty f_n(r) e^{-r} T_{tr} v \, dr \right\| \leqslant M\|v\| n(1+\varepsilon)^n \int_{1+\varepsilon}^\infty e^{(a-bn)r} \, dr$$

$$\leqslant M\|v\| \frac{n}{bn - a} (1 + \varepsilon)^n e^{(a-bn)(1+\varepsilon)} \to 0 \quad \text{as } n \to \infty,$$

since $b(1 + \varepsilon) = \varepsilon$ and $(1 + \varepsilon)e^{-\varepsilon} < 1$. A standard approximation argument now completes the proof. $\qquad\qquad\square$

We have now obtained enough necessary conditions on the generator of a strongly continuous semigroup for them to be sufficient as well.

Theorem 2.47 (Feller–Miyadera–Phillips) *A closed, densely defined operator A in V is the generator of a strongly continuous semigroup of type (M, a) if and only if*

$$\sigma(A) \subseteq \{\lambda \in \mathbb{C} : \operatorname{Re} \lambda \leqslant a\}$$

and

$$\|(\lambda I - A)^{-m}\| \leqslant M(\lambda - a)^{-m} \qquad \textit{for all } \lambda > a \textit{ and } m \in \mathbb{N}. \tag{2.4}$$

Proof Let A be the generator of a strongly continuous semigroup T of type (M, a). The spectral condition is a consequence of Theorem 2.45, and the norm inequality follows from Theorem 2.46.

For the converse, let the operator A be closed, densely defined, such that (2.4) holds and having spectrum not containing (a, ∞). Setting $A_\lambda := \lambda A(\lambda I - A)^{-1}$, note that $\{A_\lambda : \lambda \in (a, \infty)\}$ is a commuting family of bounded operators such that $A_\lambda v \to Av$ as $\lambda \to \infty$, for all $v \in \operatorname{dom} A$; see Exercise 2.48 for more details.

With $T_t^\lambda := \exp(t A_\lambda)$, the inequalities (2.4) imply $\|T_t^\lambda\| \leqslant M \exp\big(a\lambda t/(\lambda - a)\big)$ for all $\lambda > a$ and $t \in \mathbb{R}_+$, so $\limsup_{\lambda \to \infty} \|T_t^\lambda\| \leqslant M e^{at}$. Since

$$(T_t^\lambda - T_t^\mu)v = \int_0^t \frac{\mathrm{d}}{\mathrm{d}s}\big(T_s^\lambda T_{t-s}^\mu v\big)\,\mathrm{d}s = \int_0^t T_s^\lambda T_{t-s}^\mu (A_\lambda - A_\mu)v\,\mathrm{d}s,$$

if $\lambda, \mu > 2a_+ = 2\max\{a, 0\}$ and $v \in \operatorname{dom} A$ then

$$\|(T_t^\lambda - T_t^\mu)v\| \leqslant t M^2 e^{2a_+ t}\|(A_\lambda - A_\mu)v\| \to 0 \qquad \text{as } \lambda, \mu \to \infty,$$

locally uniformly in t. An approximation argument shows that $T_t u = \lim_{\lambda \to \infty} T_t^\lambda u$ exists for all $t \in \mathbb{R}_+$ and $u \in V$, and that $T = (T_t)_{t \in \mathbb{R}_+}$ is a strongly continuous one-parameter semigroup of type (M, a).

To see that the generator of T is A, note that the previous working and Lemma 2.34 imply that

$$T_t v - v = \lim_{\lambda \to \infty} T_t^\lambda v - v = \lim_{\lambda \to \infty} \int_0^t T_s^\lambda A_\lambda v\,\mathrm{d}s = \int_0^t T_s Av\,\mathrm{d}s \qquad \text{for all } v \in \operatorname{dom} A;$$

dividing by t and letting $t \to 0$ shows that the generator B of T is an extension of A. Note that (a, ∞) is not in the spectrum of B, by Theorem 2.45; it is a simple exercise to show that $(\lambda I - A)^{-1} = (\lambda I - B)^{-1}$ for $\lambda > a$, and since the ranges of these operators are the domain of A and B, the result follows. $\qquad\square$

Exercise 2.48 Let A be an unbounded operator in V, with spectrum not containing (a, ∞) and such that $\|(\lambda I - A)^{-1}\| \leqslant M(\lambda - a)^{-1}$ for all $\lambda > a$, where M and a are constants. Prove that

$$A_\lambda := \lambda A(\lambda I - A)^{-1} = \lambda^2(\lambda I - A)^{-1} - \lambda I$$

commutes with A_μ for all $\lambda, \mu > a$. Prove also that

$$\lim_{\lambda \to \infty} \lambda(\lambda I - A)^{-1}u = u \qquad \text{for all } u \in V,$$

by showing this first for the case $u \in \operatorname{dom} A$. Deduce that $A_\lambda v \to Av$ when $v \in \operatorname{dom} A$.

For contraction semigroups, we have the following refinement of Theorem 2.47.

Theorem 2.49 (Hille–Yosida) *Let A be a closed, densely defined linear operator in the Banach space V. The following are equivalent.*

(i) *A is the generator of a strongly continuous contraction semigroup.*

(ii) *$\sigma(A) \subseteq \{\lambda \in \mathbb{C} : \operatorname{Re}\lambda \leqslant 0\}$ and*

$$\|(\lambda I - A)^{-1}\| \leqslant (\operatorname{Re}\lambda)^{-1} \qquad \text{whenever } \operatorname{Re}\lambda > 0.$$

(iii) *$\sigma(A) \cap (0, \infty)$ is empty and*

$$\|(\lambda I - A)^{-1}\| \leqslant \lambda^{-1} \qquad \text{whenever } \lambda > 0.$$

Proof Note that (i) implies (ii), by Theorem 2.45, and (ii) implies (iii) trivially. That (iii) implies (i) follows from the extension of Theorem 2.47 noted in its proof. □

In practice, verifying the norm conditions in Theorems 2.47 and 2.49 may prove to be challenging. The next section introduces the concept of operator dissipativity, which is often more tractable.

2.4   The Lumer–Phillips Theorem

Throughout this subsection, V denotes a Banach space and V^* its topological dual.

Definition 2.50 For all $v \in V$, let

$$TF(v) := \{\phi \in V^* : \phi(v) = \|v\|^2 = \|\phi\|^2\}$$

be the set of *normalised tangent functionals* to v. The Hahn–Banach theorem [25, Theorem III.6] implies that $TF(v)$ is non-empty for all $v \in V$.

Exercise 2.51 Prove that if H is a Hilbert space then $TF(v) = \{\langle v|\}$ for all $v \in H$, where the Dirac functional $\langle v|$ is such that $\langle v|u := \langle v, u\rangle$ for all $u \in H$. [Recall the Riesz–Fréchet theorem from Example 2.15.]

Exercise 2.52 Prove that if $f \in V = C(K)$ and $x_0 \in K$ is such that $|f(x_0)| = \|f\|$ then setting $\phi(g) := \overline{f(x_0)}g(x_0)$ for all $g \in V$ defines a normalised tangent functional for f. Deduce that $TF(f)$ may contain more than one element.

Definition 2.53 An unbounded operator A in V is *dissipative* if and only if there exists $\phi \in TF(v)$ such that $\operatorname{Re}\phi(Av) \leqslant 0$, for all $v \in \operatorname{dom} A$. [Note that it suffices to check this condition for unit vectors only.]

Exercise 2.54 Prove that an operator A in the Hilbert space H is dissipative if and only if $\|(I + A)v\| \leqslant \|(I - A)v\|$ for all $v \in \operatorname{dom} A$.

Exercise 2.55 Suppose T is a contraction semigroup with generator A. Prove that A is dissipative.

Proof If $v \in \operatorname{dom} A$ and $\phi \in TF(v)$, then

$$\operatorname{Re}\phi(Av) = \lim_{t \to 0+} t^{-1} \operatorname{Re}\phi(T_t v - v) \leqslant \lim_{t \to 0+} t^{-1} \|\phi\| \|v\| - \|v\|^2 = 0,$$

so A is dissipative. □

We now seek to find a converse to the result of the preceding exercise.

Lemma 2.56 *The unbounded operator A in V is dissipative if and only if*

$$\|(\lambda I - A)v\| \geqslant \lambda\|v\| \qquad \text{for all } \lambda > 0 \text{ and } v \in \operatorname{dom} A. \tag{2.5}$$

If A is dissipative and $\lambda I - A$ is surjective for some $\lambda > 0$, then $\lambda \notin \sigma(A)$ and $\|(\lambda I - A)^{-1}\| \leqslant \lambda^{-1}$.

Proof Suppose first that (2.5) holds, let $v \in \operatorname{dom} A$ be a unit vector and, for all $\lambda > 0$, choose $\phi_\lambda \in TF\big((\lambda I - A)v\big)$. Then $\phi_\lambda \neq 0$, so $\psi_\lambda = \|\phi_\lambda\|^{-1}\phi_\lambda$ is well defined, and

$$\lambda \leqslant \|(\lambda I - A)v\| = \psi_\lambda(\lambda v - Av) = \lambda \operatorname{Re}\psi_\lambda(v) - \operatorname{Re}\psi_\lambda(Av).$$

Since $\operatorname{Re}\psi_\lambda(v) \leqslant 1$ and $-\operatorname{Re}\psi_\lambda(Av) \leqslant \|Av\|$, it follows that

$$\operatorname{Re}\psi_\lambda(Av) \leqslant 0 \qquad \text{and} \qquad \operatorname{Re}\psi_\lambda(v) \geqslant 1 - \lambda^{-1}\|Av\|.$$

The Banach–Alaoglu theorem [25, Theorem IV.21] implies that the unit ball of V^* is weak* compact, so the net $(\psi_\lambda)_{\lambda > 0}$ has a weak*-convergent subnet with limit in the unit ball. Hence there exists $\psi \in V^*$ such that

$$\|\psi\| \leqslant 1, \quad \operatorname{Re}\psi(Av) \leqslant 0 \quad \text{and} \quad \operatorname{Re}\psi(v) \geqslant 1.$$

In particular,

$$|\psi(v)| \leqslant \|\psi\| \leqslant 1 \leqslant \operatorname{Re}\psi(v) \leqslant |\psi(v)|,$$

so $\psi \in TF(v)$ and A is dissipative.

Conversely, if $\lambda > 0$, $v \in \operatorname{dom} A$ and $\phi \in TF(v)$ is such that $\operatorname{Re}\phi(Av) \leqslant 0$ then

$$\|v\| \|(\lambda I - A)v\| \geqslant |\phi\big((\lambda I - A)v\big)| = |\lambda\|v\|^2 - \phi(Av)| \geqslant \lambda\|v\|^2.$$

Thus (2.5) holds, and $\lambda I - A$ is injective.

If $\lambda I - A$ is also surjective, then (2.5) gives that $\|u\| \geqslant \lambda\|(\lambda I - A)^{-1}u\|$ for all $u \in V$, whence the final claim. □

Exercise 2.57 Let A be dissipative. Prove that $\lambda I - A$ is surjective for some $\lambda > 0$ if and only if $\lambda I - A$ is surjective for all $\lambda > 0$. [Hint: for a suitable choice of λ and λ_0, consider the series $R_\lambda := \sum_{n=0}^{\infty} (\lambda - \lambda_0)^n (\lambda_0 I - A)^{-(n+1)}$.]

Proof Suppose that $\lambda_0 > 0$ is such that $\lambda_0 I - A$ is surjective. It follows from Lemma 2.56 that $\|(\lambda_0 I - A)^{-1}\| \leqslant \lambda_0^{-1}$. The series

$$R_\lambda = \sum_{n=0}^{\infty} (\lambda_0 - \lambda)^n (\lambda_0 I - A)^{-(n+1)}$$

is norm convergent for all $\lambda \in (0, 2\lambda_0)$; if we can show that $R_\lambda = (\lambda I - A)^{-1}$, then the result follows.

If $C \in B(V)$ is such that $\|C\| < 1$ then $I - C$ is invertible, with $(I - C)^{-1} = \sum_{n=0}^{\infty} C^n$. Hence if $C = (\lambda_0 - \lambda)(\lambda_0 I - A)^{-1}$, then

$$R_\lambda = (\lambda_0 I - A)^{-1} (I - C)^{-1} = (I - C)^{-1} (\lambda_0 I - A)^{-1},$$

so ran $R_\lambda \subseteq \operatorname{dom}(\lambda_0 I - A) = \operatorname{dom}(\lambda I - A)$,

$$(\lambda I - A) R_\lambda = \big((\lambda - \lambda_0)I + (\lambda_0 I - A)\big) R_\lambda = \big((\lambda - \lambda_0)(\lambda_0 I - A)^{-1} + I\big)(I - C)^{-1} = I_V$$

and

$$R_\lambda(\lambda I - A) = R_\lambda\big((\lambda - \lambda_0)I + (\lambda_0 I - A)\big)$$

$$= (I - C)^{-1}\big((\lambda - \lambda_0)(\lambda_0 I - A)^{-1} + I\big) = I_{\operatorname{dom} A}.$$

$\square$

Theorem 2.58 (Lumer–Phillips) *A closed, densely defined operator A generates a strongly continuous contraction semigroup if and only if A is dissipative and $\lambda I - A$ is surjective for some $\lambda > 0$.*

Proof One implication follows from Exercise 2.57, Lemma 2.56 and Theorem 2.49. The other implication follows from Theorem 2.49 and Exercise 2.55. $\square$

Example 2.59 Let $V = L^2[0, 1]$, and let $Af := g$, where

$$\operatorname{dom} A := \Big\{ f \in V : \text{there exists } g \in V \text{ such that } f(t)$$

$$= \int_0^t g(s) \, ds \text{ for all } t \in [0, 1] \Big\}.$$

Thus $f \in \operatorname{dom} A$ if and only if $f(0) = 0$ and f is absolutely continuous on $[0, 1]$, with square-integrable derivative, and then $Af = f'$ almost everywhere. For such f,

note that

$$\mathrm{Re}\langle f, Af \rangle = \mathrm{Re} \int_0^1 \overline{f(t)} f'(t)\, \mathrm{d}t = \frac{1}{2} \int_0^t (\overline{f} f)'(t)\, \mathrm{d}t = \frac{1}{2}|f(1)|^2 \geqslant 0,$$

so $-A$ is a dissipative operator, but A is not.

Let $g \in V$ and $\lambda > 0$; we wish to find $f \in \mathrm{dom}\, A$ such that

$$(\lambda I + A)f = g \iff \lambda f + f' = g \iff f = \int (g - \lambda f).$$

We proceed by iterating this relation: given $h \in \{f, g\}$, let $h_0 := h$ and, for all $n \in \mathbb{Z}_+$, let $h_{n+1} \in V$ be such $h_{n+1}(t) = \int_0^t h_n(s)\, \mathrm{d}s$ for all $t \in [0, 1]$. Then

$$f = g_1 - \lambda \int f = g_1 - \lambda \int \int (g - \lambda f) = \cdots = \sum_{j=0}^{n-1} (-\lambda)^j g_{j+1} + (-\lambda)^n f_n$$

for all $n \in \mathbb{N}$. The series $\sum_{j=0}^\infty (-\lambda)^j g_{j+1}$ is uniformly convergent on $[0, 1]$, so defines a function $F \in \mathrm{dom}\, A$, whereas $(-\lambda)^n f_n \to 0$ as $n \to \infty$. Thus

$$(\lambda I + A)F = -\sum_{j=0}^\infty (-\lambda)^{j+1} g_{j+1} + \sum_{j=0}^\infty (-\lambda)^j g_j = g_0 = g,$$

so $\lambda I + A$ is surjective. By the Lumer–Phillips theorem, the operator $-A$ generates a contraction semigroup.

Exercise 2.60 Fill in the details at the end of Example 2.59. [Hint: with the notation of the example, show that if $h \in \{f, g\}$ then $|h_n(t)|^2 \leqslant t^n \|h\|_2^2 / n!$ for all $n \in \mathbb{N}$.]

Remark 2.61 We can explain informally why the operator A defined in Example 2.59 does not generate a semigroup, and why $-A$ does. Recall that each element of a semigroup leaves the domain of the generator invariant, by Lemma 2.34, and A would generate a left-translation semigroup, which does not preserve the boundary condition $f(0) = 0$. Moreover, $-A$ generates the right-translation semigroup, and this does preserve the boundary condition.

If we let A_0 be the restriction of A to the domain

$$\mathrm{dom}\, A_0 := \{f \in \mathrm{dom}\, A : f(1) = 0\},$$

so adding a further boundary condition, then both A_0 and $-A_0$ are dissipative, but neither generates a semigroup. We cannot solve the equation $(\lambda I \pm A_0)f = g$ for all g when subject to the constraint that $f \in \mathrm{dom}\, A_0$. [Take $g \in L^2[0, 1]$ such that $g(t) = t$ for all $t \in [0, 1]$, construct F as in Example 2.59 and note that $F(1) \neq 0$.]

Example 2.62 Recall the weak derivatives D^α and Sobolev spaces $H^k(\mathbb{R}^d)$ defined in Example 2.9, and let $2e_j \in \mathbb{Z}_+^d$ be the multi-index with 2 in the jth coordinate and 0 elsewhere. The *Laplacian*

$$\Delta := \sum_{j=1}^d \frac{\partial^2}{\partial x_j^2} = \sum_{j=1}^d D^{2e_j}$$

is a densely defined operator in $L^2(\mathbb{R}^d)$ with domain $\operatorname{dom} \Delta := H^2(\mathbb{R}^d)$. It may be shown that

$$\langle \Delta f, g \rangle_{L^2(\mathbb{R}^d)} = -\langle \nabla f, \nabla g \rangle_{L^2(\mathbb{R}^d)} \qquad \text{for all } f, g \in H^2(\mathbb{R}^d), \tag{2.6}$$

where

$$\nabla := (D^{e_1}, \ldots, D^{e_d}) : f \mapsto \left(\frac{\partial f}{\partial x_1}, \ldots, \frac{\partial f}{\partial x_d} \right);$$

consequently, the Laplacian Δ is dissipative. One way to establish (2.6) is to use the Fourier transform. Fourier-theoretic results can also be used to prove that $\lambda I - \Delta$ is surjective for all $\lambda > 0$, essentially because the map $x \mapsto 1/(\lambda + |x|^2)$ is bounded on $\mathbb{R}^d$. Thus the Laplacian generates a contraction semigroup.

Exercise 2.63 Let A be a densely defined operator on the Hilbert space H. Prove that if A is symmetric, so that

$$\langle u, Av \rangle = \langle Au, v \rangle \quad \text{for all } u, v \in \operatorname{dom} A,$$

then iA is dissipative. Deduce that if H is self-adjoint then iH and $-iH$ are the generators of contraction semigroups.

   Prove, further, that if $T = (T_t)_{t \in \mathbb{R}_+}$ has generator iH, with H self-adjoint, then T_t is unitary, so that $T_t^* T_t = I = T_t T_t^*$, for all $t \in \mathbb{R}_+$.

Proof The first part is an immediate consequence of Theorem 2.58, the Lumer–Phillips theorem, together with Exercise 2.41.

   For the next part, fix $u, v \in \operatorname{dom} H$ and $t \in \mathbb{R}_+$. If $h > 0$ then

$$h^{-1} \langle u, (T_{t+h}^* T_{t+h} - T_t^* T_t) v \rangle = \langle T_{t+h} u, h^{-1} (T_h - I) T_t v \rangle$$

$$+ \langle h^{-1} (T_h - I) T_t u, T_t v \rangle$$

$$\to \langle T_t u, iH T_t v \rangle + \langle iH T_t u, T_t v \rangle = 0$$

as $h \to 0+$, since $T_t u, T_t v \in \operatorname{dom} H$ and T is strongly continuous. A real-valued function on $\mathbb{R}_+$ is constant if it is continuous and its right derivative is identically zero, so this working shows that $T_t^* T_t = I$.

Now let $S = (S_t)_{t \in \mathbb{R}_+}$ be the strongly continuous semigroup with generator $-iH$. The previous working shows that $S_t^* S_t = I$ for all $t \in \mathbb{R}_+$, so it suffices to let $t > 0$ and prove that $S_t = T_t^*$. To see this, let $u, v \in \operatorname{dom} H$ and consider the function

$$F : [0, t] \to \mathbb{C}; \quad s \mapsto \langle u, T_{t-s}^* S_s v \rangle.$$

Working as above, it is straightforward to show that $F' \equiv 0$ on $(0, t)$, so $F(0) = F(t)$ and the result follows. $\qquad\square$

Exercise 2.64 Suppose U is a strongly continuous one-parameter semigroup on the Hilbert space H, with U_t unitary, so that $U_t^* U_t = I = U_t U_t^*$, for all $t \in \mathbb{R}_+$. Let A be the generator of U.

Prove that $U^* = (U_t^*)_{t \in \mathbb{R}_+}$ is also a strongly continuous one-parameter semigroup, with generator $-A$. Deduce that $H := iA$ is self-adjoint.

Proof The semigroup property for U^* is immediate, and strong continuity holds because

$$\|(U_t^* - I)v\|^2 = \langle (I - U_t)v, v \rangle - \langle v, (U_t - I)v \rangle \leqslant 2\|(U_t - I)v\| \, \|v\| \to 0$$

as $t \to 0+$, for any $v \in \mathsf{H}$.

Next, denote the generator of U^* by B, and let $v \in \operatorname{dom} A$. Then

$$t^{-1}(U_t^* - I)v = -U_t^* t^{-1}(U_t - I)v \to -Av \qquad \text{as } t \to 0+,$$

so $-A \subseteq B$. Since $(U^*)^* = U$, applying this argument with U replaced by U^* gives the reverse inclusion. Thus U^* has generator $B = -A$, as claimed.

Finally, let $H = iA$ and suppose first that $u, v \in \operatorname{dom} H = \operatorname{dom} A$. Then

$$\langle -iHu, v \rangle = \lim_{t \to 0+} \langle t^{-1}(U_t - I)u, v \rangle = \lim_{t \to 0+} \langle u, t^{-1}(U_t^* - I)v \rangle = \langle u, iHv \rangle,$$

so $H \subseteq H^*$. For the reverse inclusion, note that

$$U_t^* v = v + \int_0^t U_s^* A^* v \, ds \qquad \text{for all } v \in \operatorname{dom} A^*,$$

by Lemma 2.34 applied to U and properties of the adjoint. Thus $A^* \subseteq -A$, the generator of U^*, and therefore $H^* = -iA^* \subseteq iA = H$. $\qquad\square$

Remark 2.65 Exercises 2.63 and 2.64 lead to Stone's theorem, which gives a one-to-one correspondence between self-adjoint operators and strongly continuous one-parameter groups of unitary operators. This result has significant consequences for the mathematical foundations of quantum theory; see [25, Section VIII.4].

3 Classical Markov Semigroups

Throughout this section, the triple $(\Omega, \mathcal{F}, \mathbb{P})$ will denote a probability space, so that $\mathbb{P} : \mathcal{F} \to [0, 1]$ is a probability measure on the σ-algebra $\mathcal{F}$ of subsets of Ω, and E will denote a topological space, with $\mathcal{E}$ its Borel σ-algebra, generated by the open subsets.

An *E-valued random variable* is a $\mathcal{F}$-$\mathcal{E}$-measurable mapping $X : \Omega \to E$. If X is an E-valued random variable, then $\sigma(X)$ is the smallest sub-σ-algebra $\mathcal{F}_0$ of $\mathcal{F}$ such that X is $\mathcal{F}_0$-$\mathcal{E}$ measurable. More generally, if $(X_i)_{i \in I}$ is an indexed set of E-valued random variables, then $\sigma(X_i : i \in I)$ is the smallest sub-σ-algebra $\mathcal{F}_0$ of $\mathcal{F}$ such that X_i is $\mathcal{F}_0$-$\mathcal{E}$ measurable for all $i \in I$.

3.1 Markov Processes

Definition 3.1 Given a real-valued random variable X which is integrable, so that

$$\mathbb{E}\big[|X|\big] := \int_{\Omega} |X(\omega)| \, \mathbb{P}(\mathrm{d}\omega) < \infty,$$

and a sub-σ-algebra $\mathcal{F}_0$ of $\mathcal{F}$, the *conditional expectation* $\mathbb{E}[X|\mathcal{F}_0]$ is a real-valued random variable Y which is $\mathcal{F}_0$-$\mathcal{E}$ measurable and such that

$$\mathbb{E}[1_A X] = \mathbb{E}[1_A Y] \qquad \text{for all } A \in \mathcal{F}_0.$$

The choice of Y is determined *almost surely*: if Y and Z are both versions of the conditional expectation $\mathbb{E}[X|\mathcal{F}_0]$, then $\mathbb{P}(Y \neq Z) = 0$. The existence of Y is guaranteed by the Radon–Nikodým theorem.

The fact that $\mathbb{E}[X|\mathcal{F}_0]$ is determined almost surely can be recast as saying that $\mathbb{E}[\cdot|\mathcal{F}_0]$ is a linear operator from $L^1(\Omega, \mathcal{F}, \mathbb{P})$ to $L^1(\Omega, \mathcal{F}_0, \mathbb{P}|_{\mathcal{F}_0})$. In fact, the map $X \mapsto \mathbb{E}[X|\mathcal{F}_0]$ is a contraction from $L^p(\Omega, \mathcal{F}, \mathbb{P})$ onto $L^p(\Omega, \mathcal{F}_0, \mathbb{P}|_{\mathcal{F}_0})$, for all $p \in [1, \infty]$.

Remark 3.2 Let $X \in L^2(\Omega, \mathcal{F}, \mathbb{P})$. Informally, we can think of $Y := \mathbb{E}[X|\mathcal{F}_0]$ as the best guess for X given the information in $\mathcal{F}_0$. In other words, the conditional expectation Y of X with respect to $\mathcal{F}_0$ is the essentially unique choice of $\mathcal{F}_0$-measurable random variable Z which minimises the least-squares distance $\|Z - X\|_2$.

Definition 3.3 Given a topological space E, let the Banach space

$$B_b(E) := \{ f : E \to \mathbb{C} \mid f \text{ is Borel measurable and bounded} \},$$

with vector-space operations defined pointwise and supremum norm

$$\|f\| := \sup\{|f(x)| : x \in E\}.$$

Exercise 3.4 Verify that $B_b(E)$ is a Banach space. Show further that the norm $\|\cdot\|$ is submultiplicative, where multiplication of functions is defined pointwise, so that $B_b(E)$ is a *Banach algebra*. Show also that the Banach algebra $B_b(E)$ is *unital*: the multiplicative unit 1_E is such that $\|1_E\| = 1$. Show finally that the C^* *identity* holds:

$$\|f\|^2 = \|\overline{f} f\| \qquad \text{for all } f \in B_b(E),$$

where the isometric involution $f \mapsto \overline{f}$ is such that $\overline{f}(x) := \overline{f(x)}$ for all $x \in E$.

Definition 3.5 (Provisional) A *Markov process* with *state space* E is a collection of E-valued random variables $X = (X_t)_{t \in \mathbb{R}_+}$ on a common probability space such that, given any $f \in B_b(E)$,

$$\mathbb{E}[f(X_t) \mid \sigma(X_r : 0 \leqslant r \leqslant s)] = \mathbb{E}[f(X_t) \mid \sigma(X_s)]$$

for all $s, t \in \mathbb{R}_+$ such that $s \leqslant t$.

A Markov process is *time homogeneous* if, given any $f \in B_b(E)$,

$$\mathbb{E}[f(X_t) \mid X_s = x] = \mathbb{E}[f(X_{t-s}) \mid X_0 = x] \tag{3.1}$$

for all $s, t \in \mathbb{R}_+$ such that $s \leqslant t$ and $x \in E$.

Definition 3.5 is well motivated by Remark 3.2, but it is somewhat unsatisfactory; for example, what should be the proper meaning of (3.1)? To improve upon it, we introduce the following notion.

Definition 3.6 A *transition kernel* on $(E, \mathcal{E})$ is a map $p : E \times \mathcal{E} \to [0, 1]$ such that

    (i) the map $x \mapsto p(x, A)$ is Borel measurable for all $A \in \mathcal{E}$ and

    (ii) the map $A \mapsto p(x, A)$ is a probability measure for all $x \in E$.

We interpret $p(x, A)$ as the probability that the transition ends in A, given that it started at x.

Exercise 3.7 If p and q are transition kernels on $(E, \mathcal{E})$, then the *convolution* $p * q$ is defined by setting

$$(p * q)(x, A) := \int_E p(x, dy) q(y, A) \qquad \text{for all } x \in E \text{ and } A \in \mathcal{E}.$$

Prove that $p * q$ is a transition kernel. Prove also that convolution is associative: if p, q and r are transition kernels then $(p * q) * r = p * (q * r)$.

Definition 3.8 A triangular collection $\{p_{s,t} : s, t \in \mathbb{R}_+, \ s \leqslant t\}$ of transition kernels is *consistent* if $p_{s,t} * p_{t,u} = p_{s,u}$ for all $s, t, u \in \mathbb{R}_+$ with $s \leqslant t \leqslant u$; that is,

$$p_{s,u}(x, A) = \int_E p_{s,t}(x, \mathrm{d}y) p_{t,u}(y, A) \qquad \text{for all } x \in E \text{ and } A \in \mathcal{E}. \tag{3.2}$$

Equation (3.2) is the *Chapman–Kolmogorov equation*. We interpret $p_{s,t}(x, A)$ as the probability of moving from x at time s to somewhere in A at time t.

Similarly, a one-parameter collection $\{p_t : t \in \mathbb{R}_+\}$ of transition kernels is *consistent* if $p_s \star p_t = p_{s+t}$ for all $s, t \in \mathbb{R}_+$. In this case, the Chapman–Kolmogorov equation becomes

$$p_{s+t}(x, A) = \int_E p_s(x, \mathrm{d}y) p_t(y, A) \qquad \text{for all } x \in E \text{ and } A \in \mathcal{E}. \tag{3.3}$$

We interpret $p_t(x, A)$ as the probability of moving from x into A in t units of time.

Definition 3.9 A family of E-valued random variables $X = (X_t)_{t \in \mathbb{R}_+}$ on a common probability space is a *Markov process* if there exists a consistent triangular collection of transition kernels such that

$$\mathbb{E}[1_A(X_t) \mid \sigma(X_r : 0 \leqslant r \leqslant s)] = p_{s,t}(X_s, A) \quad \text{almost surely}$$

for all $A \in \mathcal{E}$ and $s, t \in \mathbb{R}_+$ such that $s \leqslant t$.

The family X is a *time-homogeneous Markov process* if there exists a consistent one-parameter collection of transition kernels such that

$$\mathbb{E}[1_A(X_t) \mid \sigma(X_r : 0 \leqslant r \leqslant s)] = p_{t-s}(X_s, A) \quad \text{almost surely}$$

for all $A \in \mathcal{E}$ and $s, t \in \mathbb{R}_+$ such that $s \leqslant t$.

The connection between time-homogeneous Markov processes and semigroups is provided by the following definition and theorem.

Definition 3.10 A *Markov semigroup* is a contraction semigroup T on $B_b(E)$ such that, for all $t \in \mathbb{R}_+$, the bounded linear operator T_t is *positive*: whenever $f \in B_b(E)$ is such that $f \geqslant 0$, that is, $f(x) \in \mathbb{R}_+$ for all $x \in E$, then $T_t f \geqslant 0$. [Note that we impose no condition with respect to continuity at the origin.]

If T_t preserves the unit, that is, $T_t 1_E = 1_E$ for all $t \in \mathbb{R}_+$, then the Markov semigroup T is *conservative*.

Remark 3.11 Positive linear maps preserve order: if T is such a map and $f \leqslant g$, in the sense that $f(x) \leqslant g(x)$ for all $x \in E$, then $Tf \leqslant Tg$. The image of a real-valued function h under a positive linear map is real valued, since if h takes real values, then $h = h^+ - h^-$, where $h^+ : x \mapsto \max\{h(x), 0\}$ and $h^- := x \mapsto \max\{-h(x), 0\}$. Consequently, positive linear maps also commute with the conjugation, in the sense that $T\overline{f} = \overline{Tf}$.

Exercise 3.12 Suppose the mapping $T : B_b(E) \to B_b(E)$ is linear and positive. Show that $|Tf|^2 \leqslant T|f|^2 \, T1_E$ for all $f \in B_b(E)$, and deduce that T is bounded, with norm $\|T\| \leqslant \|T1_E\|$.

Proof If $f \in B_b(E)$, $x \in E$ and $\lambda \in \mathbb{R}$, then

$$0 \leqslant T\big(|f - \lambda(Tf)(x)|^2\big)(x) = \lambda^2 (T1_E)(x) \, |(Tf)(x)|^2 - 2\lambda |(Tf)(x)|^2 + (T|f|^2)(x).$$

Inspecting the discriminant of this polynomial in λ gives the first claim, and the second follows because.

$$|(T_t f)(x)|^2 \leqslant (T|f|^2)(x) \, (T1_E)(x) \leqslant \|f\|^2 (T1_E)^2(x) \leqslant \|f\|^2 \|T1_E\|^2.$$

$\square$

Theorem 3.13 *Let* $p = \{p_t : t \in \mathbb{R}_+\}$ *be a family of transition kernels. Setting*

$$(T_t f)(x) := \int_E p_t(x, \mathrm{d}y) f(y) \qquad \text{for all } f \in B_b(E) \text{ and } x \in E$$

defines a bounded linear operator on $B_b(E)$ *which is positive, contractive and unit preserving. Furthermore, the family* $T = (T_t)_{t \in \mathbb{R}_+}$ *is a Markov semigroup if and only if* p *is consistent.*

Proof If $f \in B_b(E)$, $x \in E$ and $s, t \in \mathbb{R}_+$, then the Chapman–Kolmogorov equation (3.3) implies that

$$(T_{s+t} f)(x) = \int_E p_{s+t}(x, \mathrm{d}z) f(z) = \int_E \int_E p_s(x, \mathrm{d}y) p_t(y, \mathrm{d}z) f(z)$$

$$= \int_E p_s(x, \mathrm{d}y)(T_t f)(y)$$

$$= \big(T_s(T_t f)\big)(x).$$

Verifying the remaining claims is left as an exercise. $\square$

If we have more structure on the semigroup T, then it is possible to provide a converse to Theorem 3.13. This will be sketched in the following section.

3.2 Feller Semigroups

Definition 3.14 Let the topological space E be locally compact. Then

$$C_0(E) := \{f : E \to \mathbb{C} \mid f \text{ is continuous and vanishes at infinity}\} \subseteq B_b(E)$$

is a Banach space when equipped with pointwise vector-space operations and the supremum norm. [A function $f : E \to \mathbb{C}$ *vanishes at infinity* if, for all $\varepsilon > 0$, there exists a compact set $K \subseteq E$ such that $|f(x)| < \varepsilon$ for all $x \in E \setminus K$.]

Exercise 3.15 Prove that $C_0(E)$ lies inside $B_b(E)$ and is indeed a Banach space. Prove that the multiplicative unit 1_E is an element of $C_0(E)$ if and only if E is compact.

Definition 3.16 A Markov semigroup T is *Feller* if the following conditions hold:

(i) $T_t\big(C_0(E)\big) \subseteq C_0(E)$ for all $t \in \mathbb{R}_+$ and

(ii) $\lim\limits_{t \to 0+} \|T_t f - f\| = 0$ for all $f \in C_0(E)$.

Remark 3.17 If a time-homogeneous Markov process X has Feller semigroup T, then

$$\mathbb{E}\big[f(X_{t+h}) - f(X_t) \mid \sigma(X_t)\big] = (T_h f - f)(X_t) = h\,(Af)(X_t) + o(h),$$

so the generator A describes the change in X over an infinitesimal time interval.

Definition 3.18 An $\mathbb{R}^d$-valued stochastic process $X = (X_t)_{t \in \mathbb{R}_+}$ is a *Lévy process* if and only if X

(i)   has independent increments, so that $X_t - X_s$ is independent of the past σ-algebra $\sigma(X_r : 0 \leqslant r \leqslant s)$ for all $s, t \in \mathbb{R}_+$ with $s \leqslant t$,

(ii) has stationary increments, so that $X_t - X_s$ has the same distribution as $X_{t-s} - X_0$, for all $s, t \in \mathbb{R}_+$ with $s \leqslant t$ and

(iii) is continuous in probability at the origin, so $\lim\limits_{t \to 0+} \mathbb{P}\big(|X_t - X_0| \geqslant \varepsilon\big) = 0$ for all $\varepsilon > 0$.

Remark 3.19 Lévy processes are well behaved; they have cádlág modifications, and such a modification is a semimartingale, for example.

Exercise 3.20 Prove that if X is a stochastic process with independent and stationary increments, and with cádlág paths, then X is continuous at the origin in probability.

Theorem 3.21 *Every Lévy process gives rise to a conservative Feller semigroup.*

Proof (Sketch Proof) For all $t \in \mathbb{R}_+$, define a transition kernel p_t by setting

$$p_t(x, A) := \mathbb{E}[1_A(X_t - X_0 + x)] \qquad \text{for all } x \in \mathbb{R}^d \text{ and Borel } A \subseteq \mathbb{R}^d.$$

If $s \in \mathbb{R}_+$, then

$$p_t(x, A) = \mathbb{E}[1_A(X_{s+t} - X_s + x)] = \mathbb{E}[1_A(X_{s+t} - X_s + x) \mid \mathcal{F}_s], \tag{3.4}$$

where $\mathcal{F}_s := \sigma(X_r : 0 \leqslant r \leqslant s)$; the first equality holds by stationarity and the second by independence. In particular,

$$p_t(X_s, A) = \mathbb{E}[1_A(X_{s+t}) \mid \mathcal{F}_s],$$

so X is a Markov process with transition kernels $\{p_t : t \in \mathbb{R}_+\}$ if these are consistent. For consistency, we use Theorem 3.13; let T be defined as there and note that

$$(T_t f)(x) = \int_E p_t(x, dy) f(y) = \mathbb{E}[f(X_t - X_0 + x)]. \tag{3.5}$$

From the previous working, it follows that

$$(T_t f)(x) = \mathbb{E}[f(X_{s+t} - X_s + x) \mid \mathcal{F}_s],$$

and replacing x with the $\mathcal{F}_s$-measurable random variable $X_s - X_0 + x$ gives that

$$(T_{s+t} f)(x) = \mathbb{E}[f(X_{s+t} - X_0 + x)] = \mathbb{E}[(T_t f)(X_s - X_0 + x)] = \big(T_s(T_t f)\big)(x),$$

as required. Equation (3.5) also shows that T is conservative.

If $f \in C_0(\mathbb{R}^d)$, then $x \mapsto f(X_t - X_0 + x) \in C_0(\mathbb{R}^d)$ almost surely, and therefore the Dominated Convergence Theorem gives that $T_t f \in C_0(\mathbb{R}^d)$.

For continuity, let $\varepsilon > 0$ and note that $f \in C_0(\mathbb{R}^d)$ is uniformly continuous, so there exists $\delta > 0$ such that $|f(x) - f(y)| < \varepsilon$ whenever $|x - y| < \delta$. Hence

$$\|T_t f - f\| \leqslant \sup_{x \in \mathbb{R}^d} \mathbb{E}\big[|f(X_t - X_0 + x) - f(x)|\big]$$

$$= \sup_{x \in \mathbb{R}^d} \Big(\mathbb{E}\big[1_{|X_t - X_0| < \delta} |f(X_t - X_0 + x) - f(x)|\big]$$

$$+ \mathbb{E}\big[1_{|X_t - X_0| \geqslant \delta} |f(X_t - X_0 + x) - f(x)|\big]\Big)$$

$$\leqslant \varepsilon + 2\|f\| \mathbb{P}\big(|X_t - X_0| \geqslant \delta\big)$$

$$\to \varepsilon \qquad \text{as } t \to 0+.$$

$\square$

Theorem 3.22 *Let T be a conservative Feller semigroup. If the state space E is metrisable, then there exists a time-homogeneous Markov process which gives rise to T.*

Proof (Sketch Proof) For all $t \in (0, \infty)$, let

$$p_t(x, A) := (T_t 1_A)(x) \quad \text{for all } x \in E \text{ and } A \in \mathcal{E}.$$

Then p_t is readily verified to be a transition kernel.

Let μ be a probability measure on E. If $t_n \geqslant \cdots \geqslant t_1 \geqslant 0$ and $A_1, \ldots A_n \in \mathcal{E}$, then

$$p_{t_1, \ldots, t_n}(A_1 \times \cdots \times A_n) = \int_E \mu(\mathrm{d}x_0) \int_{A_1} p_{t_1}(x_0, \mathrm{d}x_1) \ldots \int_{A_n} p_{t_n - t_{n-1}}(x_{n-1}, \mathrm{d}x_n).$$

By the Chapman–Kolmogorov equation (3.3), these finite-dimensional distributions form a projective family. The Daniell–Kolmogorov extension theorem now yields a probability measure on the product space

$$\Omega := E^{\mathbb{R}_+} = \{\omega = (\omega_t)_{t \in \mathbb{R}_+} : \omega_t \in E \text{ for all } t \in \mathbb{R}_+\}$$

such the coordinate projections $X_t : \Omega \to E$; $\omega \mapsto \omega_t$ form a time-homogeneous Markov process X with associated semigroup T. $\square$

Example 3.23 (Uniform Motion) If $E = \mathbb{R}$ and $X_t = X_0 + t$ for all $t \in \mathbb{R}_+$, then

$$(T_t f)(x) = f(x + t) = \int_{\mathbb{R}} p_t(x, \mathrm{d}y) f(y) \qquad \text{for all } f \in C_0(\mathbb{R}) \text{ and } x \in \mathbb{R},$$

where the transition kernel $p_t : (x, A) \mapsto \delta_{x+t}(A)$. It follows that X gives rise to a Feller semigroup with generator A such that $Af = f'$ whenever $f \in \mathrm{dom}\, A$.

Example 3.24 (Brownian Motion) If $E = \mathbb{R}$ and X is a standard Brownian motion, then Itô's formula gives that

$$f(X_t) = f(X_0) + \int_0^t f'(X_s) \, \mathrm{d}X_s + \frac{1}{2} \int_0^t f''(X_s) \, \mathrm{d}s \qquad \text{for all } f \in C^2(\mathbb{R}).$$

It follows that the Lévy process X has a Feller semigroup with the generator A such that $Af = \frac{1}{2}f''$ for all $f \in C^2(\mathbb{R}) \cap \mathrm{dom}\, A$. [Informally,

$$t^{-1}\big(\mathbb{E}[f(X_t) \mid X_0 = x] - f(x)\big) = \frac{1}{2t} \int_0^t \mathbb{E}[f''(X_s)|X_0 = x] \, \mathrm{d}s \to \frac{1}{2} f''(x)$$

as $t \to 0+$.]

Example 3.25 (Poisson Process) If $E = \mathbb{R}$ and X is a homogeneous Poisson process with unit intensity and unit jumps, then

$$\mathbb{E}[f(X_t)|X_0 = x] = e^{-t} \sum_{n=0}^{\infty} \frac{t^n}{n!} f(x + n) \qquad \text{for all } t \in \mathbb{R}_+.$$

Hence the Lévy process X has a Feller semigroup with the bounded generator A such that $(Af)(x) = f(x + 1) - f(x)$ for all $x \in \mathbb{R}$ and $f \in C_0(\mathbb{R})$. [To see this,

note that

$$\frac{(T_t f - f)(x)}{t} = \frac{e^{-t} - 1}{t} f(x) + e^{-t} f(x+1) + O(t) \qquad \text{as } t \to 0+,$$

uniformly for all $x \in \mathbb{R}$.]

The following exercise and theorem show that it is possible to move from the non-conservative to the conservative setting, and from a locally compact state space to a compact one.

Exercise 3.26 Let $\mathcal{T}$ be a locally compact topology on E and let ∞ denote a point not in E. Prove that $\widehat{E} := E \cup \{\infty\}$ is compact when equipped with the topology

$$\widehat{\mathcal{T}} := \mathcal{T} \cup \big\{ (E \setminus K) \cup \{\infty\} : K \in \mathcal{T} \text{ is compact} \big\},$$

and that $\widehat{\mathcal{T}}$ is Hausdorff if and only if $\mathcal{T}$ is. [This is the *Alexandrov one-point compactification*.] Prove further that $C_0(E)$ has co-dimension one in $C(\widehat{E})$.

Theorem 3.27 *Let T be a Feller semigroup with locally compact state space E. If*

$$\widehat{T}_t f := f(\infty) + T_t\big(f|_E - f(\infty)\big) \qquad \text{for all } t \in \mathbb{R}_+ \text{ and } f \in B_b(\widehat{E}),$$

then $\widehat{T} = \big(\widehat{T}_t\big)_{t \in \mathbb{R}_+}$ is a conservative Feller semigroup with compact state space $\widehat{E}$.

Proof Fix $t \in \mathbb{R}_+$. The hardest step is to prove that $\widehat{T}_t$ is positive, that is, if $\lambda \in \mathbb{R}_+$ and $g \in B_b(E)$ are such that $\lambda + g(x) \geqslant 0$ for all $x \in E$, then $\lambda + (T_t g)(x) \geqslant 0$ for all $x \in E$. Note that g is real valued, and T_t maps real-valued functions to real-valued functions, by positivity. Let the function $g^- := x \mapsto \max\{-g(x), 0\}$ and note that $\lambda \geqslant g^-(x)$ for all $x \in E$. Hence

$$(T_t g^-)(x) \leqslant \|T_t g^-\| \leqslant \|g^-\| \leqslant \lambda$$

and $(T_t g)(x) \geqslant (-T_t g^-)(x) \geqslant -\lambda$, as required.

It is immediate that $\widehat{T}_t$ preserves the unit, so $\widehat{T}_t$ is contractive, by Exercise 3.12. The remaining claims are straightforward to verify. $\qquad\square$

3.3 The Hille–Yosida–Ray Theorem

As noted above, it can be difficult to show that the hypotheses of the Hille–Yosida theorem, Theorem 2.49, hold. The Lumer–Phillips theorem gives an alternative for contraction semigroups, via the notion of dissipativity. Here, we will show that the additional structure available for Feller semigroups gives another possible approach.

Throughout this subsection, E denotes a locally compact Hausdorff space. Here, a *Feller semigroup* on $C_0(E)$ means a strongly continuous contraction semigroup on $C_0(E)$ composed of positive operators. This is the restriction to $C_0(E)$ of the Feller semigroups considered above.

Let

$$C_0(E; \mathbb{R}) := \{ f : E \to \mathbb{R} \mid f \in C_0(E) \}$$

denote the real subspace of $C_0(E)$ containing those functions which take only real values.

Definition 3.28 A linear operator A in $C_0(E)$ is *real* if and only if

(i) $\overline{f} \in \operatorname{dom} A$ whenever $f \in \operatorname{dom} A$, so that the domain of A is closed under conjugation, and

(ii) $\overline{Af} = A\overline{f}$ for all $f \in \operatorname{dom} A$, so that A commutes with the conjugation.

Exercise 3.29 Show that (i) and (ii) are equivalent to

(i) $f + ig \in \operatorname{dom} A$ implies $f, g \in \operatorname{dom} A$ whenever $f, g \in C_0(E; \mathbb{R})$, and

(ii) $A\big(\operatorname{dom} A \cap C_0(E; \mathbb{R})\big) \subseteq C_0(E; \mathbb{R})$,

respectively.

Exercise 3.30 Prove that T is real whenever T is positive.

Prove further that if $T = (T_t)_{t \in \mathbb{R}_+}$ is a Feller semigroup on $C_0(E)$ and T_t is real for all $t \in \mathbb{R}_+$ then the generator A of T is real.

Proof The first claim is an immediate consequence of Remark 3.11.

For the second, suppose A is the generator of the Feller semigroup T on $C_0(E)$, with each T_t real, and let $f \in \operatorname{dom} A$. Then, since conjugation is isometric, if $t > 0$, then

$$\| t^{-1}(T_t f - t) - Af \| = \| t^{-1}(\overline{T_t f} - \overline{f}) - \overline{Af} \| = \| t^{-1}(T_t \overline{f} - \overline{f}) - \overline{Af} \|,$$

and so $\overline{f} \in \operatorname{dom} A$, with $A\overline{f} = \overline{Af}$. The result follows. $\qquad \square$

Definition 3.31 A linear operator A in $C_0(E)$ satisfies the *positive maximum principle* if, whenever $f \in \operatorname{dom} A \cap C_0(E; \mathbb{R})$ and $x_0 \in E$ are such that $f(x_0) = \| f \|$, it holds that $(Af)(x_0) \leqslant 0$.

Theorem 3.32 (Hille–Yosida–Ray) *A closed, densely defined operator A in $C_0(E)$ is the generator of a Feller semigroup on $C_0(E)$ if and only if A is real and satisfies the positive maximum principle, and $\lambda I - A$ is surjective for some $\lambda > 0$*

Proof Suppose first that A generates a Feller semigroup on $C_0(E)$. By the Lumer–Phillips theorem, Theorem 2.58, and Exercise 3.30, it suffices to prove that A satisfies the positive maximum principle. For this, let $f \in \operatorname{dom} A \cap C_0(E; \mathbb{R})$ and

$x_0 \in E$ be such that $f(x_0) = \|f\|$. Setting $f^+ := x \mapsto \max\{f(x), 0\}$, we see that

$$(T_t f)(x_0) \leqslant (T_t f^+)(x_0) \leqslant \|T_t f^+\| \leqslant \|f^+\| = f(x_0).$$

Thus

$$(Af)(x_0) = \lim_{t \to 0+} \frac{(T_t f - f)(x_0)}{t} \leqslant 0.$$

Conversely, suppose A is real and satisfies the positive maximum principle. Given any $f \in \operatorname{dom} A$, there exist $x_0 \in E$ and $\theta \in \mathbb{R}$ such that $e^{i\theta} f(x_0) = \|f\|$. The real-valued function $g := \operatorname{Re} e^{i\theta} f \in \operatorname{dom} A$, since A is real, and $\|f\| = g(x_0) \leqslant \|g\| \leqslant \|f\|$, so $\operatorname{Re}(Ae^{i\theta} f)(x_0) = (Ag)(x_0) \leqslant 0$, by the positive maximum principle. If $\lambda > 0$, then

$$\|(\lambda I - A)f\| = \|(\lambda I - A)e^{i\theta} f\| \geqslant |\lambda e^{i\theta} f(x_0) - (Ae^{i\theta} f)(x_0)|$$

$$\geqslant \operatorname{Re} \lambda e^{i\theta} f(x_0) - \operatorname{Re}(Ae^{i\theta} f)(x_0) \geqslant \lambda\|f\|,$$

so A is dissipative, by Lemma 2.56, and $\lambda I - A$ is injective. In particular, T is a strongly continuous contraction semigroup, by the Lumer–Phillips theorem.

To prove that each T_t is positive, let $\lambda > 0$ be such that $\lambda I - A$ is surjective, so invertible, let $f \in C_0(E)$ be non-negative, and consider $g = (\lambda I - A)^{-1} f \in C_0(E)$. Either g does not attain its infimum, in which case $g \geqslant 0$ because g vanishes at infinity, or there exists $x_0 \in E$ such that $g(x_0) = \inf\{g(x) : x \in E\}$. Then

$$\lambda g - Ag = (\lambda I - A)g = f \iff \lambda g - f = Ag,$$

so $\lambda g(x_0) - f(x_0) = (Ag)(x_0) \geqslant 0$, by the positive maximum principle applied to $-g$. Thus if $x \in E$, then

$$\lambda g(x) \geqslant \lambda g(x_0) \geqslant f(x_0) \geqslant 0,$$

which shows that $\lambda(\lambda I - A)^{-1}$ is positive and therefore so is $(\lambda I - A)^{-1}$. Finally, Theorem 2.46 gives that

$$T_t f = \lim_{n \to \infty} (I - tn^{-1}A)^{-n} f$$

$$= \lim_{n \to \infty} (t^{-1}n)^n (t^{-1}nI - A)^{-n} f \qquad \text{for all } f \in C_0(E), \qquad (3.6)$$

so each T_t is positive also.   □

Exercise 3.33 Prove that if the operator A is real then its resolvent $(\lambda I - A)^{-1}$ is real for all $\lambda \in \mathbb{R} \setminus \sigma(A)$. Deduce with the help of Theorem 2.46 that the Feller semigroup T is real if its generator A is.

Proof Suppose A is real and $\lambda \in \mathbb{R} \setminus \sigma(A)$. If $f \in C_0(E)$, then $f = (\lambda I - A)g$ for some $g \in C_0(E)$, and

$$\overline{f} = \overline{(\lambda I - A)g} = \lambda \overline{g} - \overline{Ag} = (\lambda I - A)\overline{g}.$$

Hence

$$\overline{(\lambda I - A)^{-1} f} = \overline{g} = (\lambda I - A)^{-1} \overline{f},$$

as required. Since conjugation is isometric, the deduction is immediate.   □

Example 3.34 Let the linear operator A be defined by setting

$$\operatorname{dom} A := \left\{ f \in C_0(\mathbb{R}) \cap C^2(\mathbb{R}) : f'' \in C_0(\mathbb{R}) \right\} \quad \text{and} \quad Af = \frac{1}{2} f''.$$

It is a familiar result from elementary calculus that A satisfies the positive maximum principle

Remark 3.35 Courrège has classified the linear operators in $C_0(\mathbb{R}^d)$ with domains containing $C_c^\infty(\mathbb{R}^d)$ which satisfy the positive maximum principle. See [3, §3.5.1] and references therein.

4   Quantum Feller Semigroups

To move beyond the classical, we need to replace the commutative domain $C_0(E)$ with the correct non-commutative generalisation. This is what we introduce in the following section.

4.1   *C* Algebras*

Definition 4.1 A *Banach algebra* is a complex Banach space and simultaneously a complex associative algebra: it has an associative multiplication compatible with the vector-space operators and the norm, which is submultiplicative. If the Banach algebra is *unital*, so that it has a multiplicative identity 1, called its *unit*, then we require the norm $\|1\|$ to be 1.

An *involution* on a Banach algebra is an isometric conjugate-linear map which reverses products and is self-inverse.

A Banach algebra with involution A is a *C* algebra* if and only if the *C* identity* holds:

$$\|a^*a\| = \|a\|^2 \qquad \text{for all } a \in \mathsf{A}.$$

Remark 4.2 The C^* identity connects the algebraic and analytic structures in a very rigid way. For example, there exists at most one norm for which an associative algebra is a C^* algebra, and $*$-homomorphisms between C^* algebras are automatically contractive [30, Proposition I.5.2].

Theorem 4.3 (Gelfand) *Every commutative C^* algebra is isometrically isomorphic to $C_0(E)$, where E is a locally compact Hausdorff space. The algebra is unital if and only if E is compact, in which case $C_0(E) = C(E)$.*

Theorem 4.4 (Gelfand–Naimark) *Any C^* algebra is isometrically $*$-isomorphic to a norm-closed $*$-subalgebra of $B(\mathsf{H})$ for some Hilbert space H, a so-called* concrete C^* algebra.

Remark 4.5 Let A be a C^* algebra. Given any $n \in \mathbb{N}$, let $M_n(\mathsf{A})$ be the complex algebra of $n \times n$ matrices with entries in A, equipped with the usual algebraic operations. By the Gelfand–Naimark theorem, we may assume that $\mathsf{A} \subseteq B(\mathsf{H})$ for some Hilbert space H, and so $M_n(\mathsf{A}) \subseteq B(\mathsf{H}^n)$, where matrices of operators act in the usual manner on column vectors with entries in H. We equip $M_n(\mathsf{A})$ with the restriction of the operator norm on $B(\mathsf{H}^n)$, and then $M_n(\mathsf{A})$ becomes a C^* algebra.

Remark 4.5 is the root of the theory of operator spaces [10, 24].

Definition 4.6 A unital concrete C^* algebra $\mathsf{A} \subseteq B(\mathsf{H})$ is a *von Neumann algebra* if and only if any of the following equivalent conditions hold.

(i) Closure in the strong operator topology: if the net $(a_i) \subseteq \mathsf{A}$ and $a \in B(\mathsf{H})$ are such that $a_i v \to av$ for all $v \in \mathsf{H}$, then $a \in \mathsf{A}$.

(ii) Closure in the weak operator topology: if the net $(a_i) \subseteq \mathsf{A}$ and $a \in B(\mathsf{H})$ are such that $\langle v, a_i v \rangle \to \langle v, av \rangle$ for all $v \in \mathsf{H}$, then $a \in \mathsf{A}$.

(iii) Equality with its bicommutant: letting

$$S' := \{a \in \mathsf{A} : ab = ba \text{ for all } b \in S\}$$

denote the commutant of $S \subseteq \mathsf{A}$, then $\mathsf{A}'' := (\mathsf{A}')' = \mathsf{A}$ [von Neumann].

(iv) Existence of a predual: there exists a Banach space A_* with $(\mathsf{A}_*)^* = \mathsf{A}$ [Sakai].

Sakai's characterisation (iv) prompts consideration of the predual of $B(\mathsf{H})$. The predual A_* is naturally a subspace of A^*, and a bounded linear functional ϕ on $B(\mathsf{H})$ is an element of $B(\mathsf{H})_*$ if and only it is σ-*weakly continuous*: there exist square-summable sequences $(u_n)_{n=1}^\infty$ and $(v_n)_{n=1}^\infty \subseteq \mathsf{H}$ such that

$$\sum_{n=1}^\infty \left(\|u_n\|^2 + \|v_n\|^2 \right) < \infty \quad \text{and} \quad \phi(T) = \sum_{n=1}^\infty \langle u_n, T v_n \rangle \qquad \text{for all } T \in B(\mathsf{H}).$$

$$(4.1)$$

This yields a fifth characterisation of von Neumann algebras.

(v) Closure in the σ-weak topology: if the net $(a_i) \subseteq$ A and $a \in B(H)$ are such that $\phi(a_i) \to \phi(a)$ for all $\phi \in B(H)_*$, then $a \in$ A.

The predual A_* consists of all those bounded linear functionals on A which are continuous in the σ-weak topology; equivalently, they are the restriction to A of elements of $B(H)_*$ as described in (4.1).

Example 4.7 Recall from Example 2.15 that $L^\infty(\Omega, \mathcal{F}, \mu) \cong \left(L^1(\Omega, \mathcal{F}, \mu)\right)^*$, and so every L^∞ space is a commutative von Neumann algebra. Furthermore, every commutative von Neumann algebra is isometrically $*$-isomorphic to $L^\infty(\Omega, \mathcal{F}, \mu)$ for some locally compact Hausdorff space Ω and positive Radon measure μ; see [30, Theorem III.1.18].

4.2 Positivity

Definition 4.8 In a C^* algebra A we have the notion of *positivity*: we write $a \geqslant 0$ if and only if there exists $b \in$ A such that $a = b^*b$. The set of positive elements in A is denoted by A_+, is closed in the norm topology and is a *cone*: it is closed under addition and multiplication by non-negative scalars. Note that a positive element is self-adjoint.

This notion of positivity agrees with that encountered previously.

Lemma 4.9 *Let $T \in B(H)$ be such that $\langle v, Tv \rangle \geqslant 0$ for all $v \in$ H. There exists a unique operator $S \in B(H)$ such that $\langle v, Sv \rangle \geqslant 0$ for all $v \in$ H, and $S^2 = T$. Furthermore, S is the limit of a sequence of polynomials in T with no constant term.*

Proof This may be established with the assistance of the Maclaurin series for the function $z \mapsto (1 - z)^{1/2}$. See [25, Theorem VI.9] for the details.   □

Corollary 4.10 *If $a \in A_+$, then there exists a unique element $a^{1/2} \in A_+$, the square root of a, such that $(a^{1/2})^2 = a$. The square root $a^{1/2}$ lies in the closed linear subspace of A spanned by the set of monomials $\{a^n : n \in \mathbb{N}\}$.*

Proof This is a straightforward exercise.   □

Exercise 4.11 Prove that $f \in C_0(E)_+$ if and only if $f(x) \geqslant 0$ for all $x \in E$. Prove also that if the C^* algebra $A \subseteq B(H)$, where H is a Hilbert space, then $a \in A_+$ if and only if $\langle v, av \rangle \geqslant 0$ for all $v \in$ H. [The existence of square roots is crucial for both parts.]

Proposition 4.12 *Let A by a C^* algebra. Then any element $a \in$ A may be written in the form $(a_1 - a_2) + i(a_3 - a_4)$, where $a_1, \ldots, a_4 \in A_+$.*

Proof The self-adjoint elements $\operatorname{Re} a := (a + a^*)/2$ and $\operatorname{Im} a := (a - a^*)/(2\mathrm{i})$ are such that $a = \operatorname{Re} a + \mathrm{i} \operatorname{Im} a$. Thus it suffices to show that any self-adjoint element of A is the difference of two positive elements.

Let $a \in \mathsf{A}$ be self-adjoint and let A_0 be the closed linear subspace of A spanned by the set of monomials $\{a^n : n \in \mathbb{N}\}$. As A_0 is a commutative C^* algebra, Theorem 4.3 gives an isometric $*$-isomorphism $j : \mathsf{A}_0 \to C_0(E)$, where E is a locally compact Hausdorff space. Then $f := j(a)$ is real valued, so

$$f^+ := x \mapsto \max\{f(x), 0\} \qquad \text{and} \qquad f^- := x \mapsto \max\{-f(x), 0\}$$

are well-defined elements of $C_0(E)_+$ such that $f = f^+ - f^-$. Hence $a = a^+ - a^-$, where $a^+ := j^{-1}(f^+)$ and $a^- := j^{-1}(f^-)$ are positive, as desired. $\qquad\square$

Remark 4.13 The proof of Proposition 4.12 shows that if $a \in \mathsf{A}$ is self-adjoint, then there exist $a^+, a^- \in \mathsf{A}_+$ such that $a = a^+ - a^-$ and $a^+ a^- = 0$.

Definition 4.14 The positive cone provides a partial order on the set of self-adjoint elements of A. Given elements $a, b \in \mathsf{A}$, we write $a \leqslant b$ if and only if $a = a^*$, $b = b^*$ and $b - a \in \mathsf{A}_+$.

This order respects the norm.

Proposition 4.15 *Let $a, b \in \mathsf{A}_+$ be such that $a \leqslant b$. Then $\|a\| \leqslant \|b\|$.*

Proof Suppose without loss of generality that $\mathsf{A} \subseteq B(\mathsf{H})$. Then $a \leqslant b \leqslant \|b\| I$, by transitivity, Exercise 4.11 and the Cauchy–Schwarz inequality. If A_0 denotes the unital commutative C^* algebra generated by the set of monomials $\{a^n : n \in \mathbb{Z}_+\}$, then Theorem 4.3 gives an isometric $*$-isomorphism $j : \mathsf{A}_0 \to C(E)$, where E is a compact Hausdorff space. Hence

$$0 \leqslant j(\|b\| I - a)(x) = \|b\| - j(a)(x) \qquad \text{for all } x \in E,$$

so $0 \leqslant j(a)(x) \leqslant \|b\|$ for all such x and $\|a\| = \|j(a)\|_\infty \leqslant \|b\|$, as claimed. $\qquad\square$

Exercise 4.16 Prove that if $a \in \mathsf{A}_+$ and $n \in \mathbb{Z}_+$, then $\|a^n\| = \|a\|^n$. [Hint: work as in the proof of Proposition 4.15.]

Definition 4.17 A linear map $\Phi : \mathsf{A} \to \mathsf{B}$ between C^* algebras is *positive* if and only if $\Phi(\mathsf{A}_+) \subseteq \mathsf{B}_+$.

Note that any algebra $*$-homomorphism is positive; this fact has been utilised in the proof of Proposition 4.15.

Corollary 4.18 *Let $\Phi : \mathsf{A} \to \mathsf{B}$ be a positive linear map between C^* algebras. Then*

(i) *the map Φ commutes with the involution, so that $\Phi(a^*) = \Phi(a)^*$ for all $a \in \mathsf{A}$, and*

(ii) *the map Φ is bounded.*

Proof Part (i) is an exercise.

For (ii), it suffices to prove that Φ is bounded on A_+; suppose otherwise for contradiction. For all $n \in \mathbb{N}$, let $a_n \in A_n$ be such that $\|a_n\| = 1$ and $\|\Phi(a_n)\| > 3^n$. If $a := \sum_{n \geqslant 1} 2^{-n} a_n \in A_+$, then $a \geqslant 2^{-n} a_n$ for all $n \in \mathbb{N}$. Hence $\Phi(a) \geqslant 2^{-n} \phi(a_n)$ and $\|\phi(a)\| \geqslant 2^{-n} \|\Phi(a_n)\| > (3/2)^n$, by Proposition 4.15, which is a contradiction for sufficiently large n. $\qquad\square$

We will now begin to investigate the generators of positive semigroups, following in the footsteps of Evans and Hanche-Olsen [12].

Theorem 4.19 *Let* $T = (T_t)_{t \in \mathbb{R}_+}$ *be a uniformly continuous one-parameter semigroup on the* C^* *algebra* A. *If* T_t *is positive for all* $t \in \mathbb{R}_+$, *then the semigroup generator* $\mathcal{L}$ *is bounded and* $*$-*preserving.*

Proof The boundedness of $\mathcal{L}$ follows immediately from Theorem 2.23, and if $a \in$ A, then

$$\mathcal{L}(a)^* = \lim_{t \to 0+} t^{-1}(T_t(a) - a)^* = \lim_{t \to 0+} t^{-1}(T_t(a^*) - a^*) = \mathcal{L}(a^*),$$

by continuity of the involution and the fact that positive maps are $*$-preserving. $\quad\square$

The following result is a variation on [12, Theorem 2]. The proof exploits an idea of Fagnola [14, Proof of Proposition 3.10].

Theorem 4.20 *Let* $\mathcal{L}$ *be a* $*$-*preserving bounded linear map on the* C^* *algebra* A. *The following are equivalent.*

(i) *If* $a, b \in A_+$ *are such that* $ab = 0$, *then* $a\mathcal{L}(b)a \geqslant 0$.

(ii) $(\lambda I - \mathcal{L})^{-1}$ *is positive for all sufficiently large* $\lambda > 0$.

(iii) $T_t = \exp(t\mathcal{L})$ *is positive for all* $t \in \mathbb{R}_+$.

Proof Suppose (i) holds; we will show that $(\lambda I - \mathcal{L})^{-1}$ is positive if $\lambda > \|\mathcal{L}\|$. It suffices to take $a \in$ A such that $(\lambda I - \mathcal{L})(a)$ is positive, and prove that $a \in A_+$. Note that a is self-adjoint, so Remark 4.13 gives b and $c \in A_+$ with $a = b - c$ and $bc = 0$. Thus (ii) holds if $c = 0$.

The condition $bc = 0$ implies that $b^{1/2}c = 0$, so (i) gives that $c\mathcal{L}(b)c \geqslant 0$. Hence

$$0 \leqslant c^*(\lambda a - \mathcal{L}(a))c = \lambda c(b - c)c - c\mathcal{L}(b)c + c\mathcal{L}(c)c \leqslant -\lambda c^3 + c\mathcal{L}(c)c,$$

and therefore $0 \leqslant \lambda c^3 \leqslant c\mathcal{L}(c)c$. It follows that $\lambda \|c\|^3 = \lambda \|c^3\| \leqslant \|\mathcal{L}\| \|c\|^3$, which holds only when $c = 0$, as required.

That (ii) and (iii) are equivalent is a consequence of Theorems 2.45 and 2.46. To see that (iii) implies (i), note that if $a, b \in A_+$ are such that $ab = 0$, then

$$0 \leqslant t^{-1}aT_t(b)a = t^{-1}a(b + t\mathcal{L}(b) + O(t))a = a\mathcal{L}(b)a + O(t) \to a\mathcal{L}(b)a$$

as $t \to 0+$. $\qquad\square$

In the quantum world, we can go beyond positivity to find a stronger notion, complete positivity, which is of great importance to the theories of open quantum systems and quantum information.

4.3   Complete Positivity

Recall from Remark 4.5 that matrix algebras over C^* algebras are also C^* algebras.

Definition 4.21 Let $n \in \mathbb{N}$. A linear map $\Phi : A \to B$ between C^* algebras is *n-positive* if and only if the ampliation

$$\Phi^{(n)} : M_n(A) \to M_n(B); \quad (a_{ij})_{i,j=1}^n \mapsto \left(\Phi(a_{ij})\right)_{i,j=1}^n$$

is positive. If Φ is *n*-positive for all $n \in \mathbb{N}$, then Φ is *completely positive*.

Remark 4.22 Choi [6] produced examples of maps which are *n*-positive but not $n + 1$-positive.

Exercise 4.23 Let $n \in \mathbb{N}$ and let $T = (T_t)_{t \in \mathbb{R}_+}$ be a one-parameter semigroup on the C^* algebra A. Prove that $T^{(n)} = (T_t^{(n)})_{t \in \mathbb{R}_+}$ is a one-parameter semigroup on $M_n(A)$, Prove further that if T is uniformly continuous, with generator $\mathcal{L}$, then $T^{(n)}$ is also uniformly continuous, with generator $\mathcal{L}^{(n)}$.

Proposition 4.24 (Paschke [23]) *Let* $A = (a_{ij})_{i,j=1}^n \in M_n(A)$, *where* A *is a* C^* *algebra. The following are equivalent.*

  (i) *The matrix* $A \in M_n(A)_+$.
 (ii) *The matrix* A *may be written as the sum of at most n matrices of the form* $(b_i^* b_j)_{i,j=1}^n$, *where* $b_1, \ldots, b_n \in A$.
(iii) *The sum* $\sum_{i,j=1}^n c_i^* a_{ij} c_j \in A_+$ *for any* $c_1, \ldots, c_n \in A$.

Proof To see that (iii) implies (i), we use the fact that any C^* algebra has a faithful representation which is a direct sum of cyclic representations [30, Theorem III.2.4]. Thus we may assume without loss of generality that $A \subseteq B(H)$ and there exists a unit vector $u \in H$ such that $\{au : a \in H\}$ is dense in H.

Given this and Exercise 4.11, let $c_1, \ldots, c_n \in A$. Then (iii) implies that

$$0 \leqslant \sum_{i,j=1}^n \langle u, c_i^* a_{ij} c_j u \rangle_H = \langle v, Av \rangle_{H^n},$$

where $v = (c_1 u, \ldots, c_n u)^T \in H^n$. Vectors of this form are dense in H^n as $c_1, \ldots, c_n$ vary over A, so the result follows by another application of Exercise 4.11.

The other implications are straightforward to verify.   □

Exercise 4.25 Let $n \in \mathbb{N}$. Use Proposition 4.24 to prove that a linear map $\Phi : \mathsf{A} \to \mathsf{B}$ between C^* algebras is n-positive if and only if

$$\sum_{i,j=1}^{n} b_i^* \Phi(a_i^* a_j) b_j \geqslant 0$$

for all $a_1, \ldots, a_n \in \mathsf{A}$ and $b_1, \ldots, b_n \in \mathsf{B}$. Deduce that any $*$-homomorphism between C^* algebras is completely positive, as is any map of the form

$$B(\mathsf{K}) \to B(\mathsf{H}); \quad a \mapsto T^* a T, \qquad \text{where } T \in B(\mathsf{H}; \mathsf{K}).$$

Theorem 4.26 *A positive linear map* $\Phi : \mathsf{A} \to \mathsf{B}$ *between* C^* *algebras is completely positive if* A *is commutative or* B *is commutative.*

Proof The first result is due to Stinespring [29] and the second to Arveson [4]. We will prove the latter.

We may suppose that $\mathsf{B} = C_0(E)$, where E is a locally compact Hausdorff space, by Theorem 4.3. If $a_1, \ldots, a_n \in \mathsf{A}$, $b_1, \ldots, b_n \in B$ and $x \in E$, then

$$\left(\sum_{i,j=1}^{n} b_i^* \Phi(a_i^* a_j) b_j \right)(x) = \sum_{i,j=1}^{n} \overline{b_i(x)} \Phi(a_i^* a_j)(x) b_j(x) = \Phi\big(c(x)^* c(x)\big)(x) \geqslant 0,$$

where $c(x) := \sum_{i=1}^{n} b_i(x) a_i \in \mathsf{A}$. Exercises 4.11 and 4.25 give the result. $\qquad \square$

Definition 4.27 A map $\Phi : \mathsf{A} \to \mathsf{B}$ between unital algebras is *unital* if $\Phi(1_\mathsf{A}) = 1_\mathsf{B}$, where 1_A and 1_B are the multiplicative units of A and B, respectively.

Theorem 4.28 (Kadison) *A* 2-*positive unital linear map* $\Phi : \mathsf{A} \to \mathsf{B}$ *between unital* C^* *algebras is such that*

$$\Phi(a)^* \Phi(a) \leqslant \Phi(a^* a) \qquad \text{for all } a \in \mathsf{A}. \tag{4.2}$$

Proof Note first that if $a \in \mathsf{A}$ then

$$A := \begin{bmatrix} 1 & a \\ a^* & a^* a \end{bmatrix} = \begin{bmatrix} 1 & a \\ 0 & 0 \end{bmatrix}^* \begin{bmatrix} 1 & a \\ 0 & 0 \end{bmatrix} \geqslant 0,$$

so

$$0 \leqslant \Phi^{(2)}(A) = \begin{bmatrix} 1 & \Phi(a) \\ \Phi(a)^* & \Phi(a^* a) \end{bmatrix}.$$

Suppose without loss of generality that $\mathsf{B} \subseteq B(\mathsf{H})$ for some Hilbert space H, and note that, by Exercise 4.11, if $u \in \mathsf{H}$ and

$$v := \begin{bmatrix} -\Phi(a)u \\ u \end{bmatrix} \in \mathsf{H}^2 \quad \text{then} \quad 0 \leqslant \langle v, \Phi^{(2)}(A)v \rangle = \langle u, (\Phi(a^*a) - \Phi(a)^*\Phi(a))u \rangle.$$

As u is arbitrary, the claim follows. $\qquad\square$

Remark 4.29 The inequality (4.2) is known as the *Kadison–Schwarz inequality*.

Exercise 4.30 Show that the inequality (4.2) holds if Φ is required only to be positive as long as a is *normal*, so that $a^*a = aa^*$. [Hint: use Theorem 4.26.]

4.4 Stinespring's Dilation Theorem

Exercise 4.25 gives two classes of completely positive maps. The following result makes clear that these are, in a sense, exhaustive.

Theorem 4.31 (Stinespring [29]) *Let* $\Phi : \mathsf{A} \to B(\mathsf{H})$ *be a linear map, where* A *is a unital* C^* *algebra and* H *is a Hilbert space. Then* Φ *is completely positive if and only if there exists a Hilbert space* K, *a unital* $*$-*homomorphism* $\pi : \mathsf{A} \to B(\mathsf{K})$ *and a bounded operator* $T : \mathsf{H} \to \mathsf{K}$ *such that*

$$\Phi(a) = T^*\pi(a)T \qquad (a \in \mathsf{A}).$$

Proof One direction is immediate. For the other, let $\mathsf{K}_0 := \mathsf{A} \otimes \mathsf{H}$ be the algebraic tensor product of A with H, considered as complex vector spaces. Define a sesquilinear form on K_0 such that

$$\langle a \otimes u, b \otimes v \rangle = \langle u, \Phi(a^*b)v \rangle_{\mathsf{H}} \qquad \text{for all } a, b \in \mathsf{A} \text{ and } u, v \in \mathsf{H}.$$

It is an exercise to check that this form is positive semidefinite, using the assumption that Φ is completely positive, and that the kernel

$$\mathsf{K}_{00} := \{x \in \mathsf{K}_0 : \langle x, x \rangle = 0\}$$

is a vector subspace of K_0. Let K be the completion of $\mathsf{K}_0/\mathsf{K}_{00} = \{[x] : x \in \mathsf{K}_0\}$. If

$$\pi(a)[b \otimes v] := [ab \otimes v] \qquad \text{for all } a, b \in \mathsf{A} \text{ and } v \in \mathsf{H},$$

then $\pi(a)$ extends by linearity and continuity to an element of $B(\mathsf{K})$, denoted in the same manner. Furthermore, the map $a \mapsto \pi(a)$ is a unital $*$-homomorphism from A to $B(\mathsf{K})$.

To conclude, let $T \in B(\mathsf{H}; \mathsf{K})$ be defined by setting $Tv = [1 \otimes v]$ for all $v \in \mathsf{H}$. It is a final exercise to verify that $\Phi(a) = T^*\pi(a)T$, as required. □

The following result extends the Kadison–Schwarz inequality, Theorem 4.28.

Corollary 4.32 *If* $\Phi : \mathsf{A} \to B(\mathsf{H})$ *is unital and completely positive then*

$$\sum_{i,j=1}^{n} \langle v_i, \big(\Phi(a_i^*a_j) - \Phi(a_i)^*\Phi(a_j)\big)v_j \rangle \geqslant 0$$

for all $n \in \mathbb{N}$, $a_1, \ldots, a_n \in \mathsf{A}$ *and* $v_1, \ldots, v_n \in \mathsf{H}$.

Proof Let π and T be as in Theorem 4.31. Then $\|T\|^2 = \|T^*\pi(1_\mathsf{A})T\| = \|\Phi(1_\mathsf{A})\| = 1$ and

$$\sum_{i,j=1}^{n} \langle v_i, \Phi(a_i^*a_j)v_j \rangle = \sum_{i,j=1}^{n} \langle Tv_i, \pi(a_i^*a_j)Tv_j \rangle = \left\| \sum_{i=1}^{n} \pi(a_i)Tv_i \right\|^2$$

$$\geqslant \left\| T^* \sum_{i=1}^{n} \pi(a_i)Tv_i \right\|^2$$

$$= \left\| \sum_{i=1}^{n} \Phi(a_i)v_i \right\|^2$$

$$= \sum_{i,j=1}^{n} \langle v_i, \Phi(a_i)^*\Phi(a_j)v_j \rangle.$$

□

Definition 4.33 A triple (K, π, T) as in Theorem 4.31 is a *Stinespring dilation* of Φ. Such a dilation is *minimal* if

$$\mathsf{K} = \overline{\mathrm{lin}}\{\pi(a)Tv : a \in \mathsf{A}, \ v \in \mathsf{H}\}.$$

Proposition 4.34 *Any unital completely positive map* $\Phi : \mathsf{A} \to B(\mathsf{H})$ *has a minimal Stinespring dilation.*

Proof One may take (K, π, T) as in Theorem 4.31 and restrict to the smallest closed subspace of K containing $\{\pi(a)Tv : a \in \mathsf{A}, \ v \in \mathsf{H}\}$. □

Exercise 4.35 Prove that the minimal Stinespring dilation is unique in an appropriate sense.

Definition 4.36 Let $(a_i) \subseteq \mathsf{A}$ be a net in the von Neumann algebra $\mathsf{A} \subseteq B(\mathsf{H})$. We write $a_i \searrow 0$ if $a_i \geqslant a_j \geqslant 0$ whenever $i \geqslant j$ and $\langle v, a_iv \rangle \to 0$ for all $v \in \mathsf{H}$.

[It follows from Vigier's theorem [22, Theorem 4.1.1.] that the decreasing net (a_i) converges in the strong operator topology to some element $a \in A_+$.]

A linear map $\Phi : A \to B(K)$ is *normal* if $a_i \searrow 0$ implies that $\langle v, \Phi(a_i)v \rangle \to 0$ for all $v \in K$.

Proposition 4.37 *Let* A *be a von Neumann algebra. If the linear map* $\Phi : A \to B(H)$ *is completely positive and normal, then the unital $*$-homomorphism* π *of Theorem 4.31 may be chosen to be normal also.*

Proof Let (K, π, T) be a minimal Stinespring dilation for Φ. If $v \in H$, $a \in A$ and the net $(a_i) \subseteq A_+$ is such that $a_i \searrow 0$, then

$$\langle \pi(a)Tv, \pi(a_i)\pi(a)Tv \rangle = \langle v, T^*\pi(a^*a_i a)Tv \rangle = \langle v, \Phi(a^*a_i a)v \rangle \to 0,$$

since $a^*a_i a \searrow 0$. It now follows by polarisation and minimality that $\pi(a_i) \searrow 0$, as required. $\qquad\square$

Proposition 4.38 *A linear map* $\Phi : A \to B(H)$ *is normal if and only if it is σ-weakly continuous.*

Proof It suffices to prove that if $(b_i) \subseteq B(K)$ is a norm-bounded net then $b_i \to 0$ in the σ-weak topology if and only if $\langle v, b_i v \rangle \to 0$ for all $v \in K$. Furthermore, by polarisation, we need only consider σ-weakly continuous functionals of the form

$$\phi : B(K) \to \mathbb{C}; \quad a \mapsto \sum_{n=1}^{\infty} \langle x_n, ax_n \rangle, \qquad \text{where } \sum_{n=1}^{\infty} \|x_n\|^2 < \infty.$$

The result now follows by a standard truncation argument. $\qquad\square$

4.5 Semigroup Generators

We will now introduce the class of quantum Feller semigroups, and proceed toward a classification of the semigroup generators for a uniformly continuous subclass. As above, we will first establish some necessary conditions that hold in greater generality.

Definition 4.39 A *quantum Feller semigroup* $T = (T_t)_{t \in \mathbb{R}_+}$ on a C^* algebra A is a strongly continuous contraction semigroup such that each T_t is completely positive.

If A is unital, with unit 1, and $T_t 1 = 1$ for all $t \in \mathbb{R}_+$ then T is *conservative*.

Exercise 4.40 Let T be a quantum Feller semigroup on a unital C^* algebra. Prove that T is conservative if and only if $1 \in \text{dom } \mathcal{L}$, with $\mathcal{L}(1) = 0$. [Hint: Theorem 2.46 may be useful.]

To begin the characterisation of the generators of these semigroups, we introduce a concept due to Evans [11].

Proposition 4.41 *Let* $\Phi : A \to B(H)$ *be a linear map on the unital concrete* C^* *algebra* $A \subseteq B(H)$. *The following are equivalent.*

(i) *If* $n \in \mathbb{N}$ *and* $a \in M_n(A)$, *then*

$$\Phi^{(n)}(a^*a) + a^*\Phi^{(n)}(1)a - \Phi^{(n)}(a^*)a - a^*\Phi^{(n)}(a) \in M_n\big(B(H)\big)_+.$$

(ii) *If* $n \in \mathbb{N}$ *and* $a_1, \ldots, a_n \in A$, *then*

$$\big(\Phi(a_i^*a_j) + a_i^*\Phi(1)a_j - \Phi(a_i^*)a_j - a_i^*\Phi(a_j)\big)_{i,j=1}^n \in M_n\big(B(H)\big)_+.$$

(iii) *If* $n \in \mathbb{N}$, $a_1, \ldots, a_n \in A$ *and* $v_1, \ldots, v_n \in H$ *are such that* $\sum_{i=1}^n a_i v_i = 0$, *then*

$$\sum_{i,j=1}^n \langle v_i, \Phi(a_i^*a_j)v_j \rangle \geqslant 0.$$

(iv) *If* $n \in \mathbb{N}$, $a_1, \ldots, a_n \in A$ *and* $b_1, \ldots, b_n \in B(H)$ *are such that* $\sum_{i=1}^n a_i b_i = 0$, *then*

$$\sum_{i,j=1}^n b_i^*\Phi(a_i^*a_j)b_j \geqslant 0.$$

Proof Given $a_1, \ldots, a_n \in A$, let $a = (a_{ij}) \in M_n(A)$ be such that $a_{1j} = a_j$ and $a_{ij} = 0$ otherwise. Then

$$\big(\Phi^{(n)}(a^*a) + a^*\Phi^{(n)}(1)a - \Phi^{(n)}(a^*)a - a^*\Phi^{(n)}(a)\big)_{ij}$$
$$= \Phi(a_i^*a_j) + a_i\Phi(1)a_j - \Phi(a_i^*)a_j - a_i^*\Phi(a_j)$$

for all $i, j = 1, \ldots, n$, so (i) implies (ii).

Conversely, let $a = (a_{ij}) \in M_n(A)$. Applying (ii) to $a_{k1}, \ldots, a_{kn}$ and then summing over k gives that

$$0 \leqslant \sum_{k=1}^n [\Phi(a_{ki}^*a_{kj}) + a_{ki}^*\Phi(1)a_{kj} - \Phi(a_{ki}^*)a_{kj} - a_{ki}^*\Phi(a_{ki}^*)]_{i,j=1}^n$$
$$= \Phi^{(n)}(a^*a) - a^*\Phi^{(n)}(1)a - \Phi^{(n)}(a^*)a - a^*\Phi^{(n)}(a).$$

Thus (ii) implies (i).

The implication from (ii) to (iii) is clear, as is that from (iii) to (iv). For the final part, let $a_1, \ldots, a_n \in A$ and $b_1, \ldots, b_n \in B(H)$, let $a_0 = 1$ and $b_0 = -\sum_{i=1}^n a_i b_i$, and note that $\sum_{i=0}^{n} a_i b_i = 0$. Hence (iv) gives that

$$0 \leqslant \sum_{i,j=0}^{n} b_i^* \Phi(a_i^* a_j) b_j = \sum_{i,j=1}^{n} b_i^* \big(\Phi(a_i^* a_j) + a_i^* \Phi(1) a_j - a_i^* \Phi(a_j) - \Phi(a_i^*) a_j \big) b_j.$$

Thus (ii) now follows from the first part of Exercise 4.25. □

Definition 4.42 A linear map $\Phi : \mathsf{A} \to B(\mathsf{H})$ on the unital C^* algebra $\mathsf{A} \subseteq B(\mathsf{H})$ is *conditionally completely positive* if and only if any of the equivalent conditions in Proposition 4.41 hold.

Exercise 4.43 Prove that the set of conditionally completely positive maps from A to $B(\mathsf{H})$ is a cone, that is, closed under addition and multiplication by non-negative scalars. Prove also that this cone contains all completely positive maps and scalar multiples of the identity map. Finally, prove that the cone is closed under pointwise weak-operator convergence: the net $\Phi_i \to \Phi$ if and only if $\langle v, \Phi_i(a)v \rangle \to \langle v, \Phi(a)v \rangle$ for all $a \in \mathsf{A}$ and $v \in \mathsf{H}$.

Exercise 4.44 Let A be as in Definition 4.42. A linear map $\delta : \mathsf{A} \to B(\mathsf{H})$ is a *derivation* if and only if

$$\delta(ab) = a\delta(b) + \delta(a)b \qquad \text{for all } a, b \in \mathsf{A}.$$

Prove that a derivation is conditionally completely positive. Prove also that the map

$$\mathsf{A} \to B(\mathsf{H}); \quad a \mapsto G^* a + aG$$

is conditionally completely positive and normal for all $G \in B(\mathsf{H})$.

Theorem 4.45 *Let T be a uniformly continuous quantum Feller semigroup on the unital C^* algebra $\mathsf{A} \subseteq B(\mathsf{H})$. The semigroup generator $\mathcal{L}$ is bounded, $*$-preserving and conditionally completely positive.*

Proof The first two claims follow immediate from Theorem 4.19. For conditional complete positivity, let $a_1, \ldots, a_n \in \mathsf{A}$ and $v_1, \ldots, v_n \in \mathsf{H}$. By Corollary 4.32, if $t > 0$, then

$$t^{-1} \sum_{i,j=1}^{n} \langle v_i, \big(T_t(a_i^* a_j) - T_t(a_i)^* T_t(a_j) \big) v_j \rangle \geqslant 0.$$

Letting $t \to 0+$ gives that

$$\sum_{i,j=1}^{n} \langle v_i, \big(\mathcal{L}(a_i^* a_j) - \mathcal{L}(a_i)^* a_j - a_i^* \mathcal{L}(a_j) \big) v_j \rangle \geqslant 0,$$

and if $\sum_{i=1}^{n} a_i v_i = 0$ then the second and third terms vanish. □

Exercise 4.46 Use Exercise 4.43 to provide an alternative proof that $\mathcal{L}$ in Theorem 4.45 is conditionally completely positive.

The following result is [11, Theorem 2.9] of Evans, who credits Lindblad [21].

Theorem 4.47 (Lindblad, Evans) *Let $\mathcal{L}$ be a $*$-preserving bounded linear map on the unital C^* algebra $\mathsf{A} \subseteq B(\mathsf{H})$. The following are equivalent.*

(i) $\mathcal{L}$ *is conditionally completely positive.*

(ii) $(\lambda I - \mathcal{L})^{-1}$ *is completely positive for all sufficiently large $\lambda > 0$.*

(iii) $T_t = \exp(t\mathcal{L})$ *is completely positive for all $t \in \mathbb{R}_+$.*

Proof The equivalence of (ii) and (iii) is given by Theorems 2.45 and 2.46, together with Exercise 4.23. The solution to Exercise 4.46 gives that (iii) implies (i); to complete the proof, it suffices to show that (i) implies (iii).

Suppose first that $\mathcal{L}(1) \leqslant 0$. Then $\mathcal{L}^{(n)}(1) \leqslant 0$ for all $n \in \mathbb{N}$, so if $a \in M_n(\mathsf{A})$ then

$$\mathcal{L}^{(n)}(a^*a) \geqslant a^*\mathcal{L}^{(n)}(a^*)a + a^*\mathcal{L}^{(n)}(a).$$

Thus if $b, c \in M_n(\mathsf{A})_+$ are such that $bc = 0$ then $b^{1/2}c = 0$ and

$$c\mathcal{L}^{(n)}(b)c \geqslant c\mathcal{L}^{(n)}(b^{1/2})b^{1/2}c + cb^{1/2}\mathcal{L}^{(n)}(b)c = 0.$$

Theorem 4.20 now gives that $T_t^{(n)} = \exp(t\mathcal{L}^{(n)})$ is positive for all $t \in \mathbb{R}_+$, so (iii) holds.

Finally, if $\mathcal{L}(1) > 0$, then the conditionally completely positive map

$$\mathcal{L}' : \mathsf{A} \to B(\mathsf{H}); \quad a \mapsto \mathcal{L}(a) - \|\mathcal{L}(1)\|a$$

is such that $\mathcal{L}'(1) \leqslant 0$, since $0 \leqslant \mathcal{L}(1) \leqslant \|\mathcal{L}(1)\|I$. It follows that $T_t' = \exp(t\mathcal{L}')$ is completely positive for all $t \in \mathbb{R}_+$, and therefore so is $T_t = \exp(\|\mathcal{L}(1)\|t)T_t'$. $\square$

Remark 4.48 Since completely positive unital linear maps between unital C^* algebras are automatically contractive, by Theorem 4.31 and the fact that $*$-homomorphisms between C^* algebras are contractive, the previous result characterises the generators of uniformly continuous conservative quantum Feller semigroups.

4.6 The Gorini–Kossakowski–Sudarshan–Lindblad Theorem

In order to provide a more explicit description of the generators of quantum Feller semigroups, we will establish some results of Lindblad and Christensen, and of Kraus. The Kraus decomposition is a key tool in quantum information theory.

Theorem 4.49 (Lindblad, Christensen) *Let $\mathcal{L}$ be a $*$-preserving bounded linear map on the von Neumann algebra* A. *Then $\mathcal{L}$ is conditionally completely positive and normal if and only if there exists a completely positive, normal map $\Psi :$ A $\to$ A and an element $g \in$ A such that*

$$\mathcal{L}(a) = \Psi(a) + g^*a + ag \qquad \text{for all } a \in \text{A}.$$

Proof The second part of Exercise 4.44 shows that $\mathcal{L}$ is conditionally completely positive or normal if and only if Ψ has the same property.

Given this, it remains to prove that if $\mathcal{L}$ is conditionally completely positive, then there exists $g \in$ A such that $a \mapsto \mathcal{L}(a) - g^*a - ag$ is completely positive. We will show this under the assumption that A $= B(\mathsf{H})$; see [14, Proof of Theorem 3.14]. The general case [7] requires considerably more work.

Given $u, v \in \mathsf{H}$, let the Dirac dyad

$$|u\rangle\langle v| : \mathsf{H} \to \mathsf{H}; \quad w \mapsto \langle v, w\rangle u.$$

Fix a unit vector $u \in \mathsf{H}$, and let $G \in B(\mathsf{H})$ be such that

$$G^* : \mathsf{H} \to \mathsf{H}; \quad v \mapsto \mathcal{L}(|v\rangle\langle u|)u - \frac{1}{2}\langle u, \mathcal{L}(|u\rangle\langle u|)u\rangle v.$$

Given $a_1, \ldots a_n \in$ A and $v_1, \ldots, v_n \in \mathsf{H}$, let $v_0 = u$ and $a_0 = -\sum_{i=1}^{n} |a_i v_i\rangle\langle u|$, so that $\sum_{i=0}^{n} a_i v_i = 0$. The conditional complete positivity of $\mathcal{L}$ implies that

$$
0 \leqslant \sum_{i,j=1}^{n} \big(\langle v_i, \mathcal{L}(a_i^* a_j)v_j\rangle - \langle v_i, \mathcal{L}(a_i^*|a_jv_j\rangle\langle u|)u\rangle - \langle u, \mathcal{L}(|u\rangle\langle a_i v_i|a_j)v_j\rangle
$$

$$
+ \langle u, \mathcal{L}(|u\rangle\langle a_i v_i||a_j v_j\rangle\langle u|)u\rangle \big)
$$

$$
= \sum_{i,j=1}^{n} \langle v_i, \mathcal{L}(a_i^* a_j)v_j\rangle - \langle v_i, \mathcal{L}(|a_i^* a_j v_j\rangle\langle u|)u\rangle - \langle u, \mathcal{L}(|u\rangle\langle a_j^* a_i v_i|)v_j\rangle
$$

$$
+ \langle u, \mathcal{L}(|u\rangle\langle u|)u\rangle\langle a_i v_i, a_j v_j\rangle
$$

$$
= \sum_{i,j=1}^{n} \langle v_i, \big(\mathcal{L}(a_i^* a_j) - G^* a_i^* a_j - a_i^* a_j G\big)v_j\rangle.
$$

The result follows. $\qquad\qquad\qquad\square$

Remark 4.50 If A is required only to be a C^* algebra, then Christensen and Evans [7] showed that Theorem 4.49 remains true if $\mathcal{L}$ and Ψ no longer required to be normal, but then g and the range of Ψ must be taken to lie in the σ-weak closure of A.

Theorem 4.51 (Kraus [18]) *Suppose* $A \subseteq B(H)$ *is a von Neumann algebra. A linear map* $\Psi : A \to B(K)$ *is normal and completely positive if and only if there exists a family of operators* $(L_i)_{i\in\mathbb{I}} \subseteq B(K; H)$ *such that*

$$\Psi(a) = \sum_{i\in\mathbb{I}} L_i^* a L_i \qquad \text{for all } a \in A,$$

with convergence in the strong operator topology. The cardinality of the index set $\mathbb{I}$ *may be taken to be no larger than* $\dim K$.

Proof If Ψ has this form, then it is completely positive and normal. The first claim is readily verified; for the second, let $a_j \searrow 0$, fix j_0 and note that $\langle u, a_j u \rangle \leqslant \langle u, a_{j_0} u \rangle$ for all $u \in H$ and $j \geqslant j_0$. Fix $\varepsilon > 0$ and $v \in K$, choose a finite set $\mathbb{I}_0 \subseteq \mathbb{I}$ such that the sum $\sum_{i\in\mathbb{I}_0} \langle L_i v, a_{j_0} L_i v \rangle > \langle v, \Psi(a_{j_0})v \rangle - \varepsilon$, and note that

$$\langle v, \Psi(a_j)v \rangle \leqslant \sum_{i\in\mathbb{I}_0} \langle L_i v, a_j L_i v \rangle + \sum_{i\in\mathbb{I}\setminus\mathbb{I}_0} \langle L_i v, a_{j_0} L_i v \rangle < \sum_{i\in\mathbb{I}_0} \langle L_i v, a_j L_i v \rangle + \varepsilon.$$

This shows that Ψ is normal, as required.

For the converse, Theorem 4.31 shows it suffices to prove that if $\pi : A \to B(K)$ is a normal unital $*$-homomorphism, then π can be written as in the statement of the theorem.

Let $(e_i)_{i\in\mathbb{I}}$ be an orthonormal basis for H, and consider the net $(I_H - \sum_{i\in\mathbb{I}_0} |e_i\rangle\langle e_i|)$, where the index $\mathbb{I}_0$ runs over all finite subsets of $\mathbb{I}$, ordered by inclusion. Since π is normal and unital, we have that $I_K = \sum_{i\in\mathbb{I}} \pi(|e_i\rangle\langle e_i|)$ in the weak-operator sense; thus, there exists some $i_0 \in \mathbb{I}$ such that $P := \pi(|e_{i_0}\rangle\langle e_{i_0}|)$ is a non-zero orthogonal projection.

Let $u \in K$ be a unit vector such that $Pu = u$, let $a \in A$, and note that

$$\|\pi(a)u\|^2 = \langle Pu, \pi(a^*a)Pu \rangle = \langle u, \pi(|e_{i_0}\rangle\langle e_{i_0}|a^*a|e_{i_0}\rangle\langle e_{i_0}|)u \rangle = \|ae_{i_0}\|^2.$$

Hence there exists a partial isometry $L_0 : K \to H$ with initial space K_0, the norm closure of $\{\pi(a)u : a \in A\}$, and final space H_0, the norm closure of $\{ae_0 : a \in A\}$, and such that $L_0\pi(a)u = ae_0$ for all $a \in A$. Note that K_0 is invariant under the action of $\pi(a)$, for all $a \in A$, so

$$\pi(a)\pi(b)u = P_0\pi(ab)u = L_0^*L_0\pi(ab)u = L_0^*abe_0 = L_0^*aL_0\pi(b)u \qquad \text{for all } b \in A.$$

Thus $\pi(a)|_{K_0} = L_0^*aL_0|_{K_0}$, and since $L_0(K_0^\perp) = \{0\}$, it follows that $\pi(a)P_0 = L_0^*aL_0$ for all $a \in A$, where $P_0 := L_0^*L_0$ is the orthogonal projection onto the initial space K_0.

Repeating this argument, but on $K_0^\perp$, there exists a partial isometry $L_1 : K \to H$ with initial projection P_1 such that $P_0P_1 = 0$ and $\pi(a)P_1 = L_1^*aL_1$ for all $a \in A$. An application of Zorn's lemma now gives the result. $\square$

Remark 4.52 With Ψ and $(L_i)_{i \in \mathbb{I}}$ as in Theorem 4.51, we may write

$$\Psi(a) = L^*(a \otimes I_{\mathsf{K}_{\mathbb{I}}})L \qquad \text{for all } a \in \mathsf{A},$$

where $\mathsf{K}_{\mathbb{I}}$ is the Hilbert space with orthonormal basis $(e_i)_{i \in \mathbb{I}}$ and $L \in B(\mathsf{K}; \mathsf{H} \otimes \mathsf{K}_{\mathbb{I}})$ is such that

$$Lv = \sum_{i \in \mathbb{I}} L_i v \otimes e_i \qquad \text{for all } v \in \mathsf{K}.$$

Exercise 4.53 Use Theorem 4.51 and the second part of Theorem 4.26 to show that every positive normal linear functional on the von Neumann algebra A has the form

$$a \mapsto \sum_{n=1}^{\infty} \langle x_n, a x_n \rangle, \qquad \text{where } \sum_{n=1}^{\infty} \|x_n\|^2 < \infty.$$

[Every bounded linear functional is the linear combination of four positive ones [22, Theorem 3.3.10], and Grothendieck [15] observed that each of these may be taken to be normal if the original is [17, Theorem 7.4.7]. Hence every normal linear functional is of the form used to define the σ-weak topology in Definition 4.6.]

Lemma 4.54 *Let T be a uniformly continuous semigroup on a von Neumann algebra with generator $\mathcal{L}$. Then $\mathcal{L}$ is normal if and only if T_t is normal for all $t \in \mathbb{R}_+$.*

Proof This holds because the limit of a norm-convergent sequence of normal maps is normal. To see this, let $\Phi_n, \Phi : \mathsf{A} \to B(\mathsf{H})$ be such that $\|\Phi_n - \Phi\| \to 0$, let the net $(a_i) \subseteq \mathsf{A}_+$ be such that $a_i \searrow 0$, and let $v \in \mathsf{H}$. Fix i_0 and note that $\|a_i\| \leqslant \|a_{i_0}\|$ whenever $i \geqslant i_0$, so

$$|\langle v, \Phi(a_i)v \rangle| \leqslant \|v\|^2 \|a_{i_0}\| \|\Phi_n - \Phi\| + |\langle v, \Phi(a_i)v \rangle| \qquad \text{for all } i \geqslant i_0.$$

The claim follows. $\qquad\square$

Theorem 4.55 (Gorini–Kossakowski–Sudarshan, Lindblad) *Let $\mathsf{A} \subseteq B(\mathsf{H})$ be a von Neumann algebra. A bounded linear map $\mathcal{L} \in B(\mathsf{A})$ is the generator of a uniformly continuous conservative quantum Feller semigroup composed of normal maps if and only if*

$$\mathcal{L}(a) = -\mathrm{i}[h, a] - \frac{1}{2}\left(L^*La - 2L^*(a \otimes I)L + aL^*L\right) \qquad \text{for all } a \in \mathsf{A},$$

where $h = h^* \in \mathsf{A}$ and $L \in B(\mathsf{H}; \mathsf{H} \otimes \mathsf{K})$ for some Hilbert space K.

Proof If $\mathcal{L}$ has this form, then it is straightforward to verify that the semigroup it generates is as claimed.

Conversely, suppose $\mathcal{L}$ is the generator of a semigroup as in the statement of the theorem. Then Theorem 4.47 gives that $\mathcal{L}$ is conditionally completely positive and $\mathcal{L}(1) = 0$. Moreover, $\mathcal{L}$ is normal, by the preceding lemma, and so Theorem 4.49 gives that

$$\mathcal{L}(a) = \Psi(a) + g^*a + ag \qquad \text{for all } a \in \mathsf{A},$$

where $\Psi : \mathsf{A} \to \mathsf{A}$ is completely positive and normal, and $g \in \mathsf{A}$. Taking $a = 1$ in this equation shows that $g^* + g = -\Psi(1)$, so $g = -\frac{1}{2}\Psi(1) + ih$ for some self-adjoint element $h \in \mathsf{A}$. The result now follows by Theorem 4.51. $\square$

The story of the previous theorem is very well told in [8]. Going beyond the case of bounded generators is the subject of much interest. See the survey [28] for some recent developments.

4.7 Quantum Markov Processes

We will conclude by giving a very brief indication of how a quantum process may be defined.

Remark 4.56 Let E be a compact Hausdorff space. If X is an E-valued random variable on the probability space $(\Omega, \mathcal{F}, \mathbb{P})$, then

$$j_X : \mathsf{A} \to \mathsf{B}; \ f \mapsto f \circ X$$

is a unital $*$-homomorphism, where $\mathsf{A} = C(E)$ and $\mathsf{B} = L^\infty(\Omega, \mathcal{F}, \mathbb{P})$.

Definition 4.57 A *non-commutative random variable* is a unital $*$-homomorphism j between unital C^* algebras.

A family $(j_t : \mathsf{A} \to \mathsf{B})_{t \in \mathbb{R}_+}$ of non-commutative random variables is a *dilation* of the quantum Feller semigroup T on A if there exists a conditional expectation $\mathbb{E}$ from B onto A such that $T_t = \mathbb{E} \circ j_t$ for all $t \in \mathbb{R}_+$.

The problem of constructing such dilations has attracted the interest of many authors, including Evans and Lewis [13], Accardi et al. [1], Vincent-Smith [31], Kümmerer [19], Sauvageot [27] and Bhat and Parthasarathy [5].

Essentially, one attempts to mimic the functional-analytic proof of Theorem 3.22. Given the appropriate analogue of an initial measure, which is a state μ on the C^* algebra A, the sesquilinear form

$$\mathsf{A}^{\otimes n} \times \mathsf{A}^{\otimes n} \to \mathbb{C}; \ (a_1 \otimes \cdots \otimes a_n, b_1 \otimes \cdots \otimes b_n) \mapsto \mu\big(T_{t_1}(a_1^* \ldots (T_{t_n - t_{n-1}}(a_n^* b_n)) \ldots b_1)\big)$$

must be shown to be positive semidefinite. The key to this is the complete positivity of the semigroup maps. There are many technical issues to be addressed; see [5] for more details.

References

1. L. Accardi, A. Frigerio, J.T. Lewis, Quantum stochastic processes. Publ. Res. Inst. Math. Sci. **18**(1), 97–133 (1982)
2. R. Alicki, K. Lendi, *Quantum Dynamical Semigroups and Applications*. Lecture Notes in Physics, vol. 717, 2nd edn. (Springer, Berlin, 2007)
3. D. Applebaum, *Lévy Processes and Stochastic Calculus*, 2nd edn. (Cambridge University Press, Cambridge, 2009)
4. W.B. Arveson, Subalgebras of C^*-algebras. Acta Math. **123**, 141–224 (1969)
5. B.V.R. Bhat, K.R. Parthasarathy, Kolmogorov's existence theorem for Markov processes in C^* algebras. Proc. Indian Acad. Sci. Math. Sci. **104**(1), 253–262 (1994)
6. M.-D. Choi, Positive linear maps on C^*-algebras. Can. J. Math. **24**(3), 520–529 (1972)
7. E. Christensen, D.E. Evans, Cohomology of operator algebras and quantum dynamical semigroups. J. Lond. Math. Soc. **20**(2), 358–368 (1979)
8. D. Chruściński, S. Pascazio, A brief history of the GKLS equation. Open Syst. Inf. Dyn. **24**(3), 1740001, 20pp. (2017)
9. E.B. Davies, *Linear Operators and Their Spectra* (Cambridge University Press, Cambridge, 2007)
10. E.G. Effros, Z.-J. Ruan, *Operator Spaces*. London Mathematical Society Monographs, vol. 23 (Oxford University Press, Oxford, 2000)
11. D.E. Evans, Conditionally completely positive maps on operator algebras. Q. J. Math. **28**(3), 271–283 (1977)
12. D.E. Evans, H. Hanche-Olsen, The generators of positive semigroups. J. Funct. Anal. **32**, 207–212 (1979)
13. D.E. Evans, J.T. Lewis, *Dilations of Irreversible Evolutions in Algebraic Quantum Theory*. Communications of the Dublin Institute for Advanced Studies Series A, vol. 24 (Dublin Institute for Advanced Studies, Dublin, 1977), v+104 pp.
14. F. Fagnola, Quantum Markov semigroups and quantum flows. Proyecciones **18**(3), 144 pp. (1999)
15. A. Grothendieck, Un résultat sur le dual d'une C^*-algèbre. J. Math. Pures Appl. (9) **36**(2), 97–108 (1957)
16. E. Hille, R.S. Phillips, *Functional Analysis and Semi-Groups*. AMS Colloquium Publications, vol. 31, third printing of the revised 1957 edition (American Mathematical Society, Rhode Island, 1974)
17. R.V. Kadison, J.R. Ringrose, *Fundamentals of the Theory of Operator Algebras II. Advanced Theory*. Graduate Studies in Mathematics, vol. 16. (American Mathematical Society, Providence, 1997)
18. K. Kraus, General state changes in quantum theory. Ann. Phys. **64**(2), 311–335 (1971)
19. B. Kümmerer, Markov dilations on $W*$-algebras. J. Funct. Anal. **63**(2), 139–177 (1985)
20. T.M. Liggett, *Continuous Time Markov Processes* (American Mathematical Society, Providence, 2010)
21. G. Lindblad, On the generators of quantum dynamical semigroups. Commun. Math. Phys. **48**(2), 119–130 (1976)
22. G.J. Murphy, *C^*-Algebras and Operator Theory* (Academic, New York, 1990)
23. W.L. Paschke, Inner product modules over B^*-algebras. Trans. Am. Math. Soc. **182**, 443–468 (1973)
24. V. Paulsen, *Completely Bounded Maps and Operator Algebras*. Cambridge Studies in Advanced Mathematics, vol. 78 (Cambridge University Press, Cambridge, 2002)
25. M. Reed, B. Simon, *Methods of Modern Mathematical Physics I. Functional Analysis*, revised and enlarged edition (Academic, New York, 1980)
26. L.C.G. Rogers, D. Williams, *Diffusions, Markov Processes and Martingales I. Foundations*, 2nd edn. (Cambridge University Press, Cambridge, 2000)

27. J.-L. Sauvageot, Markov quantum semigroups admit covariant Markov C^*-dilations. Commun. Math. Phys. **106**(1), 91–103 (1986)
28. I. Siemon, A.S. Holevo, R.F. Werner, Unbounded generators of dynamical semigroups. Open Syst. Inf. Dyn. **24**(4), 1740015, 24pp. (2017)
29. W.F. Stinespring, Positive functions on C^*-algebras. Proc. Am. Math. Soc. **6**(2), 211–216 (1955)
30. M. Takesaki, *Theory of Operator Algebras I*. Encyclopaedia of Mathematical Sciences, vol. 124 (Springer, Berlin, 2002)
31. G.F. Vincent-Smith, Dilation of a dissipative quantum dynamical system to a quantum Markov process. Proc. Lond. Math. Soc. **49**(1), 58–72 (1984)

Introduction to Non-Markovian Evolution of n-Level Quantum Systems

Dariusz Chruściński

Abstract We analyze quantum dynamical maps and the corresponding master equations beyond the celebrated quantum Markovian master equation derived by Gorini, Kossakowski, Sudarshan, and Lindblad. In the Heisenberg picture such maps are represented by completely positive and unital maps, whereas in the Schrödinger picture by completely positive and trace-preserving maps. Both time-local equations governed by time dependent generators and time non-local equations of the Nakajima-Zwanzig form governed by the corresponding memory kernels are considered. We use the Schrödinger picture to discuss time-local case and Heisenberg picture for the non-local one. These equations describe quantum non-Markovian evolution that takes into account memory effects. Our analysis is illustrated by several simple examples.

1 Introduction

Open quantum systems are of paramount importance in the study of the interaction between a quantum system and its environment [1–7] that leads to important physical processes like dissipation, decay, and decoherence. Very often to describe the evolution of a "small" system neglecting degrees of freedom of the "big" environment one applies very successful Markovian approximation leading to the celebrated quantum Markovian semigroup. This approximation usually assumes a weak coupling between the system and environment and separation of system and environment time scales (the system's degrees of freedom are "slow" and that of the environment are "fast"). A typical example is a quantum optical system where Markovian approximation is often legitimate due to the weak coupling between a system (atom) and the environment (electromagnetic field) [3]. Quantum Markovian semigroups were fully characterized by Gorini et al. [8] and independently

D. Chruściński (✉)
Institute of Physics, Faculty of Physics, Astronomy and Informatics Nicolaus Copernicus University, Torun, Poland
e-mail: darch@fizyka.umk.pl

© Springer Nature Switzerland AG 2019
D. Bahns et al. (eds.), *Open Quantum Systems*, Tutorials, Schools, and Workshops in the Mathematical Sciences, https://doi.org/10.1007/978-3-030-13046-6_2

by Lindblad [9]. Current laboratory techniques and technological progress call, however, for more refined approach taking into account non-Markovian memory effects (see recent review papers [10–13]).

A mathematical representation of the evolution of open quantum system is provided by a *quantum dynamical map*—a family of completely positive and trace-preserving maps parameterized by time. Nowadays dynamical maps define one of the basic ingredients of modern quantum theory. Being quantum channels they define at the same time one of the most fundamental objects of quantum information theory [14]. Completely positive maps in operator algebras [15–18] were invented by Stinespring in 1955 [19] and found elegant application in physics already in the 1960s with seminal papers of Kraus and collaborators (summarized in the monograph [20]).

In this paper we analyze quantum evolution beyond the Markovian approximation. One usually assumes that such map satisfies linear differential equation either in a time-local form

$$\partial_t \Lambda_t = \mathcal{L}_t \Lambda_t , \quad \Lambda_0 = \mathrm{id}, \tag{1.1}$$

or a memory kernel master equation

$$\partial_t \Lambda_t = \int_0^t K_{t-\tau} \Lambda_\tau , \quad \Lambda_0 = \mathrm{id}, \tag{1.2}$$

with suitable time-local generator $\mathcal{L}_t$ and memory kernel K_t. We analyze the properties of $\mathcal{L}_t$ and K_t which guarantee that (1.1) and (1.2) lead to completely positive maps Λ_t. Moreover, on the level of time-local equation (1.1) we analyze the property of Markovianity based on the concept of divisibility of the corresponding dynamical map. Our analysis is illustrated by several examples.

In this paper we consider mainly quantum systems living in a finite dimensional Hilbert space (n-level quantum system). It turns out, however, that several results presented in this paper can be generalized to infinite dimensional cases. Therefore, in the next introductory section we recall some well-known results which hold for the infinite dimensional case as well.

2 Preliminaries: Quantum States and Quantum Channels

We begin by introducing basic notation and terminology.

2.1 The Structure of Quantum States

In the Schrödinger picture to any quantum system one assigns a separable Hilbert space $\mathcal{H}$ and normalized vector $\psi \in \mathcal{H}$ represents (up to a phase factor) a pure state of the system. Mixed states are represented by density operators, that is, semi-positive trace-class operators $\rho \in \mathcal{T}(\mathcal{H})$ with additional normalization condition $\mathrm{Tr}\rho = 1$. Recall that $\mathcal{T}(\mathcal{H})$ defines a Banach space with the norm defined by the trace-norm $|| \cdot ||_1$. Hence density operators satisfy $||\rho||_1 = 1$. In this paper we consider mainly a quantum system living in n-dimensional Hilbert space $\mathcal{H}$. Fixing an orthonormal basis $\{e_1, \ldots, e_n\}$ in $\mathcal{H}$ any linear operator in $\mathcal{H}$ may be identified with $n \times n$ complex matrix from $M_n(\mathbb{C})$.

Clearly, a space of pure states correspond to complex projective space $\mathbb{C}\mathbb{P}^{n-1}$. Mixed states may be interpreted as convex combinations (mixtures) of pure states

$$\rho = \sum_k w_k |\psi_k\rangle \langle \psi_k| , \tag{2.1}$$

with $w_k > 0$ and $\sum_k w_k = 1$. It should be stressed that the above representation is highly non-unique. This is actually one of the distinguished features of quantum theory. To illustrate a concept of density operators let us consider the following.

Example 2.1 A 2-level system (qubit) living in $\mathbb{C}^2$. Any hermitian operator ρ may be decomposed as follows

$$\rho = \frac{1}{2} \left(\mathbb{1} + \sum_{k=1}^{3} x_k \sigma_k \right) , \tag{2.2}$$

where $\mathbf{x} = (x_1, x_2, x_3) \in \mathbb{R}^3$ and $\{\sigma_1, \sigma_2, \sigma_3\}$ are Pauli matrices. It is, therefore, clear that ρ is entirely characterized by the Bloch vector $\mathbf{x}$. This representation already guaranties that $\mathrm{Tr}\,\rho = 1$. Hence, ρ represents density operator if and only if the corresponding eigenvalues $\{\lambda_-, \lambda_+\}$ are non-negative. One easily finds $\lambda_\pm = \frac{1}{2}(1 \pm |\mathbf{x}|)$ and hence $\rho \geq 0$ if and only if $|\mathbf{x}| = \sqrt{x_1^2 + x_2^2 + x_3^2} \leq 1$. This condition defines a unit ball in $\mathbb{R}^3$ known as a Bloch ball. A state is pure if ρ defines rank-1 projector, i.e. $\lambda_- = 0$ and $\lambda_+ = 1$. It shows that pure states belong to Bloch sphere corresponding to $|\mathbf{x}| = 1$. Unfortunately, this simple geometric picture is much more complicated if $n > 2$ (see, e.g., [21, 22]).

In the Heisenberg picture one assigns to a quantum system a unital C^*-algebra $\mathfrak{A}$. Self-adjoint elements of $\mathfrak{A}$ represent quantum observables. In this approach states are represented by normalized positive functional $\omega : \mathfrak{A} \to \mathbb{C}$, that is, $\omega(a) \geq 0$ for $a \geq 0$, and $\omega(\mathbb{1}) = 1$. Very often $\mathfrak{A} = \mathcal{B}(\mathcal{H})$ for some Hilbert space $\mathcal{H}$ and in this case one relates Schrödinger and Heisenberg picture using duality relation $\mathcal{T}(\mathcal{H})^* = \mathcal{B}(\mathcal{H})$. One has $\omega(a) = \mathrm{Tr}(a\rho)$ for some density operator $\rho \in \mathcal{T}(\mathcal{H})$. States corresponding to density operators are called normal states.

2.2 Positive and Completely Positive Maps [15–18]

Consider a linear map $\Phi : \mathfrak{A} \to \mathfrak{B}$ between two unital C^*-algebras $\mathfrak{A}$ and $\mathfrak{B}$.

Definition 2.2 Φ is positive if $\Phi(x^*x) \geq 0$ for any $x \in \mathfrak{A}$. Φ is unital if $\Phi(\mathbb{1}_{\mathfrak{A}}) = \mathbb{1}_{\mathfrak{B}}$.

Any positive map Φ is necessarily Hermitian, that is, $\Phi(x)^* = \Phi(x^*)$, and $||\Phi|| = ||\Phi(\mathbb{1}_{\mathfrak{A}})||$, where

$$||\Phi|| = \sup_{x \in \mathfrak{A}} \frac{||\Phi(x)||}{||x||}. \tag{2.3}$$

Hence unital positive map satisfies $||\Phi|| = 1$.

Definition 2.3 Φ is k-positive if

$$\mathrm{id}_k \otimes \Phi : M_k(\mathbb{C}) \otimes \mathfrak{A} \to M_k(\mathbb{C}) \otimes \mathfrak{B} \tag{2.4}$$

is positive. Φ is completely positive (CP) if it is k-positive for all $k = 1, 2, \ldots$.

If $\mathcal{P}_k(\mathfrak{A}, \mathfrak{B})$ denotes a convex cone of k-positive maps, then we have

$$\mathcal{P}_1(\mathfrak{A}, \mathfrak{B}) \supset \mathcal{P}_2(\mathfrak{A}, \mathfrak{B}) \supset \ldots \supset \mathcal{P}_\infty(\mathfrak{A}, \mathfrak{B}) \tag{2.5}$$

with $\mathcal{P}_\infty(\mathfrak{A}, \mathfrak{B}) = \mathcal{P}_{\mathrm{CP}}(\mathfrak{A}, \mathfrak{B})$.

Theorem 2.4 (Stinespring) A linear map $\Phi : \mathfrak{A} \to \mathcal{B}(\mathcal{H})$ is CP if and only if there exist a Hilbert space $\mathcal{K}$ and a $*$-homomorpism $\pi : \mathfrak{A} \to \mathcal{B}(\mathcal{K})$ and a linear operator $V : \mathcal{K} \to \mathcal{H}$ such that

$$\Phi(x) = V\pi(x)V^* . \tag{2.6}$$

Moreover, Φ is unital if V is an isometry.

One proves

Proposition 2.5 ([15]) *A linear map $\Phi : \mathfrak{A} \to \mathcal{B}(\mathcal{H})$ is CP if and only if*

$$\sum_{i,j=1}^{n} \langle \psi_i | \Phi(a_i a_j^*) | \psi_j \rangle \geq 0, \tag{2.7}$$

for any $a_1, \ldots, a_n \in \mathfrak{A}$ and $\psi_1, \ldots, \psi_n \in \mathcal{H}$, where $n = 1, 2, 3, \ldots$.

Theorem 2.6 (Kraus) Any CP map $\Phi : \mathcal{B}(\mathcal{H}_1) \to \mathcal{B}(\mathcal{H}_2)$ is of the following form

$$\Phi(X) = \sum_i K_i X K_i^* , \tag{2.8}$$

where $K_i : \mathcal{H}_1 \rightarrow \mathcal{H}_2$ are bounded operators and the sum converges in the strong operator sense. Moreover, Φ is unital if in addition the following condition is satisfied $\sum_i K_i K_i^* = \mathbb{1}_{\mathcal{H}_2}$.

Consider now a finite dimensional case corresponding to matrix algebras $\mathfrak{A} = M_n(\mathbb{C})$. In this case one has

Theorem 2.7 (Choi) A linear map $\Phi : M_n(\mathbb{C}) \rightarrow \mathfrak{B}$ is CP if and only if the matrix

$$[\Phi(e_{ij})] \in M_n(\mathfrak{B}) \tag{2.9}$$

is positive, where $\{e_{ij}\}$ denote the matrix units in $M_n(\mathbb{C})$. A linear map $\Phi : M_{n_1}(\mathbb{C}) \rightarrow M_{n_2}(\mathbb{C})$ is CP if and only if it is n-positive, where $n = \min\{n_1, n_2\}$.

If $\Phi : M_n(\mathbb{C}) \rightarrow M_n(\mathbb{C})$ is a linear map, then its dual $\Phi^* : M_n(\mathbb{C}) \rightarrow M_n(\mathbb{C})$ is defined by

$$\mathrm{Tr}[A\Phi^*(B)] = \mathrm{Tr}[\Phi(A)B] , \tag{2.10}$$

for all $A, B \in M_n(\mathbb{C})$. Φ is unital if and only if Φ^* is trace-preserving.

Definition 2.8 A quantum dynamical map (Heisenberg picture) is represented by a family of CP and unital maps $\Lambda_t : \mathcal{B}(\mathcal{H}) \rightarrow \mathcal{B}(\mathcal{H})$ ($t \geq 0$). A quantum dynamical map (Schrödinger picture) is represented by a family of CP and trace-preserving (CPTP) maps $\Lambda_t^* : \mathcal{T}(\mathcal{H}) \rightarrow \mathcal{T}(\mathcal{H})$ ($t \geq 0$). One calls a CPTP map $\mathcal{E}^* : \mathcal{T}(\mathcal{H}) \rightarrow \mathcal{T}(\mathcal{H})$ a quantum channel [14].

3 Markovian Semigroup

A strongly continuous quantum dynamical semigroup is a one-parameter family $\Lambda_t : \mathcal{B}(\mathcal{H}) \rightarrow \mathcal{B}(\mathcal{H})$ for $t \geq 0$ satisfying

1. Λ_t is completely positive and unital,
2. $\Lambda_{t+s} = \Lambda_t \Lambda_s$,
3. the map $t \rightarrow \Lambda_t$ is strongly continuous,
4. $\lim_{t \rightarrow 0+} \Lambda_t(x) = x$.

Such maps possess a densely defined generator

$$\mathcal{L}(x) = \lim_{t \rightarrow 0+} \frac{\Lambda_t(x) - x}{t} . \tag{3.1}$$

The structure of the generator is not known in the general case. However, the most interesting cases are characterized due to the following theorems.

Theorem 3.1 (Lindblad [9]) If $\mathcal{L}$ is bounded, then it defines a generator of a strongly continuous semigroup if and only if

$$\mathcal{L}(x) = i[H, x] + \Phi(x) - \frac{1}{2}(\Phi(\mathbb{1})x + x\Phi(\mathbb{1})), \tag{3.2}$$

where Φ is completely positive, $H = H^*$, and $[H, x] = Hx - xH$.

Remark 3.2 In the unbounded case there are examples of generators which are not of the standard Lindblad form (cf. [23]).

Consider now finite dimensional case $\dim \mathcal{H} = n$. One considers a dynamical semigroup in the Schrödinger picture $\Lambda_t^* : M_n(\mathbb{C}) \to M_n(\mathbb{C})$ which satisfies the following master equation

$$\partial_t \Lambda_t^* = \mathcal{L}^* \Lambda_t^*, \tag{3.3}$$

with initial condition $\Lambda_{t=0} = \mathrm{id}$. The dynamical map $\Lambda_t^* = e^{t\mathcal{L}^*}$ is completely positive and trace-preserving (CPTP)

Theorem 3.3 (Gorini-Kossakowski-Sudarshan [8]) $\mathcal{L}^*$ generates dynamical semigroup in the Schrödinger picture if and only if $\mathcal{L}^*$ has the following *canonical form*:

$$\mathcal{L}^*(\rho) = -i[H, \rho] + \frac{1}{2} \sum_{i,j=1}^{n^2-1} c_{ij} \left([F_i \rho, F_j^*] + [F_i, \rho F_j^*] \right), \tag{3.4}$$

where $[c_{ij}]$ is a semi-positive definite matrix and operators F_i satisfy: $\mathrm{Tr} F_i = 0$, and $\mathrm{Tr}(F_i F_j^*) = \delta_{ij}$ for $i, j = 1, 2, \ldots, n^2 - 1$.

In the Heisenberg picture one has

$$\mathcal{L}(X) = i[H, X] + \frac{1}{2} \sum_{i,j=1}^{n^2-1} c_{ij} \left(F_i[X, F_j^*] + [F_i, X]F_j^* \right). \tag{3.5}$$

One usually calls the generator of a quantum dynamical semigroup a GKSL generator (for both Heisenberg and Schrödinger pictures).

4   Open Quantum Systems: Beyond Markovian Semigroup

The standard approach to the dynamics of open quantum systems is based on the scheme of the reduced dynamics [1, 5, 6]: one considers the unitary evolution of the composed "system + environment" system governed by the von Neumann equation

$$\partial_t \rho_{SE}(t) = -i[\mathbf{H}, \rho_{SE}(t)], \tag{4.1}$$

and defines the evolution of the reduced density operator of the system via $\rho_t :=$ $\text{Tr}_E \rho_{SE}(t)$, where Tr_E denotes the partial trace over the environmental degrees of freedom. The total Hamiltonian $\mathbf{H} = H_S + H_E + H_{\text{int}}$ obviously represents the system Hamiltonian H_S, environmental (or bath) Hamiltonian H_E, and the interaction Hamiltonian H_{int} but this splitting is not unique. If the initial system-environment state factorizes $\rho_{SE}(0) = \rho \otimes \rho_E$, then the following formula

$$\Lambda_t^*(\rho) := \text{Tr}_E[U_t \, \rho \otimes \rho_E \, U_t^*], \quad U_t = e^{-i\mathbf{H}t}, \tag{4.2}$$

defines a family of CPTP maps. A system-environment density matrix $\rho_{SE}(t)$ satisfies von Neumann equation (4.1) with total Hamiltonian $\mathbf{H}$. To find the corresponding equation for the reduced density matrix ρ_t one applies the standard Nakajima-Zwanzig projection operator technique [24] (see also [25] for more general discussion) which shows that under fairly general conditions, and initial product state the *generalized master equation* for ρ_t takes the form of the following nonlocal equation:

$$\partial_t \Lambda_t^* = \int_0^t K_{t-\tau}^* \Lambda_\tau^* d\tau \; ; \quad \Lambda_0^* = \text{id} \,, \tag{4.3}$$

where the super-operator $K_t^* : \mathcal{T}(\mathcal{H}) \to \mathcal{T}(\mathcal{H})$ encodes all dynamical properties of the system in question. The characteristic feature of Nakajima-Zwanzig equation (4.3) is the appearance of a *memory kernel*: this simply means that the rate of change of the state represented by the density operator ρ_t at time t depends on its history. It should be stressed that the structure of the memory kernel K_t is highly nontrivial. It depends upon the total Hamiltonian and the initial state $\rho_E(0)$. In practice very often it turns out that H_{int} is sufficiently small (weak interaction) and one tries various approximation schemes [1]. Approximating (4.3) is a delicate issue [26, 27]. One often applies second order Born approximation which considerably simplifies the structure of K_t. However, this approximation in general violates basic properties of the generalized master equation, for example positivity of ρ_t [28]. Due to the nontrivial structure of (4.3) one tries to replace time non-local Nakajima-Zwanzig equation by the time-local one so-called *time-convolutionless* (TCL) master equation

$$\partial_t \Lambda_t^* = \mathcal{L}_t^* \Lambda_t^* \,, \quad \Lambda_0^* = \text{id}, \tag{4.4}$$

This procedure, however, requires existence of the inverse of Λ_t^* and leads to the following formula for the generator $\mathcal{L}_t^* = (\partial_t \Lambda_t^*) \Lambda_t^{*-1}$ [29, 30].

Example 4.1 Consider the evolution of 2-level system described by the following dynamical map

$$\Lambda_t^*(\rho) = \begin{pmatrix} \rho_{11} & \cos t \rho_{12} \\ \cos t \rho_{21} & \rho_{11} \end{pmatrix}, \tag{4.5}$$

where ρ_{ij} are matrix elements of the initial density matrix ρ. One easily checks that this map is CPTP for any $t \geq 0$. Indeed, its Kraus representation reads

$$\Lambda_t^*(\rho) = \frac{1 + \cos t}{2} \rho + \frac{1 - \cos t}{2} \sigma_3 \rho \sigma_3. \tag{4.6}$$

Note, however, that for $t = (2n + 1)\pi/2$ it is not invertible and hence time-local generator $\mathcal{L}_t^*$ is singular. Suppose that (4.5) satisfies Nakajima-Zwanzig master equation and let us look for the corresponding memory. Passing to the Laplace transform domain

$$\widetilde{F}_s := \int_0^\infty F_t e^{-st} dt, \tag{4.7}$$

one finds

$$s\widetilde{\Lambda}_s^* - \mathrm{id} = \widetilde{K}_s^* \widetilde{\Lambda}_s^*, \tag{4.8}$$

and hence

$$\widetilde{K}_s^* = s\,\mathrm{id} - \widetilde{\Lambda}_s^{*-1}. \tag{4.9}$$

Using (4.5) one gets

$$K_t^*(\rho) = \frac{1}{2} k(t)(\sigma_3 \rho \sigma_3 - \rho), \tag{4.10}$$

with $k(t) = H(t)$ (Heaviside step function). Time-local generator $\mathcal{L}_t^*$ has exactly the same structure

$$\mathcal{L}_t^*(\rho) = \frac{1}{2} \gamma(t)(\sigma_3 \rho \sigma_3 - \rho), \tag{4.11}$$

but now $\gamma(t) = \tan t$ (for $t \geq 0$) is singular.

In this paper we analyze the structure of Eqs. (4.3) and (4.4). A natural question one may ask is what are the properties of memory kernel K_t^* and time-local generator $\mathcal{L}_t^*$ which guarantee that solutions to (4.3) and (4.4) are physically legitimate, that is, Λ_t^* is CPTP. Note that the formal solution of (4.4) is given by

$$\Lambda_t^* = \mathbf{T} \exp\left(\int_0^t \mathcal{L}_\tau^* d\tau \right) = \mathrm{id} + \int_0^t dt_1 \mathcal{L}_{t_1}^* + \int_0^t dt_1 \int_0^{t_1} dt_2 \mathcal{L}_{t_1}^* \mathcal{L}_{t_2}^* + \dots, \tag{4.12}$$

where $\mathbf{T}$ stands for the time ordering operator. In practice, however, it is very hard to compute T-product exponential formula defined via the infinite Dyson expansion. One might be tempted to truncate the series (4.12) and to look for an approximate solution. Note, however, that any truncation immediately spoils complete positivity of Λ_t^*. The analysis simplifies in the commutative case, i.e. when $[\mathcal{L}_t, \mathcal{L}_\tau] = 0$ for any $t, \tau \geq 0$. In this case we have a simple sufficient condition.

Proposition 4.2 *If the integral $\int_0^t \mathcal{L}_\tau^* d\tau$ provides time dependent GKSL generator for any $t > 0$, then $\mathcal{L}_t^*$ is an admissible generator.*

The converse however need not be true (cf. Sect. 6).

5 Non-Markovian Quantum Evolution

As is stresses in [13] the concept of quantum Markovianity is context dependent and there is no universal approach to quantum Markovian process/evolution/map. The name "Markovian" is borrowed from the theory of classical stochastic processes [31]: a process is Markovian if the conditional probability satisfies

$$p(x_t, t_n | x_{n-1}, t_{n-1}; \ldots; x_1, t_1) = p(x_t, t_n | x_{n-1}, t_{n-1}). \tag{5.1}$$

It implies that $p(x_t, t_n | x_{n-1}, t_{n-1})$ satisfies the celebrated Chapman-Kolmogorov equation [31]

$$p(x_3, t_3 | x_1, t_1) = \sum_{x_2} p(x_3, t_3 | x_2, t_2) p(x_2, t_2 | x_1, t_1). \tag{5.2}$$

This definition cannot be used in the quantum theory due to the lack of a proper definition of conditional probability. In the literature there are many different approaches (see recent reviews [10–13]). One of the most influential approaches is based on the following.

Definition 5.1 A dynamical map Λ_t^* is divisible if for any $t \geq s$ there exists $V_{t,s}^*$: $\mathcal{T}(\mathcal{H}) \to \mathcal{T}(\mathcal{H})$ such that

$$\Lambda_t^* = V_{t,s}^* \Lambda_s^*, \tag{5.3}$$

for any $t \geq s$. Moreover, Λ_t^* is called CP-divisible if $V_{t,s}^*$ is CPTP, and P-divisible if $V_{t,s}^*$ is positive and trace-preserving.

Note that if Λ_t^* is invertible, then it is always divisible and $V_{t,s}^* = \Lambda_t^* \Lambda_s^{*-1}$. In this paper we accept the following.

Definition 5.2 ([32, 33]) Quantum evolution represented by a dynamical map Λ_t^* is Markovian if and only if the corresponding dynamical map Λ_t^* is CP-divisible.

Actually, following [34] one calls Λ_t^* k-divisible if $V_{t,s}^*$ is k-positive ($k = 1, 2, \ldots, n$). Let us recall the following well-known result.

Proposition 5.3 ([15, 16]) *Let* $\Phi : \mathcal{B}(\mathcal{H}) \to \mathcal{B}(\mathcal{H})$ *be a unital hermitian map.*
Then Φ *is positive if and only if*

$$||\Phi(X)|| \leq ||X||, \tag{5.4}$$

for any $X = X^*$. *Equivalently let* $\Phi^* : \mathcal{T}(\mathcal{H}) \to \mathcal{T}(\mathcal{H})$ *be a hermitian trace-preserving map. Then* Φ^* *is positive if and only if*

$$||\Phi^*(X)||_1 \leq ||X||_1, \tag{5.5}$$

for any $X = X^*$.

Now, if Λ_t^* is P-divisible, then

$$\frac{d}{dt}||\Lambda_t^*(X)||_1 \leq 0, \tag{5.6}$$

for any Hermitian X and $t \geq 0$. Indeed, P-divisibility implies that $\Lambda_{t+\epsilon}^* = V_{t+\epsilon,t}^* \Lambda_t^*$ and hence

$$\frac{d}{dt}||\Lambda_t^*(X)||_1 = \lim_{\epsilon \to 0+} \frac{1}{\epsilon} \left(||\Lambda_{t+\epsilon}^*(X)||_1 - ||\Lambda_t^*(X)||_1 \right)$$

$$= \lim_{\epsilon \to 0+} \frac{1}{\epsilon} \left(||V_{t+\epsilon,t}^* \Lambda_t^*(X)||_1 - ||\Lambda_t^*(X)||_1 \right) \leq 0, \tag{5.7}$$

due to the fact that $V_{t+\epsilon,t}^*$ is positive and trace-preserving. In particular if ρ and σ are arbitrary density operators, then P-divisibility implies

$$\frac{d}{dt}||\Lambda_t^*(\rho - \sigma)||_1 \leq 0. \tag{5.8}$$

The above property has an interesting physical interpretation [35]. Given two density operators ρ_1 and ρ_2 one defines distinguishability

$$D[\rho, \sigma] = \frac{1}{2}||\rho - \sigma||_1 . \tag{5.9}$$

It is clear that $D[\rho, \sigma] = 0$, i.e. ρ and σ are indistinguishable, if and only if $\rho = \sigma$. Note that if ρ and σ are orthogonally supported, then

$$D[\rho, \sigma] = \frac{1}{2}(||\rho||_1 + ||\sigma||_1) = 1 , $$

since $||\rho||_1 = 1$ for any density matrix ρ. In this case ρ and σ are perfectly distinguishable. Hence $0 \leq D[\rho, \sigma] \leq 1$. Now, the authors of [35] call the quantity $\frac{d}{dt}||\Lambda_t^*(\rho - \sigma)||_1$ an *information flow* and the condition (5.8) shows that the information flows from the system into the environment. Hence the *backflow* of information from the environment into the systems marks non-Markovian evolution or the presence of memory effects [35].

Theorem 5.4 ([36]) Let Λ_t^* be a dynamical map in $\mathcal{T}(\mathcal{H})$. If Λ_t^* is invertible for all $t > 0$, then it is CP-divisible if and only if

$$\frac{d}{dt}||[\text{id} \otimes \Lambda_t^*](X)||_1 \leq 0, \tag{5.10}$$

for any Hermitian X in $\mathcal{B}(\mathcal{H} \otimes \mathcal{H})$ and $t \geq 0$.

Proof If Λ_t^* is CP-divisible, then $\Lambda_t^* = V_{t,s}^* \Lambda_s$, with CPTP maps $V_{t,s}^*$. Hence

$$\frac{d}{dt}||[\text{id} \otimes \Lambda_t^*](X)||_1 = \lim_{\epsilon \to 0+} \frac{1}{\epsilon} \left(||[\text{id} \otimes \Lambda_{t+\epsilon}^*](X)||_1 - ||[\text{id} \otimes \Lambda_t^*](X)||_1 \right)$$

$$= \lim_{\epsilon \to 0+} \frac{1}{\epsilon} \left(||[\text{id} \otimes V_{t+\epsilon,t}^*][\text{id} \otimes \Lambda_t^*](X)||_1 \right.$$

$$\left. -||[\text{id} \otimes \Lambda_t^*](X)||_1 \right) \leq 0,$$

due to $||[\text{id} \otimes V_{t+\epsilon,t}^*](Y)||_1 \leq ||Y||_1$. Now, suppose that (5.10) is satisfied. Since $\text{id} \otimes \Lambda_t^*$ is invertible, there always exists $V_{t,s}^* = \Lambda_t^* \Lambda_s^{*-1}$. We show that $V_{t,s}^*$ is CPTP. The above calculation shows that

$$||[\text{id} \otimes V_{t+\epsilon,t}^*][\text{id} \otimes \Lambda_t^*](X)||_1 \leq ||[\text{id} \otimes \Lambda_t^*](X)||_1$$

for any $X = X^* \in \mathcal{B}(\mathcal{H} \otimes \mathcal{H})$. Now, since Λ_t^* is invertible, it implies that $||[\text{id} \otimes V_{t+\epsilon,t}^*](Y)||_1 \leq ||Y||_1$ for any $Y = Y^* \in \mathcal{B}(\mathcal{H} \otimes \mathcal{H})$ and due to Proposition 5.3 the map $V_{t,s}^*$ is CPTP. $\square$

Theorem 5.5 Suppose that Λ_t^* satisfies time-local master equation (4.4). If Λ_t^* is invertible, it is CP-divisible if and only if the time-local generator $\mathcal{L}_t^*$ has the following standard form

$$\mathcal{L}_t^*(\rho) = -i[H(t), \rho] + \frac{1}{2} \sum_{\alpha} \gamma_\alpha(t) \left([V_\alpha(t), \rho V_\alpha^*(t)] + [V_\alpha(t)\rho, V_\alpha^*(t)] \right), \tag{5.11}$$

with $\gamma_\alpha(t) \geq 0$.

Suppose now that Λ_t^* is invertible and it is defined by

$$\Lambda_t^* = \exp\left(\int_0^t M_u^* du \right) = e^{\mathbb{M}_t^*}, \tag{5.12}$$

where $\mathbb{M}_t^* := \int_0^t M_u^* du$ defines GKSL generator for any $t \geq 0$. It is clear that Λ_t is a legitimate dynamical map. Using Snider-Wilcox formula [37]

$$\partial_t \, e^{\mathbb{M}_t^*} = \int_0^1 ds \, e^{s \, \mathbb{M}_t^*} M_t^* e^{(1-s) \, \mathbb{M}_t^*}, \tag{5.13}$$

one finds the following formula for the corresponding time-local generator $\mathcal{L}_t^*$

$$\mathcal{L}_t^* = \partial_t \Lambda_t^* \, \Lambda_t^{*-1} = \left(\int_0^1 ds \, e^{s \, \mathbb{M}_t^*} M_t^* e^{(1-s) \, \mathbb{M}_t^*} \right) e^{-\mathbb{M}_t^*} = \int_0^1 ds \, e^{s \, \mathbb{M}_t^*} M_t^* e^{-s \, \mathbb{M}_t^*}, \tag{5.14}$$

where we used the fact that Λ_t^* is invertible and

$$\Lambda_t^{*-1} = e^{-\mathbb{M}_t^*}. \tag{5.15}$$

Note that in the commutative case, that is, $[M_t^*, M_\tau^*] = 0$, one finds $\mathcal{L}_t^* = M_t^*$. However, in general case formula (5.14) provides highly nontrivial relation between GKSL generator $\mathbb{M}_t^*$ and a legitimate time-local generator $\mathcal{L}_t^*$. Note that the above construction guaranties that Λ_t^* is divisible and

$$V_{t,s}^* = e^{\mathbb{M}_t^*} e^{-\mathbb{M}_s^*}. \tag{5.16}$$

If $V_{t,s}^*$ is CP, then Λ_t^* is CP-divisible and equivalently $\mathcal{L}_t^*$ defined in (5.14) is of GKSL form. Again, in the commutative case

$$V_{t,s}^* = e^{\mathbb{M}_t^* - \mathbb{M}_s^*} = \exp \left(\int_s^t \mathcal{L}_u^* du \right). \tag{5.17}$$

and in general

$$V_{t,s}^* = e^{\mathbb{M}_t^*} e^{-\mathbb{M}_s^*} = \mathbf{T} \exp \left(\int_s^t \mathcal{L}_u^* du \right). \tag{5.18}$$

6 CP- vs. P-Divisibility for Random Unitary Qubit Evolution

To illustrate the concepts of P- and CP-divisibility let us consider a qubit evolution governed by the following time-local generator

$$\mathcal{L}_t^*(\rho) = \frac{1}{2} \sum_{k=1}^3 \gamma_k(t)(\sigma_k \rho \sigma_k - \rho). \tag{6.1}$$

The corresponding solution for Λ_t^* reads

$$\Lambda_t^*(\rho) = \sum_{\alpha=0}^{3} p_\alpha(t)\sigma_\alpha\rho\sigma_\alpha, \tag{6.2}$$

where $\sigma_0 = \mathbb{1}$, and

$$p_\alpha(t) = \frac{1}{4}\sum_{\beta=0}^{3} H_{\alpha\beta}\lambda_\beta(t), \tag{6.3}$$

with $H_{\alpha\beta}$ being a Hadamard matrix

$$H = \begin{pmatrix} 1 & 1 & 1 & 1 \\ 1 & 1 & -1 & -1 \\ 1 & -1 & 1 & -1 \\ 1 & -1 & -1 & 1 \end{pmatrix}, \tag{6.4}$$

and $\lambda_\beta(t)$ are time-dependent eigenvalues of Λ_t^*

$$\Lambda_t^*(\sigma_\alpha) = \lambda_\alpha(t)\sigma_\alpha, \tag{6.5}$$

defined as follows $\lambda_0(t) = 1$ and

$$\begin{aligned} \lambda_1(t) &= \exp(-\Gamma_2(t) - \Gamma_3(t)], \\ \lambda_2(t) &= \exp(-\Gamma_1(t) - \Gamma_3(t)), \\ \lambda_3(t) &= \exp(-\Gamma_1(t) - \Gamma_2(t)), \end{aligned} \tag{6.6}$$

with $\Gamma_k(t) = \int_0^t \gamma_k(\tau)d\tau$. Now, the map (6.2) is CP iff $p_\alpha(t) \geq 0$ which is equivalent to the following set of conditions for λs [38, 39]

$$1 + \lambda_1(t) + \lambda_2(t) + \lambda_3(t) \geq 0, \tag{6.7}$$

and

$$\lambda_i(t) + \lambda_j(t) \leq 1 + \lambda_k(t), \tag{6.8}$$

and $\{i, j, k\}$ run over the cyclic permutations of $\{i, j, k\}$. Now, the map Λ_t^* is invertible if and only if $\Gamma_k(t) < \infty$. Being invertible it is CP-divisible if and only if $\gamma_k(t) \geq 0$. Finally, invertible CPTP map Λ_t^* is P-divisible if and only if [38]

$$\gamma_1(t) + \gamma_2(t) \geq 0, \quad \gamma_2(t) + \gamma_3(t) \geq 0, \quad \gamma_3(t) + \gamma_1(t) \geq 0. \tag{6.9}$$

Authors of [39] consider an interesting case corresponding to $\gamma_1(t) = \gamma_2(t) = 1$, and $\gamma_3(t) = -\tanh t$. Note that $\gamma_3(t)$ is always negative, however, the conditions (6.8) are satisfied and the map Λ_t^* is CPTP. Clearly, it is not CP-divisible but it is P-divisible due to the fact that conditions (6.9) are satisfied.

7 Quantum Jump Representation of the Markovian Semigroup

In this section we present a suitable representation of dynamical semigroup which will be used later for the construction of admissible memory kernels. Any GKSL generator (Heisenberg picture) $\mathcal{L} : \mathcal{B}(\mathcal{H}) \to \mathcal{B}(\mathcal{H})$ may be represented as follows

$$\mathcal{L} = \Phi - Z, \tag{7.1}$$

where Φ is CP and $Z : \mathcal{B}(\mathcal{H}) \to \mathcal{B}(\mathcal{H})$ is defined as

$$Z(X) = -i(CX - XC^*), \tag{7.2}$$

with $C \in \mathcal{B}(\mathcal{H})$ given by $C = H + \frac{i}{2}\Phi(\mathbb{1})$. The map Λ_t is unital if and only if $\mathcal{L}(\mathbb{1}) = 0$ which implies

$$\Phi(\mathbb{1}) = Z(\mathbb{1}).$$

Now, let us denote by N_t a solution of the following equation

$$\partial_t N_t = -Z N_t , \quad N_{t=0} = \text{id}. \tag{7.3}$$

One finds

$$N_t(X) = e^{-Zt} X = e^{iCt} X e^{-iC^* t}, \tag{7.4}$$

and hence $N_t = e^{-Zt}$ defines a semi-group of CP maps. Note, however, that it is not unital. Interestingly, one has

Proposition 7.1 *The map N_t satisfies*

$$\partial_t N_t(\mathbb{1}) \leq 0, \tag{7.5}$$

for any $t \geq 0$.

Proof One has

$$\partial_t N_t(\mathbb{1}) = e^{iCt}(iC - iC^*)e^{-iC^* t} = -N_t(\Phi(\mathbb{1})) \leq 0, \tag{7.6}$$

due to the fact that N_t is CP and $\Phi(\mathbb{1}) \geq 0$.   □

Remark 7.2 In the Schrödinger picture the map $N_t^* : \mathcal{T}(\mathcal{H}) \to \mathcal{T}(\mathcal{H})$ satisfies

$$\partial_t \mathrm{Tr}[N_t^*(\rho)] \leq 0, \tag{7.7}$$

for any $\rho \geq 0$. A CP map $\mathcal{E}^* : \mathcal{T}(\mathcal{H}) \to \mathcal{T}(\mathcal{H})$ such that

$$\mathrm{Tr}[\mathcal{E}^*(\rho)] \leq \mathrm{Tr}\rho,$$

is often called a quantum operation [14].

Theorem 7.3 The solution to (1.1) may be represented as follows

$$\Lambda_t = N_t * (\mathrm{id} + Q_t + Q_t * Q_t + Q_t * Q_t * Q_t + \ldots). \tag{7.8}$$

where $A_t * B_t := \int_0^t A_\tau B_{t-\tau} d\tau$, and $Q_t = \Phi N_t$.

Proof Passing to the Laplace transform of (3.3) and (7.3) one finds

$$\widetilde{\Lambda}_s = \frac{1}{s - \Phi + Z}, \quad \widetilde{N}_s = \frac{1}{s + Z} \tag{7.9}$$

and hence

$$\widetilde{\Lambda}_s = \widetilde{N}_s \frac{1}{\mathrm{id} - \Phi\widetilde{N}_s}. \tag{7.10}$$

Now, introducing $\widetilde{Q}_s := \Phi\widetilde{N}_s$ one obtains

$$\widetilde{\Lambda}_s = \widetilde{N}_s \sum_{k=0}^{\infty} \widetilde{Q}_s^n, \tag{7.11}$$

with $Q_t^{*n} := Q_t * \ldots * Q_t$ (n factors). It implies the formula (7.8) in the time domain. $\qquad\square$

Remark 7.4 Note that the series $\sum_{k=0}^{\infty} \widetilde{Q}_s^n$ is convergent. To prove it we show that $||\widetilde{Q}_s|| < 1$. One has

$$||\widetilde{Q}_s|| = ||\widetilde{Q}_s(\mathbb{1})|| = ||\Phi \, \widetilde{N}_s(\mathbb{1})||.$$

Now

$$\widetilde{N}_s(\mathbb{1}) = \frac{1}{s + Z(\mathbb{1})} = \frac{1}{s - \Phi(\mathbb{1})},$$

which is defined for $\operatorname{Re} s > ||\Phi(\mathbb{1})||$. One has

$$||\Phi \, \widetilde{N}_s(\mathbb{1})|| = \left|\left| \int_0^\infty e^{-st} \Phi \, N_t(\mathbb{1}) dt \right|\right| \leq \int_0^\infty |e^{-st}| \, ||\Phi \, N_t(\mathbb{1})|| \, dt \leq \frac{1}{\operatorname{Re} s} ||\Phi(\mathbb{1})||,$$

due to $N_t(\mathbb{1}) \leq \mathbb{1}$. Finally, $||\widetilde{Q}_s|| < 1$ for $\operatorname{Re} s > ||\Phi(\mathbb{1})||$. $\square$

Note that formulae (7.9) allow also for another representation, that is, instead of (7.10) one equivalently has

$$\widetilde{\Lambda}_s = \frac{1}{\mathrm{id} - \widetilde{N}_s} \, \widetilde{N}_s, \tag{7.12}$$

and hence introducing $P_t := N_t \Phi$ one finds the following representation

$$\Lambda_t = \left(\sum_{k=0}^\infty P_t^{*n} \right) * N_t. \tag{7.13}$$

Using the definition of the convolution formula (7.13) may be rewritten as follows

$$\Lambda_t = \sum_{k=1}^\infty \int_0^t dt_k \int_0^{t_k} dt_{k-1} \dots \int_0^{t_2} dt_1 N_{t-t_k} \Phi \, N_{t_k - t_{k-1}} \Phi \, \dots \Phi \, N_{t_2 - t_1}. \tag{7.14}$$

Remark 7.5 Representations (7.8) and (7.13) are often called a *quantum jump* representation of the dynamical map Λ_t and the CP map Φ is interpreted as quantum jump.

Remark 7.6 Representations (7.8) and (7.13) are complementary to the standard exponential representation of Markovian semigroup

$$\Lambda_t = e^{t\mathcal{L}} = \sum_{k=0}^\infty \frac{t^k}{k!} \mathcal{L}^k. \tag{7.15}$$

Note that (7.15) immediately implies that Λ_t is unital but complete positivity is not evident. On the other hand, both (7.8) and (7.13) imply that Λ_t is CP but now the preservation of unity is not evident. It shows that complete positivity and unitality (or trace preservation in the Schrödinger picture) are complementary properties.

8   Memory Kernel Master Equation

In this section we generalize the quantum jump representation of the Markovian semigroup to the solution of the memory kernel master equation (4.3). Any memory kernel K_t has the following general structure

$$K_t = \Phi_t - Z_t, \tag{8.1}$$

where maps $\Phi_t, Z_t : \mathcal{B}(\mathcal{H}) \to \mathcal{B}(\mathcal{H})$ are Hermitian and satisfy

$$\Phi_t(\mathbb{1}) = Z_t(\mathbb{1}).$$

This condition guaranties that Λ_t is unital.

Theorem 8.1 ([40]) Let $\{N_t, Q_t\}$ be a pair of CP maps in $\mathcal{B}(\mathcal{H})$ such that

1. $N_{t=0} = \mathrm{id}$,
2. $Q_t(\mathbb{1}) + \partial_t N_t(\mathbb{1}) = 0$,
3. $\|\tilde{Q}_s\| < 1$.

Then the following map

$$\Lambda_t = N_t * \sum_{n=0}^{\infty} Q_t^{*n}, \tag{8.2}$$

defines a legitimate dynamical map.

Proof Condition (3) guarantees that the series

$$\tilde{\Lambda}_s = \tilde{N}_s \sum_{k=0}^{\infty} \tilde{Q}_s^n = \tilde{N}_s \frac{1}{\mathrm{id} - \tilde{Q}_s},$$

is convergent and hence (8.2) defines a CP map. Condition (1) implies that $\Lambda_{t=0} = N_{t=0} = \mathrm{id}$. Finally, condition (2) implies that the map Λ_t is unital. Indeed, passing the Laplace transform domain one finds

$$\tilde{Q}_s(\mathbb{1}) + s\tilde{N}_s(\mathbb{1}) = \mathbb{1}, \tag{8.3}$$

which is equivalent to $\tilde{\Lambda}_s(\mathbb{1}) = \frac{1}{s}\mathbb{1}$. $\square$

Remark 8.2 Note that

$$\partial_t N_t(\mathbb{1}) = -Q_t(\mathbb{1}) \leq 0, \tag{8.4}$$

since Q_t is CP. Hence, the dual map N_t^* is trace non-increasing (quantum operation).

Theorem 8.1 may be immediately generalized as follows

Corollary 8.3 *Let $\{N_t, Q_t\}$ be a pair of k-positive maps in $\mathcal{B}(\mathcal{H})$ such that*

1. $N_{t=0} = \mathrm{id}$,
2. $Q_t(\mathbb{1}) + \partial_t N_t(\mathbb{1}) = 0$,
3. $\|\tilde{Q}_s\| < 1$.

Then the map $\Lambda_t = N_t * \sum_{n=0}^{\infty} Q_t^{*n}$ is k-positive and unital.

In the same way one proves the following.

Proposition 8.4 *Let $\{N_t, P_t\}$ be a pair of CP maps such that*

1. $N_{t=0} = \mathrm{id}$,
2. $P_t(\mathbb{1}) + \partial_t N_t(\mathbb{1}) = 0$,
3. $\|\tilde{P}_s\| < 1$.

Then the following map

$$\Lambda_t = \sum_{n=0}^{\infty} P_t^{*n} * N_t, \tag{8.5}$$

defines a legitimate dynamical map.

In this case one has in the time domain

$$\Lambda_t = \sum_{k=1}^{\infty} \int_0^t dt_k \int_0^{t_k} dt_{k-1} \ldots \int_0^{t_2} dt_1 \, P_{t-t_k} P_{t_k-t_{k-1}} \ldots P_{t_3-t_2} N_{t_2-t_1}, \tag{8.6}$$

which generalizes (7.14).

Suppose now that $\{N_t, Q_t\}$ satisfy assumptions of Theorem 8.1 (i.e., conditions (1)–(3)). Moreover, let us assume that $\tilde{N}_s$ is invertible. Then one proves the following

Theorem 8.5 The operator $K_t = \Phi_t - Z_t$, where

$$\tilde{\Phi}_s = \tilde{Q}_s \tilde{N}_s^{-1}, \quad Z_s = \frac{\mathrm{id} - s\tilde{N}_s}{\tilde{N}_s}, \tag{8.7}$$

defines a legitimate memory kernel.

Proof Indeed, one has

$$\tilde{\Lambda}_s = \frac{1}{s - \tilde{\Phi}_s + \tilde{Z}_s}, \quad \tilde{N}_s = \frac{1}{s + \tilde{Z}_s} \tag{8.8}$$

which generalizes (7.9). Hence, the representation (8.2) easily follows. □

Remark 8.6 This shows that knowing $\{N_t, Q_t\}$ one may construct a legitimate memory kernel. Following [40] we call $\{N_t, Q_t\}$ a *legitimate pair*.

To illustrate the above construction let us consider the following.

Example 8.7 Let

$$N_t = \left(1 - \int_0^t f(\tau)d\tau\right)\mathrm{id}, \tag{8.9}$$

where the function $f : \mathbb{R}_+ \to \mathbb{R}$ satisfies:

$$f(t) \geq 0, \quad \int_0^\infty f(\tau)d\tau \leq 1.$$

Moreover, let $Q_t = f(t)\mathcal{E}$, where $\mathcal{E}$ is an arbitrary unital CP map. Then one finds the following formula for the memory kernel

$$K_t = \kappa(t)(\mathcal{E} - \mathrm{id}), \tag{8.10}$$

where the function $\kappa(t)$ is defined in terms of $f(t)$ as follows

$$\widetilde{\kappa}(s) = \frac{s\widetilde{f}(s)}{1 - \widetilde{f}(s)}. \tag{8.11}$$

In particular taking $f(t) = \gamma e^{-\gamma t}$ one finds $K_t = \delta(t)\mathcal{L}$, with

$$\mathcal{L} = \gamma(\mathcal{E} - \mathrm{id}), \tag{8.12}$$

being the standard GKSL generator.

Now, we show that conditions (1)–(3) from Theorem 8.1 are sufficient but not necessary, that is, formula (8.2) may give rise to legitimate CPTP map even if these conditions are not satisfied. Indeed, consider a CP unital map $\mathcal{E}$ such that $\mathcal{E}\mathcal{E} = \mathcal{E}$. Then one can easily find

$$\Lambda_t = N_t + \int_0^t f(\tau)d\tau \, \mathcal{E} = N_t + \int_0^t Q_\tau d\tau. \tag{8.13}$$

Indeed, one has

$$\sum_{k=1}^\infty \widetilde{Q}_s^n = \sum_{k=1}^\infty \widetilde{f}^n(s)\mathcal{E} = \frac{\widetilde{f}(s)}{1 - \widetilde{f}(s)}\mathcal{E}$$

and hence

$$\widetilde{\Lambda}_s = \widetilde{N}_s \sum_{k=0}^\infty \widetilde{Q}_s^n = \widetilde{N}_s + \frac{1}{s}\widetilde{f}(s)\mathcal{E} = \widetilde{N}_s + \frac{1}{s}\widetilde{Q}_s,$$

which reproduces (8.13) in the time domain. I shows that the condition $f(t) \geq 0$ is not necessary. One needs only $\int_0^t f(\tau)d\tau \in [0, 1]$. Hence $Q_t = f(t)\mathcal{E}$ need not be CP and still (8.2) defines CPTP map.

Example 8.8 (Quantum Semi-Markov Evolution [41–43]) Consider the following pair $\{N_t, Q_t\}$ such that Q_t is CP and satisfies

$$\int_0^t Q_\tau(\mathbb{1})d\tau \leq \mathbb{1}. \tag{8.14}$$

Define a CP map N_t by

$$N_t(X) = \sqrt{\mathbf{g}_t}\, X \sqrt{\mathbf{g}_t}, \tag{8.15}$$

where $\mathbf{g}_t := \mathbb{1} - \int_0^t Q_\tau(\mathbb{1})d\tau$. It is clear that N_t is CP, $N_0 = $ id, and $Q_t(\mathbb{1}) + \partial_t N_t(\mathbb{1}) = 0$. Finally, for Re $s > 0$ one has

$$||\widetilde{Q}_s|| = ||\widetilde{Q}_s(\mathbb{1})|| = ||\int_0^\infty e^{-st} Q_t(\mathbb{1})dt|| < ||\int_0^t Q_t(\mathbb{1})|| < 1$$

and hence $\{N_t, Q_t\}$ satisfies all conditions of Theorem 8.1. We stress that Λ_t in this case is defined entirely in terms of a single CP map Q_t satisfying additional condition (8.14).

For other approaches see also [44–46].

Proposition 8.9 *For any pair of functions $\{N_t, Q_t\}$ satisfying conditions 1)-3) the corresponding dynamical map (8.2) satisfies*

$$\partial_t \Lambda_t = \int_0^t \mathbb{K}_{t-\tau} \Lambda_\tau d\tau + \partial_t N_t \,, \quad \Lambda_0 = \text{id}, \tag{8.16}$$

where the new memory kernel $\mathbb{K}_t$ is defined by

$$\widetilde{\mathbb{K}}_s = s \widetilde{N}_s \widetilde{Q}_s \widetilde{N}_s^{-1}, \tag{8.17}$$

provided $\widetilde{N}_s$ is invertible.

Note that if $[N_t, Q_\tau] = 0$, then (8.17) reduces to

$$\widetilde{\mathbb{K}}_s = s \widetilde{Q}_s, \tag{8.18}$$

and implies the following relation in the time domain

$$\mathbb{K}_t = \delta(t)\, Q_0 + \partial_t Q_t, \tag{8.19}$$

with $\delta(t)$ denoting Dirac δ-distribution. In this case the corresponding memory kernel master equation has the following form

$$\partial_t \Lambda_t = Q_0 \Lambda_t + \int_0^t \partial_\tau Q_\tau \Lambda_{t-\tau} d\tau + \partial_t N_t \,, \quad \Lambda_0 = \text{id}, \tag{8.20}$$

and it incorporates three terms: local generator Q_0, memory kernel $\partial_\tau Q_\tau$, and the inhomogeneous term $\partial_t N_t$.

9 Conclusions

We analyzed the evolution of a quantum system represented by a quantum dynamical map beyond Markovian semigroup. Due to the celebrated Gorini, Kossakowski, Sudarshan, and Lindblad the structure of Markovian semigroup is fully characterized on the level of generators. Interestingly, beyond Markovian semigroup the problem is still open. Both time-local and memory kernel master equations were analyzed. In the case of time-local description we introduced the notion of divisibility and defined quantum evolution to be Markovian if the corresponding dynamical map is CP-divisible. On the level of memory kernel master equation we introduced a class of maps generated by the so-called legitimate pairs. Interestingly, this class describes many examples considered recently in the literature including semi-Markov evolution [42] and collision models [47].

Acknowledgement This paper was partially supported by the National Science Centre project 2018/30/A/ST2/00837.

References

1. H.-P. Breuer, F. Petruccione, *The Theory of Open Quantum Systems* (Oxford University Press, Oxford, 2007)
2. U. Weiss, *Quantum Dissipative Systems* (World Scientific, Singapore, 2000)
3. H.J. Carmichael, *An Open Systems Approach to Quantum Optics* (Springer, Berlin, 2009); H.J. Carmichael, *Statistical Methods in Quantum Optics I: Master Equations and Fokker-Planck Equations* (Springer, Berlin, 1999)
4. R. Alicki, K. Lendi, *Quantum Dynamical Semigroups and Applications* (Springer, Berlin, 1987)
5. Á. Rivas, S. Hulega, *Open Quantum Systems. An Introduction*. Springer Briefs in Physics (Springer, Berlin, 2011)
6. E.B. Davies, *Quantum Theory of Open Systems* (Academic, London, 1976)
7. F. Haake, *Statistical Treatment of Open Systems by Generalized Master Equations*. Springer Tracts in Modern Physics, vol. 66 (Springer, Berlin, 1973)
8. V. Gorini, A. Kossakowski, E.C.G. Sudarshan, J. Math. Phys. **17**, 821 (1976)
9. G. Lindblad, Commun. Math. Phys. **48**, 119 (1976)
10. Á. Rivas, S.F. Huelga, M.B. Plenio, Rep. Prog. Phys. **77**, 094001 (2014)
11. H.-P. Breuer, E.-M. Laine, J. Piilo, B. Vacchini, Rev. Mod. Phys. **88**, 021002 (2016)
12. I. de Vega, D. Alonso, Rev. Mod. Phys. **89**, 015001 (2017)
13. L. Li, M.J.W. Hall, H.M. Wiseman, Phys. Rep. **759**, 1–51 (2018). arXiv:1712.08879
14. M.A. Nielsen, I.L. Chuang, *Quantum Computation and Quantum Information* (Cambridge University Press, Cambridge, 2000); M. Wilde, *Quantum Information Theory* (CUP, Cambridge, 2013)

15. V. Paulsen, *Completely Bounded Maps and Operator Algebras* (Cambridge University Press, Cambridge, 2002)
16. E. Størmer, *Positive Linear Maps of Operator Algebras*. Springer Monographs in Mathematics (Springer, Berlin, 2013)
17. R.V. Kadison, J.R. Ringrose, *Fundamentals of the Theory of Operator Algebras*. Graduate Studies in Mathematics, vols. 1–2 (American Mathematical Society, Providence, 1997)
18. R. Bhatia, *Positive Definite Matrices*. Princeton Series in Applied Mathematics (Princeton University Press, Princeton, 2007)
19. W.F. Stinespring, Proc. Am. Math. Soc. **6**, 211 (1955)
20. K. Kraus, *States, Effects and Operations: Fundamental Notions of Quantum Theory*. Lecture Notes in Physics (Springer, Berlin, 1983)
21. I. Bengtsson, K. Życzkowski, *Geometry of Quantum States: An Introduction to Quantum Entanglement* (Cambridge University Press, Cambridge, 2007)
22. D. Chruściński, A. Jamiołkowski, *Geometric Phases in Classical and Quantum Mechanics* (Birkhäuser, Boston, 2004)
23. I. Siemon, A.S. Holevo, R.F. Werner, Open Syst. Inf. Dyn. **24**, 1740015 (2017)
24. S. Nakajima, Prog. Theor. Phys. **20**, 948 (1958); R. Zwanzig, J. Chem. Phys. **33**, 1338 (1960)
25. H. Grabert, *Projection Operator Techniques in Nonequilibrium Statistical Mechanics*. Springer Tracts in Modern Physics, vol. 95, ed. by G. Höhler (Springer, Berlin, 1982)
26. S.M. Barnett, S. Stenholm, Phys. Rev. A **64**, 033808 (2001)
27. A. Shabani, D.A. Lidar, Phys. Rev. A **71**, 020101 (R) (2005)
28. F. Benatti, R. Floreanini, Int. J. Mod. Phys. B **19**, 3063 (2005)
29. P. Hänggi, H. Thomas, Z. Phys. B Condens. Matter **26**, 85 (1977); H. Grabert, P. Talkner, P. Hänggi, Z. Phys. B Condens. Matter **26**, 389 (1977)
30. M.J.W. Hall, J. Phys. A **41**, 205302 (2008); B. Vacchini, Phys. Rev. Lett. **117**, 230401 (2016); D. Chruściński, A. Kossakowski, Phys. Rev. Lett. **104**, 070406 (2010)
31. N.G. van Kampen, *Stochastic Processes in Physics and Chemistry*, 3rd edn. (North Holland, Amsterdam, 2007)
32. M.M. Wolf, J. Eisert, T.S. Cubitt, J.I. Cirac, Phys. Rev. Lett. **101**, 150402 (2008)
33. Á. Rivas, S.F. Huelga, M.B. Plenio, Phys. Rev. Lett. **105**, 050403 (2010)
34. D. Chruściński, S. Maniscalco, Phys. Rev. Lett. **112**, 120404 (2014)
35. H.-P. Breuer, E.-M. Laine, J. Piilo, Phys. Rev. Lett. **103**, 210401 (2009)
36. D. Chruściński, A. Kossakowski, Á. Rivas, Phys. Rev. A **83**, 052128 (2011)
37. R.F. Snider, J. Math. Phys. **5**, 1580 (1964); R.M. Wilcox, J. Math. Phys. **8**, 962 (1967)
38. D. Chruściński, F.A. Wudarski, Phys. Lett. A **377**, 1425 (2013); D. Chruściński, F.A. Wudarski, Phys. Rev. A **91**, 012104 (2015)
39. M.J.W. Hall, J.D. Cresser, L. Li, E. Andersson, Phys. Rev. A **89**, 042120 (2014)
40. D. Chruściński, A. Kossakowski, Phys. Rev. A **94**, 020103(R) (2016)
41. A.A. Budini, Phys. Rev. A **69**, 042107 (2004); A.A. Budini, Phys. Rev. E **89**, 012147 (2014); A.A. Budini, Phys. Rev. A **88**, 012124 (2013)
42. H.-P. Breuer, B. Vacchini, Phys. Rev. Lett. **101**, 140402 (2008); Phys. Rev. E **79**, 041147 (2009); D. Chruściński, A. Kossakowski, EPL **97**, 20005 (2012); B. Vacchini, Phys. Rev. A **87**, 030101 (2013); A.A. Budini, Phys. Rev. A **88**, 029904 (2013); B. Vacchini, A. Smirne, E.-M. Laine, J. Piilo, H.-P. Breuer, New J. Phys. **13**, 093004 (2011); K. Siudzińska, D. Chruściński, Phys. Rev. A **96**, 022129 (2017)
43. D. Chruściński, A. Kossakowski, Phys. Rev. A **95**, 042131 (2017)
44. S. Maniscalco, Phys. Rev. A **72**, 024103 (2005); S. Maniscalco, F. Petruccione, Phys. Rev. A **73**, 012111 (2006)
45. F.A. Wudarski, P. Należyty, G. Sarbicki, D. Chruściński, Phys. Rev. A **91**, 042105 (2015)
46. J. Wilkie, Phys. Rev. E **62**, 8808 (2000); J. Wilkie, Y.M. Wong, J. Phys. A **42**, 015006 (2009)
47. F. Ciccarello, G.M. Palma, V. Giovannetti, Phys. Rev. A **87**, 040103(R) (2013); B. Vacchini, Phys. Rev. A **87**, 030101(R) (2013); Int. J. Quantum Inf. **12**, 1461011 (2014); S. Campbell, F. Ciccarello, G.M. Palma, B. Vacchini, Phys. Rev. A **98**, 012142 (2018)

Aspects of Micro-Local Analysis and Geometry in the Study of Lévy-Type Generators

Niels Jacob and Elian O. T. Rhind

Abstract Generators of Feller processes are pseudo-differential operators with negative definite symbols, thus they are objects of micro-local analysis. Continuous negative definite functions (and symbols) give often raise to metrics and these metrics are important to understand, for example, transition functions of certain Feller processes. In this survey we outline some of the more recent results and ideas while at the same time we long to introduce into the field.

1 Introduction

Stochastic processes are in general not objects associated with micro-local analysis and the relations between (differential) geometry and diffusions is a relatively recent subject of mathematical investigations. For non-diffusion Markov processes, e.g. general Feller processes with discontinuous paths, relations with (differential) geometry are essentially unexplored.

However, since the work of Courrège [17] in 1966 we know that generators of Feller semigroups, i.e. Feller processes, are pseudo-differential operators but their symbols are in general quite "exotic". For fixed space coordinates they must be a characteristic exponent of a Lévy process, i.e. satisfy a Lévy–Khinchine formula, which is equivalently to say that they are a continuous negative definite function. Hence they are in general neither smooth, nor do they admit some type of homogeneity decomposition—in other words, they do not fit into any "classical" symbol class. An invariantly defined principal symbol does in general not exist, hence transferring results from micro-local analysis to these operators is a problem, eventually micro-local analysis is an analysis of objects defined on the co-tangent bundle.

The author Niels Jacob wrote the appendix jointly with James Harris.

N. Jacob (✉) · E. O. T. Rhind
Swansea University, Swansea, Wales, UK
e-mail: n.jacob@swansea.ac.uk; elianrhind@googlemail.com

© Springer Nature Switzerland AG 2019   77
D. Bahns et al. (eds.), *Open Quantum Systems*, Tutorials, Schools, and Workshops
in the Mathematical Sciences, https://doi.org/10.1007/978-3-030-13046-6_3

The Lévy–Khinchine formula allows a representation of these generators as integro-differential operators and such operators had been studied intensively. In a few cases, e.g. Komatsu [47] or Kochubei [46], aspects of classical pseudo-differential operator theory were incorporated in the sense that symbols were assumed to belong to some "classical" symbol classes while being also negative definite in the co-variable.

It seems that in [35] for the first time general continuous negative definite symbols were suggested to be the point of departure for constructing and studying Markov processes. The monograph [38–40] summarizes these studies until ca. 2002, a more recent survey is given by Schilling and coauthors in [12]. Since some time the first named author and some of his (former) students make some attempts to extend "classical" ideas from micro-local analysis and the analysis in metric measure spaces to pseudo-differential operators with negative definite symbols, i.e. generators of Feller processes. We are far away from a satisfactory theory, many problems do so far resist approaches to transfer methods and results from established theories, whether from micro-local analysis or the theory of (local) metric measure spaces. However, some first results indicate that much more should be possible. Hence in front of us we have a field worth to be investigated with maybe some new, fresh ideas. By introducing such a topic to PhD students or postdocs it is possible to add to their education and maybe to raise interest and to stimulate some research. In this spirit our paper is written: explaining (partly new) concepts, discussing (some) existing results, establishing the context to other fields, and indicating some open research problems.

In Sect. 2 we set the scene by introducing strongly continuous and positivity preserving contraction semigroups on some function spaces and identifying their generators as pseudo-differential operators with negative definite symbols. Recall that $\psi : \mathbb{R}^n \longrightarrow \mathbb{C}$ is a continuous negative definite function if continuous, $\psi(0) \geq 0$ and $\xi \mapsto e^{-t\psi(\xi)}$, $t > 0$, is a positive definite function in the sense of Bochner. A symbol $q(x, \xi)$ is called a **negative definite symbol** if for every x the function $\xi \mapsto q(x, \xi)$ is a continuous negative definite function. In order to exhibit the main ideas and difficulties the translation invariant case is often sufficient, i.e. the case of symbols $\psi(\xi)$ and we often will concentrate on such symbols.

Hilbert space techniques are rather powerful and in our context this leads to Dirichlet spaces. In Sect. 3 we show that an analysis in the associated extended Dirichlet space is sometimes a more natural approach, for example in relation to Nash-type inequalities, or more generally to functional inequalities. It might be that for transient semigroups investigations in the corresponding extended Dirichlet space may lead to better or more sharp results. The semigroups we are interested in allow a kernel representation and often these kernels have a density with respect to the Lebesgue measure. Thus we are interested to study these densities. It is now common to make in such a study a distinction between the diagonal behaviour and the off-diagonal decay. In the case of diffusions geometric interpretations using the underlying Riemannian or sub-Riemannian geometry are natural and successful. In [44] it was suggested to try such an approach also for the non-diffusion case.

In Sect. 4 we handle for translation invariant semigroups (they correspond to Lévy processes) the diagonal terms of densities. For this we use the fact that in many cases the square root of a real-valued continuous negative definite function induces a metric on $\mathbb{R}^n$ and we can express the diagonal term as in the case of diffusions as a volume term with respect to this metric (and the Lebesgue measure). This metric is measuring distances between co-variables. The doubling property plays a crucial role, however it is not always satisfied. In general the geometry with respect to this metric causes some difficulties: metric balls are in general not convex and they are quite anisotropic. We added an appendix (written jointly with J. Harris) where we discussed some properties of these metric balls. Finally, in Sect. 4, we introduce subordination in the sense of Bochner as a tool to construct examples.

In Sect. 5 we turn to off-diagonal estimates. Here our results are rather modest. In [44] we conjectured that the off-diagonal terms always decay "exponentially" with respect to a (square of a) time dependent metric, but so far we have no proof, but already in [44] we could provide non-trivial (classes of) examples. It seems that the conjecture might be too general and some surprises may wait for us. One surprise is that in some cases we can associate with a given Lévy process an additive process the symbol of which induces a time dependent metric. The density of this additive process has a diagonal term controlled by this time dependent metric while the off-diagonal decay is controlled by the metric induced by the Lévy process we started with. Moreover, the off-diagonal term of the Lévy process is controlled by the metric induced by the additive process. Thus we have to use two metrics to control the density. This type of duality is interesting and deserves further investigations, we refer to [13] and [14].

While following Sects. 3–5 the reader will have developed some feeling for the problems we want to approach and their difficulties when put in the context of classical micro-local analysis. In Sect. 6 we indicate in more detail where these difficulties are by outlining what we expect to achieve when dealing with the classical situation and why we cannot transfer methods and techniques in a straightforward way.

As a function on the co-tangent bundle every symbol can be viewed as a Hamilton function and it is well known that the study of the corresponding Hamilton dynamics may contribute much to our understanding of the corresponding pseudo-differential operator. For this reason we started to look at the Hamilton dynamics associated with some negative definite symbols. Here the situation is quite similar as in micro-local analysis: in general, we cannot expect classical techniques to work. For example, when switching to the corresponding Lagrange function we need a (partial) Legendre transform, hence C^1-regularity and convexity in the co-variable is required. On the other hand, when studying the Feynman–Kac formula or related spectral problems it is desirable to know the behaviour of the associated "classical" dynamical system associated with a Schrödinger operator, say of type $\psi(D) + V(q)$. Section 7 gives first results of such an investigation, in the forthcoming thesis of the second named author much more results will be discussed.

Section 8 returns to pseudo-differential operators and following [36] we introduce some classes of operators which are perturbations of constant coefficient operators, i.e. generators of Lévy processes. We provide these results and ideas here in order to indicate how in principle previously obtained results for translation invariant operators can be extended by employing perturbation techniques. This section serves more reference purposes on the one hand side, on the other hand it can be seen as an invitation to add results for some state space dependent symbols, we refer also to [75].

Pseudo-differential operators with negative definite symbols are studied because they (may) generate Markov processes. Following [37], in particular R. Schilling could work out that in many cases the symbol can be obtained in pure probabilistic terms. Moreover in some pioneering papers he could demonstrate that the symbol obtains a lot of probabilistic interesting information about the process, i.e. it is not only a natural object, but it is also quite useful from the probabilistic point of view. In our final section we have collected some of these results, partly for reasons of "completeness", partly however to raise the expectation that micro-local analysis will contribute to our understanding of Feller processes.

A final remark of the first named author: A typical probabilist does not learn much about pseudo-differential operators and micro-local analysis, and a typical analyst working with pseudo-differential operators rarely works with stochastic processes, symbol classes as strange as ours do not belong to their world. Still I believe that both worlds belong together and their relations deserve more attention. Many of the known results about which we could report here are due to my (former) PhD students W. Hoh, R. Schilling, Victorya Knopova, B. Böttcher, Sandra Landwehr, K. Evans, Y. Zhuang, Ran Zhang, L. Bray, J. Harris and E. Rhind, and others contributed more indirectly. I consider it as a privilege to have had the opportunity to work with so many highly talented young mathematicians from different countries and cultures and to help them pursuing their careers. They all contributed much to our field of interest.

Finally, I wish to express my gratitude to Professor Ingo Witt as well as to Professor Dorothea Bahns and Professor Anke Pohl for inviting me to deliver these lectures during the workshop and to contribute to this volume. The financial support for N.J. and E.R. while attending the workshop is gratefully appreciated.

2   Auxiliary Results

This section serves to provide the reader with some background knowledge used in the main text. We believe that most of the readers will know some of the material, but parts might be less familiar. Given the mixed audience we do have and must have in mind we also feel the need to supply a coherent presentation of background material not least to fix notations. Our standard reference will be [38], a further text we want to refer to is [12]. Since we are dealing with some common material we prefer to keep the references on a minimal scale. Standard notations such as L^p ($\mathbb{R}^n$), etc. are taken for granted to be known and they will coincide with those in [38].

Let $(X, \|.\|)$ be a Banach space. A family $(T_t)_{t \geq 0}$ of linear operators $T_t : X \longrightarrow X$ is called a **strongly continuous contraction semigroup** (of linear operators on X) if

$$T_t \circ T_s = T_{t+s} , T_0 = \mathrm{id} ; \tag{2.1}$$

$$\|T_t\| \leq 1 ; \tag{2.2}$$

$$\lim_{t \to 0} \|T_t u - u\| = 0 ; \tag{2.3}$$

where $\|T_t\|$ denotes the operator norm of T_t. Note that the normalization $T_0 = \mathrm{id}$ is not always used.

A strongly continuous contraction semigroup on $(C_\infty (\mathbb{R}^n) , \|.\|_\infty)$, the space of all continuous functions vanishing at infinity equipped with the sup-norm, is called a **Feller semigroup** if it is also positivity preserving, i.e.

$$u \geq 0 \quad \text{implies} \quad T_t u \geq 0 . \tag{2.4}$$

A consequence of the Riesz representation theorem is the existence of a kernel $p_t(x, \mathrm{d}y)$ which allows the representation

$$T_t u(x) = \int_{\mathbb{R}^n} u(y) \, p_t(x, \mathrm{d}y) . \tag{2.5}$$

In the case that $p_t(x, \mathrm{d}y)$ admits a density with respect to the Lebesgue measure $\lambda^{(n)}$ we write $p_t(x, \mathrm{d}y) = p_t(x, y)\lambda^{(n)}(\mathrm{d}y)$ and we have

$$T_t u(x) = \int_{\mathbb{R}^n} u(y) p_t(x, y)\mathrm{d}y . \tag{2.6}$$

One of the main objectives is to study the density $p_t(x, y)$, more precisely the **transition function** $(t, x, y) \mapsto p_t(x, y)$.

We call a strongly continuous contraction semigroup $(T_t)_{t \geq 0}$ on $L^p (\mathbb{R}^n)$, $1 \leq p < \infty$, an **L^p-sub-Markovian semigroup** if

$$0 \leq u \leq 1 \text{ a.e.} \quad \text{implies} \quad 0 \leq T_t \leq 1 \text{ a.e.} , \tag{2.7}$$

where a.e. (almost everywhere) refers to the Lebesgue measure (or in the case a space $L^p(\Omega, \mu)$ is considered with respect to the measure μ). Often we start with a contraction semigroup of operators defined on a dense subset of $C_\infty (\mathbb{R}^n)$ which is also dense in $L^p (\mathbb{R}^n)$, for example the test functions $C_0^\infty (\mathbb{R}^n)$ or the Schwartz space $S (\mathbb{R}^n)$. Hence we may extend by continuity this semigroup to $C_\infty (\mathbb{R}^n)$ as well as to $L^p (\mathbb{R}^n)$. In such a situation we will in general not introduce separate notations for these extensions.

For a strongly continuous contraction semigroup on a Banach space $(X, \|.\|)$, we introduce the **generator** $(A, D(A))$ by

$$D(A) := \left\{ u \in X \; \middle| \; \lim_{t \to 0} \frac{T_t u - u}{t} \text{ exists as strong limit} \right\} \tag{2.8}$$

and

$$Au := \lim_{t \to 0} \frac{T_t u - u}{t} \; , \tag{2.9}$$

i.e.

$$\lim_{t \to 0} \left\| \frac{T_t u - u}{t} - Au \right\| = 0 \; . \tag{2.10}$$

The generator is always densely defined, unique, and a closed operator. For a given strongly continuous contraction semigroup $(T_t)_{t \geq 0}$, the **resolvent** is defined for $\lambda > 0$ by

$$R_\lambda u := \int_0^\infty e^{-\lambda t} T_t u \, dt \; , \quad u \in X \; . \tag{2.11}$$

It follows that

$$R_\lambda u = (\lambda - A)^{-1} u \; . \tag{2.12}$$

The central result for generators is the Hille–Yosida theorem which we give in the version of Phillips and Lumer:

Theorem 2.1 *A linear closed operator $(A, D(A))$ on a Banach space $(X, \|.\|)$ is the generator of a strongly continuous contraction semigroup $(T_t)_{t \geq 0}$ on X if and only if $D(A) \subset X$ is dense, A is a dissipative operator in the sense that $\|(\lambda - A)u\| \geq \lambda \|u\|$ for all $\lambda > 0$ and $u \in D(A)$, and for some $\lambda > 0$ we have $R(\lambda - A) = X$.*

Note that the range condition $R(\lambda - A) = X$ is equivalent to the statement that for every $f \in X$ there exists $u \in D(A)$ such that $\lambda u - Au = f$, i.e. verifying the range condition is equivalent to solve for all $f \in X$ the equation $\lambda u - Au = f$. In most interesting cases Theorem 2.1 cannot be applied. The more applicable version is

Theorem 2.2 *A linear operator $(A, D(A))$ on a Banach space $(X, \|.\|)$ is closable and its closure is the generator of a strongly continuous contraction semigroup if and only if it is densely defined and dissipative and for some $\lambda > 0$ we have $R(\lambda - A) = X$, i.e. the range of $\lambda - A$ is dense in X.*

The generator $(A, D(A))$ of a Feller semigroup satisfies the **positive maximum principle**: For $u \in D(A)$ we have

$$u(x_0) = \sup_{x \in \mathbb{R}^n} u(x) \geq 0 \quad \text{implies} \quad (Au)(x_0) \leq 0 . \tag{2.13}$$

The positive maximum principle implies the dissipativity on $C_\infty (\mathbb{R}^n)$ and we have

Theorem 2.3 (Hille–Yosida–Ray) *A linear operator $(A, D(A))$ on $C_\infty (\mathbb{R}^n)$ is closable and its closure is the generator of a Feller semigroup if and only if $D(A)$ is dense in $C_\infty (\mathbb{R}^n)$, $(A, D(A))$ satisfies the positive maximum principle and for some $\lambda > 0$ we have $\overline{R(\lambda - A)} = C_\infty (\mathbb{R}^n)$.*

We will give below, see (2.41) and (2.42), a characterization of operators satisfying the positive maximum principle, hence of generators of Feller semigroups.

On $L^2 (\mathbb{R}^n)$ we have a natural Hilbert space structure and except when discussing spectral problems we will consider $L^2 (\mathbb{R}^n)$ as a Hilbert space over $\mathbb{R}$. (A standard complexification procedure will link the real L^2-space with the complex L^2-space.) We need

Definition 2.4 A closed bilinear form $(\mathcal{E}, \mathcal{F})$ is called a **symmetric Dirichlet form** on $L^2 (\mathbb{R}^n)$ if its domain $\mathcal{F} \subset L^2 (\mathbb{R}^n)$ is dense, $\mathcal{E} : \mathcal{F} \times \mathcal{F} \longrightarrow \mathbb{R}$ is a symmetric, non-negative bilinear form, and for every $u \in \mathcal{F}$ it follows that $((0 \vee u) \wedge 1) \in \mathcal{F}$ and

$$\mathcal{E}((0 \vee u) \wedge 1, (0 \vee u) \wedge 1) \leq \mathcal{E}(u, u) . \tag{2.14}$$

Here we use $a \vee b := \max(a, b)$ and $a \wedge b := \min(a, b)$. Note that τ is called a **normal contraction** on $H \subset L^2 (\mathbb{R}^n)$ if $\tau u \in H$ for $u \in H$ and

$$|(\tau u)(x) - (\tau u)(y)| \leq |u(x) - u(y)| \quad \text{and} \quad |(\tau u)(x)| \leq |u(x)| . \tag{2.15}$$

The mapping $u \mapsto (0 \vee u) \wedge 1$ is a normal contraction on $\mathcal{F}$ and we can replace in (2.14) this special normal contraction by any other normal contraction.

As a densely defined closed bilinear form $(\mathcal{E}, \mathcal{F})$ admits a densely defined generator $(A, D(A))$ which is a closed operator, $D(A) \subset \mathcal{F}$, and for $u \in D(A)$, $v \in \mathcal{F}$ we have

$$\mathcal{E}(u, v) = (-Au, v)_{L^2} . \tag{2.16}$$

The operator $(A, D(A))$ is in fact self-adjoint and the generator of a sub-Markovian L^2-semigroup $(T_t)_{t \geq 0}$ which can be obtained by using the spectral theorem as

$$T_t u = e^{At} u . \tag{2.17}$$

Generators of symmetric Dirichlet forms are self-adjoint **Dirichlet operators** in the sense that

$$\int_{\mathbb{R}^n} (Au)(u-1)^+ \, dx \leq 0 \qquad (2.18)$$

for all $u \in D(A)$. Here $u^+ = u \vee 0$. In fact, symmetric Dirichlet forms, symmetric L^2-sub-Markovian semigroups and self-adjoint Dirichlet operators are in 1–1 correspondence. Note that $(T_t)_{t \geq 0}$ is called symmetric if every $T_t, t \geq 0$, is a symmetric operator on $L^2(\mathbb{R}^n)$.

Given a symmetric Dirichlet form on $L^2(\mathbb{R}^n)$ we can introduce the scalar products

$$\mathcal{E}_\lambda(u, v) = \mathcal{E}(u, v) + \lambda(u, v)_{L^2}, \quad \lambda > 0, \qquad (2.19)$$

and for $\lambda > 0$ these scalar products are all equivalent to each other. Moreover, $\mathcal{F}$ equipped with $\mathcal{E}_1$ is a Hilbert space and we call $(\mathcal{F}, \mathcal{E}_1)$ the **Dirichlet space** associated with the Dirichlet form $(\mathcal{E}, \mathcal{F})$. In the next section we will discuss in more detail the extended Dirichlet space $(\mathcal{F}_e, \mathcal{E})$.

Before examining the structure of Feller generators we need some facts from Fourier analysis, in particular in relation to convolution semigroups of sub-probability measures. On the Schwartz space $S(\mathbb{R}^n)$ we define the **Fourier transform** by

$$\hat{u}(\xi) := (Fu)(\xi) := (2\pi)^{-n/2} \int_{\mathbb{R}^n} e^{-ix \cdot \xi} u(x) \, dx \qquad (2.20)$$

which is a bi-continuous linear bijection from $S(\mathbb{R}^n)$ onto itself with **inverse Fourier transform**

$$\left(F^{-1} u \right)(y) = (2\pi)^{-n/2} \int_{\mathbb{R}^n} e^{i\eta \cdot y} u(\eta) \, d\eta . \qquad (2.21)$$

The Fourier transform and its inverse have natural continuous extensions from $S(\mathbb{R}^n)$ to the space $S'(\mathbb{R}^n)$ of tempered distributions and therefore they are defined for bounded continuous functions as well as bounded Borel measures. With our normalization the **Plancherel theorem** reads as

$$\|\hat{u}\|_{L^2} = \|u\|_{L^2} \qquad (2.22)$$

and the **Riemann–Lebesgue lemma** is

$$\|\hat{u}\|_\infty \leq (2\pi)^{-n/2} \|u\|_{L^1} . \qquad (2.23)$$

For convolutions in $S(\mathbb{R}^n)$ or of bounded measures we have the **convolution theorem**

$$(u * v)^\wedge(\xi) = (2\pi)^{n/2}\hat{u}(\xi)\hat{v}(\xi) \tag{2.24}$$

and whenever $(u \cdot v)^\wedge$ is defined

$$(u \cdot v)^\wedge(\xi) = (2\pi)^{-n/2}(\hat{u} * \hat{v})(\xi) . \tag{2.25}$$

The Fourier transform of a bounded Borel measure is given by

$$\hat{\mu}(\xi) = (2\pi)^{-n/2} \int_{\mathbb{R}^n} e^{-ix\cdot\xi} \mu(dx) \tag{2.26}$$

and it is a **positive definite function**, i.e. for all $N \in \mathbb{N}$ and all $\xi^1, \ldots, \xi^N \in \mathbb{R}^n$ the matrix $\left(\hat{\mu}\left(\xi^k - \xi^l\right)\right)_{k,l=1,\ldots,N}$ is positive semi-definite, i.e.

$$\sum_{k,l=1}^{N} \hat{\mu}\left(\xi^k - \xi^l\right) \lambda_k \overline{\lambda_l} \geq 0 \tag{2.27}$$

for all $\lambda_1, \ldots, \lambda_N \in \mathbb{C}$.

Bochner's theorem states that the continuous positive definite functions on $\mathbb{R}^n$ are in 1–1-correspondence to the Fourier transforms of bounded Borel measures on $\mathbb{R}^n$.

A convolution semigroup $(\mu_t)_{t\geq 0}$ is a family of sub-probability measures on $\mathbb{R}^n$ (with the Borel σ-field) which is vaguely continuous, i.e.

$$\lim_{t\to 0} \int_{\mathbb{R}^n} u \, d\mu_t = u(0) = \int_{\mathbb{R}^n} u \, d\varepsilon_0 \tag{2.28}$$

for all $u \in C_0(\mathbb{R}^n)$, the continuous functions on $\mathbb{R}^n$ with compact support, and satisfies

$$\mu_t * \mu_s = \mu_{t+s} . \tag{2.29}$$

Using the convolution theorem and Bochner's theorem we can prove that for a convolution semigroup $(\mu_t)_{t\geq 0}$ exists a unique continuous function $\psi : \mathbb{R}^n \longrightarrow \mathbb{C}$ such that

$$\hat{\mu}_t(\xi) = (2\pi)^{-n/2} e^{-t\psi(\xi)} . \tag{2.30}$$

The function ψ is a **continuous negative definite function**, a key notion in our essay. The following equivalent assertions define continuous negative definite functions $\psi : \mathbb{R}^n \longrightarrow \mathbb{C}$:

(i) $\psi(0) \geq 0$ and $(2\pi)^{-n/2}e^{-t\psi}$ is for all $t > 0$ positive definite;
(ii) for all $N \in \mathbb{N}$ and all $\xi^1, \ldots, \xi^N \in \mathbb{R}^n$ the matrix

$$\left(\psi\left(\xi^k\right) + \overline{\psi\left(\xi^l\right)} - \psi\left(\xi^k - \xi^l\right) \right)_{k,l=1,\ldots,N} \tag{2.31}$$

   is positive semi-definite;
(iii) there exists a constant $c \geq 0$, a vector $d \in \mathbb{R}^n$, a positive semi-definite real matrix $(q_{kl})_{k,l=1,\ldots,n}$, $q_{kl} = q_{lk}$, and a measure ν on $\mathbb{R}^n \setminus \{0\}$ integrating the function $y \mapsto 1 \wedge |y|^2$ (and called the **Lévy measure** of ψ) such that

$$\psi(\xi) = c + \mathrm{i}d \cdot \xi + \sum_{k,l=1}^{n} q_{kl}\xi_k\xi_l \tag{2.32}$$

$$+ \int_{\mathbb{R}^n \setminus \{0\}} \left(1 - e^{\mathrm{i}y\cdot\xi} - \frac{\mathrm{i}y \cdot \xi}{1 + |y|^2} \right) \nu(\mathrm{d}y) \,.$$

Note that (2.32) implies immediately that the differentiability of ψ is determined by the absolute moments of the Lévy measure, in particular continuous negative definite functions need not be differentiable. The formula (2.32) is called the **Lévy–Khinchine formula**. Here are some examples of continuous negative definite functions

$$\psi(\xi) = c \,, \quad c \geq 0 \,,$$

$$\psi(\xi) = \mathrm{i}d \cdot \xi \,, \quad i = \sqrt{-1}, \, d \in \mathbb{R}^n \,,$$

$$\psi(\xi) = \sum q_{kl}\xi_k\xi_l \quad \left(q_{kl} = q_{lk}, \, (q_{kl})_{k,l=1,\ldots,n} \text{ positive semi-definite} \right) \,,$$

$$\psi(\xi) = 1 - e^{-\mathrm{i}a\cdot\xi} \,,$$

$$\psi(\xi) = |\xi|^{2\alpha} \,, \quad 0 < \alpha \leq 1 \,,$$

$$\psi(\xi) = \left(|\xi|^2 + m \right)^{\alpha} - m \,, \quad 0 < \alpha \leq 1 \,,$$

$$\psi(\xi) = (\mathrm{i}\xi)^{\alpha} \,, 0 < \alpha < 1 \,, \xi \in \mathbb{R} \,,$$

$$\psi(\xi) = \ln \cosh \xi \,, \quad \xi \in \mathbb{R} \,,$$

more examples are discussed in the following sections.

Since continuous negative definite functions are of utmost importance for our investigations, here are some of their properties:

(i) The set of all continuous negative definite functions form a convex cone which is closed under uniform convergence on compact sets.
(ii) We have

$$\psi(\xi) = \overline{\psi(-\xi)} \; ; \tag{2.33}$$

$$\mathrm{Re}\ \psi(\xi) \geq 0 \; ; \tag{2.34}$$

$$\sqrt{|\psi(\xi + \eta)|} \leq \sqrt{|\psi(\xi)|} + \sqrt{|\psi(\eta)|} \; ; \tag{2.35}$$

and

$$\frac{1 + |\psi(\xi)|}{1 + |\psi(\eta)|} \leq 2\,(1 + |\psi(\xi - \eta)|) \quad \text{(Peetre's inequality)} \; . \tag{2.36}$$

(iii) If $\psi : \mathbb{R}^n \longrightarrow \mathbb{R}$ is a continuous negative definite function such that $\psi(\xi) = 0$ if and only if $\xi = 0$ then a metric is given on $\mathbb{R}^n$ by $\psi^{1/2}(\xi - \eta)$.
(iv) If ψ is a continuous negative definite function so are $\mathrm{Re}\ \psi$ and $\psi(.) - \psi(0)$.
(v) If $\psi_j : \mathbb{R}^{n_j} \longrightarrow \mathbb{C}$, $j = 1, 2$, are continuous negative definite functions, then $\psi : \mathbb{R}^n \longrightarrow \mathbb{C}$, $n = n_1 + n_2$, defined by $\psi(\xi, \eta) = \psi_1(\xi) + \psi_2(\eta)$, $(\xi, \eta) \in \mathbb{R}^n = \mathbb{R}^{n_1} \times \mathbb{R}^{n_2}$, is also a continuous negative definite function.
(vi) The zeroes of ψ form a closed subgroup of $(\mathbb{R}^n, +)$.

Let $(\mu_t)_{t \geq 0}$ be a convolution semigroup on $\mathbb{R}^n$. We define on $S(\mathbb{R}^n)$ the operators

$$(T_t u)(x) := (u * \mu_t)(x) = \int_{\mathbb{R}^n} u(x + y)\, \mu_t(dy) \tag{2.37}$$

which are contractions on $C_\infty(\mathbb{R}^n)$ as well as on $L^2(\mathbb{R}^n)$. In fact on $C_\infty(\mathbb{R}^n)$ these operators form a Feller semigroup and on $L^2(\mathbb{R}^n)$ they form an L^2-sub-Markovian semigroup. On $S(\mathbb{R}^n)$ we also obtain a representation as a pseudo-differential operator

$$T_t u(x) = (2\pi)^{-n/2} \int_{\mathbb{R}^n} e^{ix \cdot \xi} e^{-t\psi(\xi)} \hat{u}(\xi)\, d\xi \; . \tag{2.38}$$

We can use (2.38) to calculate on $S(\mathbb{R}^n)$ the Feller generator as well as the L^2-generator of $(T_t)_{t \geq 0}$ and we find

$$\lim_{t \to 0} \frac{T_t u(x) - u(x)}{t} = -\psi(D)u(x) := -(2\pi)^{-n/2} \int_{\mathbb{R}^n} e^{ix \cdot \xi} \psi(\xi) \hat{u}(\xi)\, d\xi \; . \tag{2.39}$$

Using the Lévy–Khinchine formula we obtain now

$$- \psi(D)u(x) = -cu(x) + \sum_{j=1}^{n} d_j \frac{\partial u}{\partial x_j}(x) + \sum_{k,l=1}^{n} q_{kl} \frac{\partial^2 u}{\partial x_k \partial x_l}(x) \tag{2.40}$$

$$+ \int_{\mathbb{R}^n \setminus \{0\}} \left(u(x+y) - u(x) - \sum_{j=1}^{n} \frac{y_j \frac{\partial u}{\partial x_j}(x)}{1 + |y|^2} \right) \nu(dy) \, .$$

The right-hand side of (2.40) is defined for all $u \in C_b^2(\mathbb{R}^n)$ and for $u \in C_b^2(\mathbb{R}^n) \cap C_\infty(\mathbb{R}^n)$ the operator $-\psi(D)$ satisfies the positive maximum principle: If $u(x_0) = \sup_{x \in \mathbb{R}^n} u(x) \geq 0$, then grad $u(x_0) = 0$ and we find

$$-\psi(D)u(x_0) = -cu(x_0) + \sum_{k,l=1}^{n} q_{kl} \frac{\partial^2 u}{\partial x_k \partial x_l}(x_0)$$

$$+ \int_{\mathbb{R}^n \setminus \{0\}} (u(x_0 + y) - u(x_0)) \, \nu(dy) \leq 0 \, ,$$

since $c \geq 0$, at x_0 the term $\sum_{k,l=1}^{n} q_{kl} \frac{\partial^2 u}{\partial x_k \partial x_l}(x_0) \leq 0$ since u has at x_0 a local maximum, and of course $u(x_0 + y) - u(x_0) \leq 0$ by the definition of x_0.

From (2.40) we may derive more general operators satisfying the positive maximum principle:

$$Au(x) := -c(x)u(x) + \sum_{j=1}^{n} d_j(x) \frac{\partial u}{\partial x_j}(x) + \sum_{k,l=1}^{n} q_{kl}(x) \frac{\partial^2 u}{\partial x_k \partial x_l}(x) \tag{2.41}$$

$$+ \int_{\mathbb{R}^n \setminus \{0\}} \left(u(x+y) - u(x) - \sum_{j=1}^{n} \frac{y_j \frac{\partial u}{\partial x_j}(x)}{1 + |y|^2} \right) \nu(x, dy) \, ,$$

and using the Lévy–Khinchine formula we get for $u \in S(\mathbb{R}^n)$

$$Au(x) = -(2\pi)^{-n/2} \int_{\mathbb{R}^n} e^{ix \cdot \xi} q(x, \xi) \hat{u}(\xi) \, d\xi \tag{2.42}$$

where

$$q(x, \xi) = c(x) + i \sum_{j=1}^{n} d_j(x)\xi_j + \sum_{k,l=1}^{n} q_{kl}(x)\xi_k\xi_l \tag{2.43}$$

$$+ \int_{\mathbb{R}^n \setminus \{0\}} \left(1 - e^{iy \cdot \xi} - \frac{iy \cdot \xi}{1 + |y|^2} \right) \nu(x, dy) \, ,$$

here $c(.)$, $d_j(.)$ and $q_{kl}(.) = q_{lk}(.)$ are continuous functions, $c(x) \geq 0$, furthermore we require $\sum_{k,l=1}^{n} q_{kl}(x)\xi_k\xi_l \geq 0$ for all $x \in \mathbb{R}^n$ and $\xi \in \mathbb{R}^n$, and $\nu(x, dy)$ is a kernel such that for every $x \in \mathbb{R}^n$ the function $y \mapsto 1 \wedge |y|^2$ is integrable with respect to $\nu(x, dy)$. Thus, operators defined by (2.41) or (2.42) with a symbol $q(x, \xi)$ which is with respect to ξ a continuous negative definite function are candidates for Feller generators. Indeed, a theorem due to Courrège [17] states that every linear operator from $C_0^\infty (\mathbb{R}^n)$ to $C_b (\mathbb{R}^n)$ satisfying the positive maximum principle is of type (2.41), hence of type (2.42).

The operators (2.38), (2.39) or (2.42) are pseudo-differential operators. In our essay we call any operator of type

$$h(x, D)u(x) = (2\pi)^{-n/2} \int_{\mathbb{R}^n} e^{ix \cdot \xi} h(x, \xi)\hat{u}(\xi) \, d\xi \tag{2.44}$$

where $h : \mathbb{R}^n \times \mathbb{R}^n \longrightarrow \mathbb{C}$ is a continuous function such that for $x \in \mathbb{R}^n$ fixed the function $\xi \mapsto h(x, \xi)$ has at most power growth a **pseudo-differential operator** and h is called the **symbol** of $h(x, D)$. We need to look at pseudo-differential operators having a symbol which is with respect to ξ a continuous negative definite function and this means that for our purposes in general the standard theory of pseudo-differential operators is not applicable.

Continuous negative definite functions enter into our discussion also from quite a different point of view: they are the **characteristic exponents** of Lévy processes. By definition a **Lévy process** with state space $\mathbb{R}^n$ is a stochastic process $(X_t)_{t \geq 0}$ with stationary and independent increments which is stochastically continuous, i.e. if $(\Omega, \mathcal{A}, P)$ is the underlying probability space for the random variables $X_t : \Omega \longrightarrow \mathbb{R}^n$ we have

$$P_{X_t - X_s} = \mu_{t-s}, \quad 0 < s < t ; \tag{2.45}$$

$$X_{t_1} - X_{t_0}, \ldots, X_{t_N} - X_{t_{N-1}} \text{ are independent random variables} \tag{2.46}$$

for any choice $\quad 0 \leq t_0 < t_1 < \ldots < t_N ;$

$$\lim_{s \to 0} P \left(\{ |X_{t+s} - X_t| \geq \varepsilon \} \right) = 0 \quad \text{for every} \quad \varepsilon > 0 . \tag{2.47}$$

The distributions

$$\mu_t := P_{X_t - X_0}, \quad t \geq 0 , \tag{2.48}$$

form a convolution semigroup and we have

$$\mathbb{E}^x (u (X_t)) = T_t u(x) \tag{2.49}$$

for $u \in C_\infty (\mathbb{R}^n)$ where $(T_t)_{t \geq 0}$ is the operator semigroup associated with $(\mu_t)_{t \geq 0}$. In fact (2.49) holds for bounded Borel functions, and if we choose as $u = \chi_B$ the characteristic function of a Borel set B we find

$$P_t(x, B) = T_t \chi_B(x) = P^x \{ X_t \in B \} , \qquad (2.50)$$

i.e. $P_t(x, B)$ gives the probability to be at time t in the set B if at time 0 we were at the point x. In the case where $P_t(x, \mathrm{d}y)$ has a density we have of course

$$P_t(x, B) = \int_B P_t(x, y) \, \mathrm{d}y . \qquad (2.51)$$

3 Translation Invariant Extended Dirichlet Spaces

In this paragraph we recollect some material on translation invariant extended Dirichlet spaces. We follow closely [39], however we emphasize that the presentation in [39] depends much on [6, 28] and partly [69], and indirectly on [7] and [21].

Let $(T_t)_{t \geq 0}$ be a symmetric L^2-sub-Markovian semigroup on $L^2(\mathbb{R}^n)$. It is known that $(T_t)_{t \geq 0}$ extends from $L^2(\mathbb{R}^n) \cap L^p(\mathbb{R}^n)$ to $L^p(\mathbb{R}^n)$ as sub-Markovian semigroup, $1 \leq p < \infty$, and under suitable regularity conditions it induces a Feller semigroup on $C_\infty(\mathbb{R}^n)$. For the operator

$$S_t u := \int_0^t T_s u \, \mathrm{d}s , \quad \mathrm{t} > 0 , \qquad (3.1)$$

we have

$$\|S_t u\|_{L^p} \leq t \|u\|_{L^p} . \qquad (3.2)$$

Moreover, S_t is positivity preserving and for $t < t'$ it follows for every $u \geq 0$ a.e. that $S_t u \leq S_{t'} u$ a.e.

Definition 3.1 The **potential operator** G associated with $(T_t)_{t \geq 0}$ is defined for $u \in L^1(\mathbb{R}^n)$, $u \geq 0$ a.e., by

$$Gu(x) := \lim_{N \to \infty} S_N u(x) = \sup_{N \in \mathbb{N}} S_N u(x) \leq \infty . \qquad (3.3)$$

Remark 3.2 Denoting by $(R_\lambda)_{\lambda > 0}$ the resolvent of $(T_t)_{t \geq 0}$ we find almost everywhere

$$Gu = \sup_{\lambda > 0} R_\lambda u = \lim_{\lambda \to 0} R_\lambda u . \qquad (3.4)$$

Definition 3.3 If $Gu < \infty$ a.e. for all $u \in L^1(\mathbb{R}^n)$, $u \geq 0$ a.e., then we call $(T_t)_{t\geq 0}$ **transient**. In the case that for all $u \in L^1(\mathbb{R}^n)$, $u \geq 0$ a.e., we have $Gu(x) \in \{0, \infty\}$ for almost all $x \in \mathbb{R}^n$ we call $(T_t)_{t\geq 0}$ **recurrent**.

Theorem 3.4 *Let $(T_t)_{t\geq 0}$ be a symmetric L^2-sub-Markovian semigroup with the corresponding Dirichlet form $(\mathcal{E}, D(\mathcal{E}))$. The semigroup is transient if and only if there exists a bounded function $g \in L^1(\mathbb{R}^n)$ which is strictly positive and satisfies for all $u \in D(\mathcal{E})$*

$$\int |u|g\,\mathrm{d}x \leq \mathcal{E}^{1/2}(u, u) . \tag{3.5}$$

Definition 3.5 Let $(\mathcal{E}, D(\mathcal{E}))$ be a symmetric Dirichlet space on $L^2(\mathbb{R}^n)$, i.e. $D(\mathcal{E}) \subset L^2(\mathbb{R}^n)$. The **extended Dirichlet space** $\mathcal{F}_e$ (or $(\mathcal{F}_e, \mathcal{E}_e)$) associated with $(\mathcal{E}, D(\mathcal{E}))$ is the family of all measurable functions $u : \mathbb{R}^n \longrightarrow \mathbb{R}$, $|u| < \infty$ a.e., such that there exists a sequence $(u_k)_{k\in\mathbb{N}}$, $u_k \in D(\mathcal{E})$, which converges almost everywhere to u and which is a Cauchy sequence with respect to $\mathcal{E}$, i.e. $\mathcal{E}(u_k - u_l, u_k - u_l) \to 0$ as $k, l \to \infty$.

For $u \in \mathcal{F}_e$ we call a sequence satisfying the conditions of Definition 3.5 an **approximating sequence** for u.

Theorem 3.6 *Let $(\mathcal{E}, D(\mathcal{E}))$, $D(\mathcal{E}) \subset L^2(\mathbb{R}^n)$, be a symmetric Dirichlet space with associated sub-Markovian semigroup $(T_t)_{t\geq 0}$. For every $u \in \mathcal{F}_e$ and every approximating sequence $(u_k)_{k\in\mathbb{N}}$ for u the limit $\lim_{k\to\infty}\mathcal{E}(u_k, u_k)$ exists and is independent of the choice of $(u_k)_{k\in\mathbb{N}}$. Moreover we have $D(\mathcal{E}) = \mathcal{F}_e \cap L^2(\mathbb{R}^n)$. In particular we can extend $\mathcal{E}$ to $\mathcal{F}_e$ by*

$$\mathcal{E}(u, u) := \lim_{k\to\infty}\mathcal{E}(u_k, u_k) . \tag{3.6}$$

Corollary 3.7 *For $u \in \mathcal{F}_e$ we have*

$$\lim_{t\to 0}\mathcal{E}(T_t u - u, T_t u - u) = 0 \tag{3.7}$$

and

$$\mathcal{E}(T_t u, T_t u) \leq \mathcal{E}(u, u) . \tag{3.8}$$

The following theorem summarizes the basic facts about extended Dirichlet spaces, see [28], Theorem 1.5.3, or [39], Theorem 3.5.46.

Theorem 3.8 *Let* $(\mathcal{E}, D(\mathcal{E}))$ *be a symmetric transient Dirichlet space,* $D(\mathcal{E}) \subset$ $L^2(\mathbb{R}^n)$, *i.e. the semigroup* $(T_t)_{t \geq 0}$ *associated with* $(\mathcal{E}, D(\mathcal{E}))$ *is transient. For the extended Dirichlet space* $\mathcal{F}_e$ *with scalar product* $\mathcal{E}$ *the following holds:*

  (i) $(\mathcal{F}_e, \mathcal{E})$ *is a Hilbert space.*
 (ii) *There exists a bounded strictly positive function* $g \in L^1(\mathbb{R}^n)$ *such that*

$$\int |u(x)| g(x) \, dx \leq \mathcal{E}^{1/2}(u, u) \tag{3.9}$$

   holds for all $u \in \mathcal{F}_e$.
 (iii) $\mathcal{F}_e \cap L^2(\mathbb{R}^n)$ *is dense in* $L^2(\mathbb{R}^n)$ *and* $\mathcal{F}_e$.
 (iv) *For every normal contraction* τ *and all* $u \in \mathcal{F}_e$ *we have* $\tau u \in \mathcal{F}_e$ *and*

$$\mathcal{E}(\tau u, \tau u) \leq \mathcal{E}(u, u) . \tag{3.10}$$

In addition we have $\mathcal{F}_e \cap L^2(\mathbb{R}^n) = D(\mathcal{E})$. *Conversely, suppose that* $(H, \mathcal{E})$ *satisfies (i)–(iv). Then* $(H, \mathcal{E})$ *is the extended Dirichlet space of the transient, symmetric Dirichlet form* $(\mathcal{E}, H \cap L^2(\mathbb{R}^n))$.

We need also the notion of the abstract potential operator in the sense of K. Yosida, see [74], and we give the definition here just for the case of a symmetric sub-Markovian semigroup on $L^2(\mathbb{R}^n)$.

Definition 3.9 Let $(T_t)_{t \geq 0}$ be a symmetric sub-Markovian semigroup on $L^2(\mathbb{R}^n)$ with generator $(A, D(A))$, resolvent $(R_\lambda)_{\lambda > 0}$ and potential operator $(G, D(G))$ where

$$D(G) := \left\{ u \in L^2(\mathbb{R}^n) \,\Big|\, \lim_{N \to \infty} \int_0^N T_t u \, dt \ \text{exists in } L^2(\mathbb{R}^n) \right\} . \tag{3.11}$$

The **abstract potential operator** or the **resolvent at zero** $(R_0, D(R_0))$ associated with $(T_t)_{t \geq 0}$ is the operator R_0 defined on

$$D(R_0) := \left\{ u \in L^2(\mathbb{R}^n) \,\Big|\, \lim_{\lambda \to 0} R_\lambda u \ \text{exists in } L^2(\mathbb{R}^n) \right\} \tag{3.12}$$

and we set

$$R_0 u = \lim_{\lambda \to 0} R_\lambda u , \quad u \in D(R_0) . \tag{3.13}$$

We have the following.

Proposition 3.10 *Let* $(T_t)_{t \geq 0}$ *be a symmetric transient sub-Markovian semigroup on* $L^2(\mathbb{R}^n)$ *with generator* $(A, D(A))$. *In this case the corresponding potential operator and the abstract potential operator coincide as* L^2-*operators and we have* $A = -G^{-1}$ *as well as* $G = -A^{-1}$.

With these general preparations we can start to study sub-Markovian semigroups on $L^2(\mathbb{R}^n)$ associated with convolution semigroups of measures, or equivalently with real-valued continuous negative definite functions.

Let $(\mu_t)_{t\geq 0}$ be a symmetric convolution semigroup on $\mathbb{R}^n$ with associated negative definite function $\psi : \mathbb{R}^n \longrightarrow \mathbb{R}$. Later on we will require that $\xi = 0 \in \mathbb{R}^n$ is a zero of ψ, in fact we will assume $\psi(\xi) = 0$ if and only if $\xi = 0$. It follows then that all the measures μ_t are probability measures. We also assume that μ_t has no Gaussian part which implies for the **Lévy-Khinchine representation** of ψ

$$\psi(\xi) = \int_{\mathbb{R}^n \setminus \{0\}} \left(1 - e^{iy\cdot\xi}\right) \nu(dy) = \int_{\mathbb{R}^n \setminus \{0\}} (1 - \cos y \cdot \xi) \, \nu(dy) \qquad (3.14)$$

with a symmetric **Lévy measure** ν integrating the function $y \mapsto 1 \wedge |y|^2$. The relation between $(\mu_t)_{t\geq 0}$ and ψ is given by

$$\hat{\mu}_t(\xi) = (2\pi)^{-n/2} e^{-t\psi(\xi)} . \qquad (3.15)$$

Now we can define on $L^p(\mathbb{R}^n)$, $1 \leq p < \infty$, or $C_\infty(\mathbb{R}^n)$ the operators

$$T_t^\psi u(x) = \int_{\mathbb{R}^n} u(x - y) \, \mu_t(dy) . \qquad (3.16)$$

The convolution theorem implies further

$$\left(T_t^\psi u\right)^\wedge (\xi) = (2\pi)^{n/2} \hat{u}(\xi)\hat{\mu}_t(\xi) = \hat{u}(\xi)e^{-t\psi(\xi)} \qquad (3.17)$$

which yields the pseudo-differential operator representation of T_t^ψ as

$$T_t^\psi u(x) = (2\pi)^{-n/2} \int_{\mathbb{R}^n} e^{ix\cdot\xi} e^{-t\psi(\xi)} \hat{u}(\xi) \, d\xi . \qquad (3.18)$$

The L^2-generator of $\left(T_t^\psi\right)_{t\geq 0}$ is $\left(A^\psi, D\left(A^\psi\right)\right)$ with

$$D(A^\psi) = H^{\psi,2}(\mathbb{R}^n) \qquad (3.19)$$

where with

$$\|u\|_{\psi,s}^2 = \int_{\mathbb{R}^n} (1 + \psi(\xi))^s |\hat{u}(\xi)|^2 \, d\xi \qquad (3.20)$$

we have for $s \geq 0$

$$H^{\psi,s}(\mathbb{R}^n) = \left\{u \in L^2(\mathbb{R}^n) \,\Big|\, \|u\|_{\psi,s} < \infty\right\}, \qquad (3.21)$$

and

$$A^{\psi} u(x) = -\psi(D)u(x) = (2\pi)^{-n/2} \int_{\mathbb{R}^n} e^{ix \cdot \xi}(-\psi(\xi))\hat{u}(\xi) \, d\xi \; . \tag{3.22}$$

Moreover for the resolvent $(R_\lambda)_{\lambda>0}$ we have the representation

$$R_\lambda u(x) = (2\pi)^{-n/2} \int_{\mathbb{R}^n} e^{ix \cdot \xi} \frac{1}{\lambda + \psi(\xi)} \hat{u}(\xi) \, d\xi \; . \tag{3.23}$$

Note that the condition $\psi(0) = 0$ implies that $(T_t^{\psi})_{t \geq 0}$ is a conservative semigroup, i.e. $T_t^{\psi} 1 = 1$ for all $t > 0$. Here we define $T_t^{\psi} 1$ by (3.16).

The next result is taken in the formulation of [39], Theorem 3.5.51.

Theorem 3.11 *Let $(\mu_t)_{t \geq 0}$ be a symmetric convolution semigroup of probability measures on $\mathbb{R}^n$ with corresponding continuous negative definite function ψ : $\mathbb{R}^n \longrightarrow \mathbb{R}$ and symmetric Dirichlet form $(\mathcal{E}^{\psi}, D(\mathcal{E}^{\psi})) = (\mathcal{E}^{\psi}, H^{\psi,1}(\mathbb{R}^n))$, where*

$$\mathcal{E}^{\psi}(u, v) = \int_{\mathbb{R}^n} \psi(\xi)\hat{u}(\xi)\overline{\hat{v}(\xi)} \, d\xi \; , \tag{3.24}$$

and operator semigroup $\left(T_t^{\psi}\right)_{t \geq 0}$ with T_t^{ψ} as in (3.16). The following statements are equivalent

(i) $\left(T_t^{\psi}\right)_{t \geq 0}$ is transient;

(ii) for every compact set $K \subset \mathbb{R}^n$ we have

$$\mathcal{K}(K) := \int_0^\infty \mu_t(K) \, dt < \infty \; ; \tag{3.25}$$

(iii) for all $u \in C_0(\mathbb{R}^n)$, $u \geq 0$, it follows that

$$\int_0^\infty \left(T_t^{\psi} u, u\right)_0 \, dt < \infty \; ; \tag{3.26}$$

(iv) $\frac{1}{\psi} \in L_{\text{loc}}^1(\mathbb{R}^n)$.

The equivalence of (i) and (iv) is the one of importance to us. First we state a result which is mainly due to C. Berg and G. Forst, see [6].

Theorem 3.12 *Let $\left(T_t^{\psi}\right)_{t \geq 0}$ be a transient symmetric L^2-Markovian semigroup associated with the convolution semigroup $(\mu_t)_{t \geq 0}$ and the corresponding negative definite function ψ. In this case the potential operator G and the abstract potential operator R_0 are both defined as L^2-operators and coincide. Moreover R_0 is densely defined.*

We want to note that in the situation of Theorem 3.12 the operator R_λ, $\lambda > 0$, admits the representation

$$R_\lambda u = \varrho_\lambda * u \tag{3.27}$$

with

$$\varrho_\lambda = \int_0^\infty e^{-\lambda t} \mu_t \, dt \tag{3.28}$$

where

$$\varrho_\lambda(v) = \langle \varrho_\lambda, v \rangle = \int_{\mathbb{R}^n} v(x) \varrho_\lambda \, (dx)$$

$$= \int_0^\infty e^{-\lambda t} \int_{\mathbb{R}^n} v(x) \, \mu_t \, (dx) \, dt$$

$$= \int_0^\infty e^{-\lambda t} \langle \mu_t, v \rangle \, dt .$$

It follows that $(T_t^\psi)_{t \geq 0}$ is transient if and only if $(\mu_t)_{t \geq 0}$ is transient in the sense that

$$\mathcal{K}(u) := \lim_{\lambda \to 0} \varrho_\lambda(u) , \quad u \in C_0\left(\mathbb{R}^n\right) , \tag{3.29}$$

exists. We call $\mathcal{K}$ the **potential kernel** associated with $(\mu_t)_{t \geq 0}$.

Example 3.13 Let $\psi : \mathbb{R}^n \longrightarrow \mathbb{R}$ be a continuous negative definite function such that $\psi(\xi) = 0$ if and only if $\xi = 0$ and assume that $1/\psi \in L^1_{\text{loc}}(\mathbb{R}^n)$. Hence the corresponding semigroup $\left(T_t^\psi\right)_{t \geq 0}$ and Dirichlet form $(\mathcal{E}^\psi, H^{\psi,1}(\mathbb{R}^n))$ are transient. Recall that

$$\mathcal{E}^\psi(u, v) = \int_{\mathbb{R}^n} \psi(\xi)\hat{u}(\xi)\overline{\hat{v}(\xi)} \, d\xi . \tag{3.30}$$

The corresponding extended Dirichlet space is given by

$$\mathcal{F}_e^\psi := H_e^{\psi,1}(\mathbb{R}^n) := \left\{ u \in \mathcal{S}'(\mathbb{R}^n) \,\bigg|\, u \in L^1_{\text{loc}}(\mathbb{R}^n) \text{ and } \hat{u}\psi^{\frac{1}{2}} \in L^2(\mathbb{R}^n) \right\} \tag{3.31}$$

and the extended form $\mathcal{E}_e^\psi$ is again given (and often denoted) by $\mathcal{E}^\psi$. Note that $\mathcal{F}_e^\psi$ is a subspace of the weighted L^2-space $L^2(\mathbb{R}^n; \psi\lambda^{(n)})$ with norm

$$\|u\|^2_{L^2(\mathbb{R}^n; \psi\lambda^{(n)})} = \int_{\mathbb{R}^n} |\hat{u}(\xi)|^2 \psi(\xi) \, d\xi . \tag{3.32}$$

The Lévy–Khinchine representation gives further

$$\int_{\mathbb{R}^n} |\hat{u}(\xi)|^2 \psi(\xi)\, d\xi = \frac{1}{2} \int_{\mathbb{R}^n} \left(\int_{\mathbb{R}^n \setminus \{0\}} (u(x+y) - u(x))^2 \, \nu(dy) \right) dx \ . \tag{3.33}$$

In general it is not known whether for some $p \geq 1$ we can prove that $H_e^{\psi,1} (\mathbb{R}^n) \subset L^p (\mathbb{R}^n)$. However it can be shown, compare [21] that $C_0 (\mathbb{R}^n) \cap H_e^{\psi,1} (\mathbb{R}^n)$ is dense in $H_e^{\psi,1} (\mathbb{R}^n)$ and in $C_0 (\mathbb{R}^n)$. It follows further that in this case we achieve (3.5) with a strictly positive function $g \in C_\infty (\mathbb{R}^n) \cap L^1 (\mathbb{R}^n)$.

Example 3.14 Consider $\psi_\alpha(\xi) = |\xi|^\alpha$, $0 < \alpha < 2$, with corresponding operator semigroup

$$T_t^{\psi_\alpha} u(x) = T_t^{(\alpha)} u(x) = (2\pi)^{-n/2} \int_{\mathbb{R}^n} e^{ix \cdot \xi} e^{-t|\xi|^\alpha} \hat{u}(\xi)\, d\xi \ . \tag{3.34}$$

The generator of $\left(T_t^{(\alpha)} \right)_{t \geq 0}$ is given by $(-(-\Delta)^{\alpha/2}, H^\alpha (\mathbb{R}^n))$ where for $s \in \mathbb{R}$

$$H^s (\mathbb{R}^n) = \left\{ u \in \mathcal{S}' (\mathbb{R}^n) \ \middle| \ \|u\|_s^2 < \infty \right\} \tag{3.35}$$

and

$$\|u\|_s^2 = \int_{\mathbb{R}^n} \left(1 + |\xi|^2 \right)^s |\hat{u}(\xi)|^2 \, d\xi \ . \tag{3.36}$$

Since

$$\int_{B_R(0)} \frac{1}{|\xi|^\alpha}\, d\xi = c_n \int_0^R r^{n-1-\alpha}\, dr$$

it follows that $1/\psi_\alpha \in L_{\mathrm{loc}}^1 (\mathbb{R}^n)$ if and only if $\alpha < n$, i.e. $\left(T_t^{(\alpha)} \right)_{t \geq 0}$ is transient for all $n \geq 2$ which we will assume in the following. The corresponding extended Dirichlet space is given by $(H_e^{\alpha/2} (\mathbb{R}^n), \mathcal{E}^\alpha)$ where

$$H_e^{\alpha/2} (\mathbb{R}^n) = \left\{ u \in \mathcal{S}' (\mathbb{R}^n) \ \middle| \ u \in L_{\mathrm{loc}}^1 (\mathbb{R}^n) \text{ and } |.|^{\alpha/2} \hat{u} \in L^2 (\mathbb{R}^n) \right\} \tag{3.37}$$

and

$$\mathcal{E}^\alpha (u, v) = \int_{\mathbb{R}^n} |\xi|^\alpha \hat{u}(\xi) \overline{\hat{v}(\xi)}\, d\xi \tag{3.38}$$

$$= \tilde{c}_{\alpha,n} \int_{\mathbb{R}^n} \int_{\mathbb{R}^n} \frac{(u(x) - u(y))(v(x) - v(y))}{|x - y|^{n-\alpha}}\, dx\, dy \ ,$$

where

$$\tilde{c}_{\alpha,n} = \frac{2^{\alpha-1}\alpha\Gamma\left(\frac{n+\alpha}{2}\right)}{\pi^{n/2}\Gamma\left(\frac{2-\alpha}{2}\right)}.$$

For the potential operator $G^{(\alpha)}$ we find for $n \geq 2$ that

$$G^{(\alpha)}u(x) = \int_0^\infty T_t^{(\alpha)}u(x)\mathrm{d}t = (2\pi)^{-n/2}\int_{\mathbb{R}^n} \mathrm{e}^{\mathrm{i}x\cdot\xi}\frac{1}{|\xi|^\alpha}\hat{u}(\xi)\,\mathrm{d}\xi , \qquad (3.39)$$

and with the Riesz-kernel $R^{(\alpha)}(x) = c_{\alpha,n}|x|^{-n+\alpha}$ we have

$$G^{(\alpha)} = R^{(\alpha)} * u =: R^\alpha u , \qquad (3.40)$$

and

$$c_{\alpha,n} = \frac{\Gamma\left(\frac{n-\alpha}{2}\right)}{2^\alpha \pi^{n/2}\Gamma\left(\frac{\alpha}{2}\right)} .$$

This leads to the equivalent characterization of $H_e^{\alpha/2}(\mathbb{R}^n)$ as

$$H_e^{\alpha/2}(\mathbb{R}^n) = \left\{ u \in L_{\mathrm{loc}}^1(\mathbb{R}^n) \,\Big|\, u = R^{\alpha/2}f , \ f \in L^2(\mathbb{R}^n) \right\}. \qquad (3.41)$$

Of importance for the following is now.

Theorem 3.15 (Sobolev's Inequality) *Let* $1 < q < p < \infty, 0 < \alpha < n$ *and* $1/p = 1/q - \alpha/n > 0$. *Then we have*

$$\left\| R^{(\alpha)}u \right\|_{L^p} \leq c_{\alpha,p,q,n} \|u\|_{L^q} . \qquad (3.42)$$

As consequences we obtain

Corollary 3.16 *For* $n \geq 2$ *and* $0 < \alpha < 2$ *we can embed* $H_e^{\alpha/2}(\mathbb{R}^n)$ *into* $L^p(\mathbb{R}^n)$ *where* $p = \frac{2n}{n-\alpha}$ *and the estimate*

$$\|u\|_{L^p}^2 \leq c\mathcal{E}^{(\alpha)}(u,u) \qquad (3.43)$$

holds for all $u \in H_e^{\alpha/2}(\mathbb{R}^n)$.

Corollary 3.17 *Let* $\psi : \mathbb{R}^n \longrightarrow \mathbb{R}$ *be a continuous negative definite function such that for some* $c_0 > 0$ *and* $0 < \alpha < 2$ *we have for all* $\xi \in \mathbb{R}^n$

$$c_0|\xi|^\alpha \leq \psi(\xi) . \qquad (3.44)$$

If $n \geq 2$, then the semigroup $\left(T_t^{\psi} \right)_{t \geq 0}$ is transient and the extended Dirichlet space $H_e^{\psi,1} \left(\mathbb{R}^n \right)$ is continuously embedded into $L^p \left(\mathbb{R}^n \right)$, $p = \frac{2n}{n-\alpha}$, and we have the estimate

$$\|u\|_{L^p}^2 \leq \tilde{c} \mathcal{E}^{\psi}(u, u) \tag{3.45}$$

for all $u \in H_e^{\psi,1} \left(\mathbb{R}^n \right)$.

Example 3.18 This example will be taken up at several occasions later on. Consider the continuous negative definite function $\psi_{ER} : \mathbb{R}^2 \longrightarrow \mathbb{R}$

$$\psi_{ER}(\xi, \eta) = \left(|\xi|^{\alpha_1} + |\eta|^{\beta_1} \right)^{\gamma_1} + \left(|\xi|^{\alpha_2} + |\eta|^{\beta_2} \right)^{\gamma_2} \tag{3.46}$$

for $1 < \alpha_1, \alpha_2 < 2$, $1 < \beta_1, \beta_2 < 2$, $0 < \gamma_1, \gamma_2 < 1$, $\alpha_1 \gamma_1 = \beta_2 \gamma_2$, $\alpha_2 \gamma_2 = \beta_1 \gamma_1$, $\alpha_1 \gamma_1 > \alpha_2 \gamma_2$ and $\alpha_i \gamma_i > 1$, $\beta_i \gamma_i > 1$ for $i = 1, 2$. Then for all $(\xi, \eta) \in \mathbb{R}^2$

$$\kappa_0 \left(|\xi|^2 + |\eta|^2 \right)^{\frac{\alpha_1 \gamma_1}{2}} \leq \psi_{ER}(\xi, \eta) \tag{3.47}$$

for some $\kappa_0 > 0$. Since $n = 2$ we can apply Corollary 3.17 to the associated semigroup $\left(T_t^{\psi_{ER}} \right)_{t \geq 0}$ which is transient and the extended Dirichlet space $H_e^{\psi_{ER},1} \left(\mathbb{R}^2 \right)$ is continuously embedded into $L^p \left(\mathbb{R}^2 \right)$, $p = \frac{4}{2 - \alpha_1 \gamma_1}$. Hence, by (3.45) we have

$$\|u\|_{L^{\frac{4}{2-\alpha_1\gamma_1}}}^2 \leq \tilde{c} \mathcal{E}^{\psi_{ER}}(u, u) \tag{3.48}$$

for all $u \in H_e^{\psi_{ER},1} \left(\mathbb{R}^2 \right)$.

Note that for ψ_{ER} we have also the upper bound

$$\psi_{ER}(\xi, \eta) \leq \kappa_1 \left(\left(|\xi|^2 + |\eta|^2 \right)^{\frac{\alpha_1 \gamma_1}{2}} + \left(|\xi|^2 + |\eta|^2 \right)^{\frac{\alpha_2 \gamma_2}{2}} \right) \tag{3.49}$$

for some $\kappa_1 > 0$.

4   On Diagonal Terms of Transition Functions

Let $\psi : \mathbb{R}^n \longrightarrow \mathbb{R}$ be a continuous negative definite function associated with the convolution semigroup $(\mu_t)_{t \geq 0}$ and such that $\psi(\xi) = 0$ if and only if $\xi = 0$. Then a metric on $\mathbb{R}^n$ is given by

$$d_{\psi}(\xi, \eta) = \psi^{1/2}(\xi - \eta) \tag{4.1}$$

and we assume that d_ψ induces on $\mathbb{R}^n$ the Euclidean topology, which is the case if and only if $\liminf_{|\xi|\to\infty}\psi(\xi) > 0$, see [44]. In some situations it will be convenient to consider instead d_ψ for $t > 0$ the family

$$d_{\psi,t}(\xi,\eta) = \sqrt{t\psi(\xi-\eta)}\,. \tag{4.2}$$

In addition we assume that $e^{-t\psi(\cdot)} \in L^1(\mathbb{R}^n)$ for all $t > 0$. This assumption implies that the associated semigroup of operators $\left(T_t^\psi\right)_{t\geq 0}$ admits a representation

$$T_t^\psi u(x) = \int_{\mathbb{R}^n} p_t^\psi(x-y)u(y)\,\mathrm{d}y \tag{4.3}$$

where

$$p_t^\psi(x) = \frac{1}{(2\pi)^n}\int_{\mathbb{R}^n} e^{ix\cdot\xi}e^{-t\psi(\xi)}\,\mathrm{d}\xi\,. \tag{4.4}$$

We call the function $(t,x) \mapsto p_t(x) := p_t^\psi(x)$ the transition density associated with $(T_t^\psi)_{t\geq 0}$ or $(\mu_t)_{t\geq 0}$, respectively. Clearly, $p_t(.)$ is nothing but the density of μ_t with respect to the Lebesgue measure $\lambda^{(n)}$. Since by our assumption μ_t is a probability measure we find for $t > 0$ fixed that

$$p_t \in L^1(\mathbb{R}^n) \cap C_\infty(\mathbb{R}^n) \subset \bigcap_{p\geq 1} L^p(\mathbb{R}^n) \cap C_\infty(\mathbb{R}^n) \tag{4.5}$$

and

$$\hat{p}_t(\xi) = (2\pi)^{-n/2}e^{-t\psi(\xi)}\,. \tag{4.6}$$

Our general aim is to get a good understanding of p_t^ψ and hence of the operators T_t^ψ, $A^\psi = -\psi(D)$ and R_λ^ψ. In this chapter we concentrate on the term $p_t^\psi(0)$, i.e. on

$$p_t^\psi(0) = \frac{1}{(2\pi)^n}\int_{\mathbb{R}^n} e^{-t\psi(\xi)}\,\mathrm{d}\xi\,. \tag{4.7}$$

The following result is taken from Varopoulos et al. [71], see also Theorem 3.6.1 in [39], and formulated for the case $\mathbb{R}^n$ only.

Theorem 4.1 *Let $(T_t)_{t\geq 0}$ be a symmetric Markovian semigroup on $L^2(\mathbb{R}^n)$ with corresponding Dirichlet form $(\mathcal{E}, D(\mathcal{E}))$. Further let $p > 2$ and $N := \frac{2p}{p-2} > 2$. The following estimates are equivalent*

$$\|u\|_{L^p}^2 \leq c\,\mathcal{E}(u,u)\,, \quad u \in D(\mathcal{E})\,; \tag{4.8}$$

$$\|u\|_{L^2}^{2+4/N} \le c\,\mathcal{E}(u,u)\|u\|_{L^1}^{4/N}, \quad u \in D(\mathcal{E}) \cap L^1\left(\mathbb{R}^n\right); \tag{4.9}$$

$$\|T_t\|_{L^1-L^\infty} \le c' t^{-N/2} \quad \text{for all } t > 0. \tag{4.10}$$

From (4.3) it follows that

$$\left\|T_t^\psi\right\|_{L^1-L^\infty} = p_t^\psi(0). \tag{4.11}$$

In light of Theorem 4.1 it is now clear why estimates for $\mathcal{E}^\psi$ are of importance to control $p_t^\psi(0)$ or more generally p_t^ψ. Combining Theorem 4.1 and Corollary 3.17 we get

Theorem 4.2 *For $n \ge 2$ let $\psi : \mathbb{R}^n \longrightarrow \mathbb{R}$ be a continuous negative definite function satisfying our standard assumptions. We assume additionally that with some $c_0 > 0$ and $0 < \alpha < 2$ we have for all $\xi \in \mathbb{R}^n$*

$$c_0|\xi|^\alpha \le \psi(\xi). \tag{4.12}$$

Then the estimate

$$p_t^\psi(0) \le C_\psi t^{-n/\alpha} \tag{4.13}$$

holds for all $t > 0$.

Proof By Corollary 3.17 we have

$$\|u\|_{L^p}^2 \le c\,\mathcal{E}^\psi(u,u) \tag{4.14}$$

for all $u \in D(\mathcal{E})$ with $p = \frac{2n}{n-\alpha}$. This gives for N as defined in Theorem 4.1 the value $N = N_\psi = \frac{2n}{\alpha}$. $\qquad\square$

Note that for $\psi_\alpha(\xi) = |\xi|^\alpha, 0 < \alpha < 2$, we can obtain estimate (4.13) by a direct calculation using the homogeneity of ψ_α. In this case we obtain equality with C_{ψ_α} given by

$$C_{\psi_\alpha} = \frac{\Gamma\left(\frac{n-\alpha}{2}\right)}{2^\alpha \pi^{n/2} \Gamma\left(\frac{\alpha}{2}\right)}. \tag{4.15}$$

This opens the way to derive for the general case discussed in Theorem 4.2 the estimate (4.13) directly from

$$e^{-t\psi(\xi)} \le e^{-c_0 t|\xi|^\alpha},$$

but the approach suggested by using Theorem 4.1 is still available when $\psi(\xi)$ is replaced by a continuous negative definite symbol $q(x, \xi)$ while in that case the direct comparison of symbols cannot be used for a "simple" calculation.

Example 4.3 Taking up ψ_{ER} from Example 3.18, we can investigate its corresponding associated transition density using Theorem 4.2. Recalling that ψ_{ER} satisfies an estimate of the form (4.12), namely (3.47), we can deduce from (4.13) that

$$p_t^{\psi_{ER}}(0) \leq C_{\psi_{ER}} t^{-2/\alpha_1\gamma_1}$$

for all $t > 0$.

Using the upper bound (3.49) for ψ_{ER} we first note that

$$p_t^{\psi_{ER}}(0) = \frac{1}{(2\pi)^2} \int_{\mathbb{R}^2} e^{-t\psi_{ER}(\xi,\eta)} \, d\xi d\eta$$

$$\geq \frac{1}{(2\pi)^2} \int_{\mathbb{R}^2} e^{-t\kappa_1\left((|\xi|^2+|\eta|^2)^{\frac{\alpha_1\gamma_1}{2}}+(|\xi|^2+|\eta|^2)^{\frac{\alpha_2\gamma_2}{2}}\right)} \, d\xi d\eta$$

which implies with $\tilde{\psi}(\xi, \eta) = \left(|\xi|^2 + |\eta|^2\right)^{\frac{\alpha_1\gamma_1}{2}} + \left(|\xi|^2 + |\eta|^2\right)^{\frac{\alpha_2\gamma_2}{2}}$ that

$$p_t^{\psi_{ER}}(0) \geq p_{t\kappa_1}^{\tilde{\psi}}(0) \, .$$

We want to investigate the integral (4.7) by using integration with respect to the distribution function, see [64], Theorem 13.11, for a general formulation of the result.

Theorem 4.4 ([44]) *For a continuous negative definite function* $\psi : \mathbb{R}^n \longrightarrow \mathbb{R}$ *satisfying our standard assumptions we have*

$$\int_{\mathbb{R}^n} e^{-t\psi(\xi)} \, d\xi = \int_0^\infty \lambda^{(n)}\left(B^{d_\psi}\left(0, \sqrt{\frac{r}{t}}\right)\right) e^{-r} \, dr \qquad (4.16)$$

where

$$B^{d_\psi}(x, r) := \{y \in \mathbb{R}^n \mid d_\psi(x, y) < r\} \qquad (4.17)$$

is the open ball with centre $x \in \mathbb{R}^n$ *and radius* $r > 0$ *with respect to the metric* d_ψ.

Introducing the volume functions

$$V_\psi(r) := \lambda^{(n)}\left(B^{d_\psi}(0, r)\right) \qquad (4.18)$$

and

$$\tilde{V}_\psi(r) = V_\psi\left(\sqrt{r}\right) \tag{4.19}$$

we find, see [30] or [31],

Corollary 4.5 *For* $\psi : \mathbb{R}^n \longrightarrow \mathbb{R}$ *as in Theorem 4.4 we have*

$$p_t^\psi(0) = (2\pi)^{-n} t \mathcal{L}\left(\tilde{V}_\psi\right)(t), \tag{4.20}$$

where $\mathcal{L}$ denotes the Laplace transform.

Proof We need only observe that

$$\int_0^\infty \lambda^{(n)}\left(B^{d_\psi}\left(0, \sqrt{\frac{r}{t}}\right)\right) e^{-r} \, dr = t \int_0^\infty \lambda^{(n)}\left(B^{d_\psi}\left(0, \sqrt{\varrho}\right)\right) e^{-t\varrho} \, d\varrho$$

$$= t \mathcal{L}\left(\tilde{V}_\psi\right)(t).$$

$\square$

This corollary allows us to find for some cases the function $t \mapsto p_t^\psi(0)$ explicitly. However in most cases we depend on estimates. For this we introduce

Definition 4.6 Let d be a metric on $\mathbb{R}^n$. We call $\left(\mathbb{R}^n, d, \lambda^{(n)}\right)$ a metric measure space with **doubling property** if for every $r > 0$ and all $x \in \mathbb{R}^n$ there exists a constant $\gamma > 0$ independent of r and x such that

$$\lambda^{(n)}\left(B_{2r}(x)\right) \leq \gamma \lambda^{(n)}\left(B_r(x)\right), \tag{4.21}$$

where $B_r(x)$ denotes the open ball with centre x and radius r with respect to d.

If $\left(\mathbb{R}^n, d, \lambda^{(n)}\right)$ is a metric measure space with volume doubling, then the volume function $r \mapsto V_d(x, r) := \lambda^{(n)}\left(V_d(x, r)\right)$ has at most power growth. In particular if d is a translation invariant metric with the volume doubling property, then the growth of $r \mapsto \lambda^{(n)}\left(V_d(x, r)\right)$ is independent of x and has at most power growth.

Note that if $\left(\mathbb{R}^n, d_\psi, \lambda^{(n)}\right)$ has the doubling property then it is a metric space of homogeneous type in the sense of Coifman and Weiss [16].

Theorem 4.7 ([44]) *Suppose that the continuous negative definite function $\psi :$ $\mathbb{R}^n \longrightarrow \mathbb{R}$ satisfying our standard assumptions and in addition assume that the metric measure space $\left(\mathbb{R}^n, d_\psi, \lambda^{(n)}\right)$ has the volume doubling property. Then we have*

$$p_t^\psi(0) \asymp \tilde{V}_\psi\left(\frac{1}{\sqrt{t}}\right), \quad t > 0. \tag{4.22}$$

Here we used the notation $a \asymp b$, $0 < a < b$, for the two estimates

$$c_0 \leq \frac{b}{a} \leq c \tag{4.23}$$

to hold for some $0 < c_0 \leq c$.

Example 4.8 It is not hard to see that we can improve (3.47) to

$$\kappa_2 \left(\left(|\xi|^2 + |\eta|^2 \right)^{\frac{\alpha_1 \gamma_1}{2}} + \left(|\xi|^2 + |\eta|^2 \right)^{\frac{\alpha_2 \gamma_2}{2}} \right) \leq \psi_{ER}(\xi, \eta) \tag{4.24}$$

and from (4.24) combined with (3.49) we obtain the estimates

$$\tilde{\kappa}_0 h(\xi, \eta) \leq \psi_{ER}^{1/2}(\xi, \eta) \leq \tilde{\kappa}_1 h(\xi, \eta)$$

where

$$h(\xi, \eta) = \max \left(\left(|\xi|^{\frac{\gamma_1 \alpha_1}{2}} + |\eta|^{\frac{\gamma_1 \alpha_1}{2}} \right), \left(|\xi|^{\frac{\gamma_2 \alpha_2}{2}} + |\eta|^{\frac{\gamma_2 \alpha_2}{2}} \right) \right) .$$

By $h(\xi_1 - \xi_2, \eta_1 - \eta_2)$ a metric having the doubling property is induced on $\mathbb{R}^2$ and using the concrete formulae for the volumes of the corresponding balls, we can deduce the doubling property for ψ_{ER}, see [55].

While our main interest is the study of generators of Feller semigroups i.e. pseudo-differential operators with negative definite symbols, it is worth mentioning that studying the case of L^2-operator semigroups $\left(T_t^g \right)_{t \geq 0}$ defined on $\mathcal{S}(\mathbb{R}^n)$ by

$$T_t^g u(x) = (2\pi)^{-n/2} \int_{\mathbb{R}^n} e^{ix \cdot \xi} e^{-tg(\xi)} \hat{u}(\xi) \, d\xi \tag{4.25}$$

for a continuous function $g : \mathbb{R}^n \longrightarrow \mathbb{R}$, $g \geq 0$, $e^{-tg} \in L^1(\mathbb{R}^n)$ will lead to some further insights for our problem. In [30], [31] we have discussed some results for $\left(T_t^g \right)_{t \geq 0}$.

In order to expand our reservoir of examples we need to introduce subordination in the sense of Bochner and we refer to [38] as well as [67] as reference.

Definition 4.9 A real-valued function $f \in C((0, \infty))$ is called a **Bernstein function** if

$$f(x) \geq 0 \quad \text{and} \quad (-1)^k \frac{d^k f(x)}{dx^k} \leq 0 \tag{4.26}$$

holds for all $k \in \mathbb{N}$ and $x \in (0, \infty)$.

Theorem 4.10 *If f is a Bernstein function, then there exist constants $a, b \geq 0$ and a measure μ on $(0, \infty)$ integrating $s \mapsto \frac{s}{1+s}$ such that*

$$f(x) = a + bx + \int_{0^+}^{\infty} \left(1 - e^{-xs}\right) \mu(ds), \quad x > 0. \tag{4.27}$$

Furthermore, given a Bernstein function f then there exists a convolution semigroup $(\eta_t)_{t \geq 0}$ of sub-probability measures supported on $[0, \infty)$ such that

$$\mathcal{L}(\eta_t)(x) = \int_0^{\infty} e^{-sx} \eta_t(ds) = e^{-tf(x)} \tag{4.28}$$

for all $x > 0$ and $t > 0$.

Remark 4.11 The integrability condition on μ in (4.27) yields that $\mu|_{(0,1]}$ integrates the function $s \mapsto s$ and that $\mu|_{[1,\infty)}$ integrates the function $s \mapsto 1$. Since for $z \in \mathbb{C}$, $\mathrm{Re}\, z \geq 0$ and $s > 0$ we have the estimates $|1 - e^{-sz}| \leq s|z|$ and $|1 - e^{-zs}| \leq 2$ it follows that (4.27) admits a continuous extension to the half-plane $\mathrm{Re}\, z \geq 0$, i.e. f admits an extension to $\mathrm{Re}\, z \geq 0$.

The fundamental result is

Theorem 4.12 *If $\psi : \mathbb{R}^n \longrightarrow \mathbb{C}$ is a continuous negative definite function with associated convolution semigroup $(\mu_t)_{t \geq 0}$ and if f is a Bernstein function with associated convolution semigroup $(\eta_t)_{t \geq 0}$, then $f \circ \psi$ is again a continuous negative definite function and for the associated convolution semigroup $\left(\mu_t^f\right)_{t \geq 0}$ we find (as a vague integral) the representation*

$$\mu_t^f = \int_0^{\infty} \mu_s \, \eta_t(ds), \tag{4.29}$$

i.e. for all $\varphi \in C_0(\mathbb{R}^n)$ we have

$$\int_{\mathbb{R}^n} \varphi(x) \, \mu_t^f(dx) = \int_0^{\infty} \int_{\mathbb{R}^n} \varphi(x) \, \mu_s(dx) \, \eta_t(ds). \tag{4.30}$$

Definition 4.13 We call $\left(\mu_t^f\right)_{t \geq 0}$ the convolution semigroup **subordinate** (in the sense of Bochner) to $(\mu_t)_{t \geq 0}$ with respect to $(\eta_t)_{t \geq 0}$.

Subordination extends to operator semigroups. Let $(T_t)_{t \geq 0}$ be a strongly continuous contraction semigroup on the Banach space $(X, \|.\|)$. The family of operators T_t^f, $t \geq 0$, defined by

$$T_t^f u = \int_0^{\infty} T_s u \, \eta_t(ds) \tag{4.31}$$

is called the semigroup subordinate to $(T_t)_{t\geq 0}$ with respect to $(\eta_t)_{t\geq 0}$, compare [38], Theorem 4.3.1.

Suppose that $\psi : \mathbb{R}^n \longrightarrow \mathbb{R}$ is a continuous negative definite function satisfying our standard assumptions and assume that f is a Bernstein function such that $f \circ \psi$ also satisfies our standard assumptions. In this case we have

$$p_t^{f\circ\psi}(0) = (2\pi)^{-n} t \mathcal{L}\left(\tilde{V}_{f\circ\psi}\right)(t) . \tag{4.32}$$

In addition if $\frac{1}{f\circ\psi} \in L^1_{\mathrm{loc}}(\mathbb{R}^n)$, then the Dirichlet form $\mathcal{E}^{f\circ\psi}$ and therefore the semigroup $\left(T_t^{f\circ\psi}\right)_{t\geq 0}$ is transient.

Let $(T_t)_{t\geq 0}$ be a strongly continuous contraction semigroup on a Banach space $(X, \|.\|)$ with generator $(A, D(A))$. Let f be a Bernstein function with representation (4.27) and corresponding convolution semigroup $(\eta_t)_{t\geq 0}$. Then $D(A)$ is an operator core for the generator $\left(A^f, D\left(A^f\right)\right)$ of $\left(T_t^f\right)_{t\geq 0}$ and on $D(A)$ we have

$$A^f u = -au + bAu + \int_{0+}^{\infty} (T_s u - u)\, \mu\,(ds) , \tag{4.33}$$

see [62] or [67], Theorem 13.6.

For the following it is convenient to rewrite (4.9) as

$$\|u\|_{L^2}^2 h\left(\|u\|_{L^2}^2\right) \leq \mathcal{E}(u, u) \tag{4.34}$$

for all $u \in D(\mathcal{E}) \cap L^1(\mathbb{R}^n)$ with $\|u\|_{L^1} = 1$.

The following result due to Schilling and Wang [66] extends a result of Bendikov and Maheux [4] and in [72] relations to the work of Tomisaki [70] were discussed.

Theorem 4.14 *Let $(T_t)_{t\geq 0}$ be a symmetric sub-Markovian semigroup on $L^2(\mathbb{R}^n)$ and assume that $T_t|_{L^2\cap L^1}$ extends to an L^1-contraction, i.e. $\|T_t u\|_{L^1} \leq \|u\|_{L^1}$. If the L^2-generator $(A, D(A))$ of $(T_t)_{t\geq 0}$ satisfies with an increasing function $h :$ $(0, \infty) \longrightarrow (0, \infty)$*

$$\|u\|_{L^2}^2 h\left(\|u\|_{L^2}^2\right) \leq (Au, u)_{L^2} \tag{4.35}$$

for all $u \in D(A)$ such that $\|u\|_{L^1} = 1$, then we have

$$\frac{1}{2}\|u\|_{L^2}^2 f\left(h\left(\frac{\|u\|_{L^2}^2}{2}\right)\right) \leq \left(A^f u, u\right)_{L^2} \tag{4.36}$$

for all $u \in D\left(A^f\right)$, $\|u\|_{L^1} = 1$.

In [67] plenty of examples of Bernstein functions are provided together with more detailed information. We mention here just a few

$$f(s) = s^\alpha , \quad 0 < \alpha < 1 ,$$

$$f(s) = (s + 1)^\alpha - 1 , \quad 0 < \alpha < 1 ,$$

$$f(s) = \sqrt{s} \left(1 - e^{-\sqrt{s}} \right) ,$$

$$f(s) = \ln(1 + s) ,$$

$$f(s) = \sqrt{s} \arctan \sqrt{s} ,$$

$$f(s) = \frac{s}{1 + s} \left(\frac{s^\alpha}{\sin \alpha \pi} - \cot(\alpha \pi) - \frac{1}{\pi} \ln s \right) ,$$

$$f(s) = \mathrm{arsinh}(s) ,$$

$$f(s) = \mathrm{arcosh}(s + 1) ,$$

$$f(s) = s^{1-\nu} e^s \Gamma(\nu, s) .$$

Example 4.15 We consider again the continuous negative definite function from Example 3.18. Clearly $\psi_{ER}(\xi, \eta) = 0$ if and only if $\xi = \eta = 0$ and by Theorem 4.4 it follows when taking (3.47) and (4.16) into account that

$$p_t^{\psi_{ER}}(0) = \int_0^\infty \lambda^{(2)} \left(B^{d_{\psi_{ER}}} \left(0, \sqrt{\frac{r}{t}} \right) \right) e^{-r} \, dr \tag{4.37}$$

$$= (2\pi)^{-2} t \mathcal{L} \left(\tilde{V}_{\psi_{ER}} \right)(t) .$$

From Example 4.8 we know that $\left(\mathbb{R}^2, d_{\psi_{ER}}, \lambda^{(2)} \right)$ has the volume doubling property and by Theorem 4.7 we obtain the following bounds

$$p_t^{\psi_{ER}}(0) \asymp \tilde{V}_{\psi_{ER}} \left(\frac{1}{\sqrt{t}} \right) , \quad t > 0 . \tag{4.38}$$

5   Some Remarks on Off-Diagonal Results

Still we work within the frame of Chap. 4, i.e. we start with a continuous negative definite function ψ satisfying our standard assumptions. Hence the associated operator semigroup $(T_t^\psi)_{t \geq 0}$ has a representation

$$T_t^\psi u(x) = \int_{\mathbb{R}^n} p_t^\psi (x - y) u(y) \, dy \tag{5.1}$$

where

$$p_t^\psi(x) = (2\pi)^{-n} \int_{\mathbb{R}^n} e^{ix\cdot\xi} e^{-t\psi(\xi)} \, d\xi \, . \tag{5.2}$$

The aim is to understand the behaviour of $x \mapsto p_t^\psi(x)$. This is of course a "classical" topic and we do not intend to summarize important developments here. However a few words are in order. The seminal work of Davies [18] and [19], see also [20], introduced a method to obtain for second order elliptic operators on a Riemannian manifold off-diagonal estimates from diagonal estimates. In the much quoted paper [15] an attempt was made to transfer this method to general (non-local) Dirichlet forms. In this paper much use is made of the opérateur carré du champ associated with the corresponding Dirichlet form which is natural since the studies of Hörmander-type operators or more general second order sub-elliptic differential operators have revealed that for local operators the associated opérateur carré du champ contains much information, for example about the heat kernel. For this we refer in particular to the monograph [1], the already mentioned monograph [71], and to the work of Wang [72] and [73]. However, eventually [15] does not allow to find concrete estimates for concrete non-local Dirichlet forms. This is much due to the fact that for non-local operators the opérateur carré du champ is not anymore a sum of products of derivations and Leibniz' rule as well as the chain rule are not available. Recall that for a symmetric second order sub-elliptic differential operator in divergence form $L(x, D) = \sum_{k,l=1}^n \frac{\partial}{\partial x_k} a_{kl}(x) \frac{\partial}{\partial x_l}$ the corresponding opérateur carré du champ is given by

$$\Gamma(u, v)(x) = \sum_{k,l=1}^n a_{kl}(x) \frac{\partial u}{\partial x_k}(x) \frac{\partial v}{\partial x_l}(x) \, , \tag{5.3}$$

whereas for the Dirichlet form $\mathcal{E}^\psi$ with $\psi(\xi) = \int_{\mathbb{R}^n \setminus \{0\}} (1 - \cos y \cdot \xi) \, \nu(dy)$ the opérateur carré du champ has the form

$$\Gamma^\psi(u, v)(x) = \frac{1}{2} \int_{\mathbb{R}^n \setminus \{0\}} (u(x + y) - u(x))(v(x + y) - v(x)) \, \nu(dy) \, . \tag{5.4}$$

In [51] Meyer indicated how we can write (5.4) as an infinite sum of squares of operators $\left(\text{having Hörmander-type operators } \sum_{l=1}^N X_l^2 u \text{ in mind with opérateur} \right.$ carré du champ $\left. \sum_{l=1}^N X_l u X_l v \right)$ but in his representation the operators are again not derivations. As was pointed out in [44] Meyer's proof has similarities with the proof of Schoenberg's theorem given in [5]. Schoenberg's theorem reads for our purpose as follows, compare with [44],

Theorem 5.1 *Let ψ be a continuous negative definite function and d_ψ the corresponding metric. Then the metric space $(\mathbb{R}^n, d_\psi)$ can be isometrically embedded into some Hilbert space $(\mathcal{H}, \langle ., . \rangle_\mathcal{H})$. Conversely, if $J : (\mathbb{R}^n, d) \longrightarrow (\mathcal{H}, \langle ., . \rangle_\mathcal{H})$ is an isometric embedding of the metric space $(\mathbb{R}^n, d)$ into some Hilbert space $(\mathcal{H}, \langle ., . \rangle_\mathcal{H})$ such that $\langle J(x), J(y) \rangle_\mathcal{H} = \langle J(x - y), J(0) \rangle_\mathcal{H}$, then $d = d_\psi$ for some continuous negative definite function $\psi : \mathbb{R}^n \longrightarrow \mathbb{R}$.*

A possible proof in case that the Lévy measure ν has a nice density $\tilde{n}$ and using the opérateur carré du champ starts with

$$\Gamma^\psi(u, v)(x) = \frac{1}{2} \int_{\mathbb{R}^n \setminus \{0\}} (u(x + y) - u(x))(v(x + y) - v(x)) \, \tilde{n}(y) dy \qquad (5.5)$$

and the observation that on

$$C_\psi := \{u \in C^2(\mathbb{R}^n) \mid \Gamma(u, u)(0) < \infty\} / \{u \in C^2(\mathbb{R}^n) \mid \Gamma(u, u)(0) = 0\} \qquad (5.6)$$

a scalar product is given by

$$\langle u, v \rangle_\mathcal{H} := \Gamma(u, v)(0). \qquad (5.7)$$

The space $\mathcal{H}$ is introduced as the completion of C_ψ with respect to $\langle ., . \rangle_\mathcal{H}$ and the Lévy-Khinchine formula yields with $e_\xi(x) = e^{ix \cdot \xi}$ that $\Gamma(e_\xi, e_\xi)(0) = \psi(\xi)$.

This observation in mind, as well as some examples (see below), it was conjectured that in general we can write or at least estimate $p_t^\psi(.)$ according to

$$p_t^\psi(x - y) \asymp p_t^\psi(0) e^{-\delta_{\psi,t}^2(x,y)} \qquad (5.8)$$

with $p_t^\psi(0) \asymp t\mathcal{L}\left(\tilde{V}_\psi\right)(t)$ and a suitable metric $\delta_{\psi,t}$ on $\mathbb{R}^n$. This approach differs from that of many other authors as it is much more geometric in spirit and uses as principal data the characteristic exponent ψ and not the Lévy measure.

Here are two examples for (5.8):

$$p_t^C(x - y) = \pi^{-\frac{n+1}{2}} \Gamma\left(\frac{n + 1}{2}\right) \frac{t}{(t^2 + |x - y|^2)^{\frac{n+1}{2}}} \qquad (5.9)$$

$$= c_1 p_t^C(0) e^{-\delta_{C,t}^2(x,y)},$$

where

$$\delta_{C,t}^2(x, y) = -\ln\left(1 + t^2|x - y|^2\right), \qquad (5.10)$$

and for $n = 1$

$$p_t^M(x - y) = \frac{2^{t-1}}{\pi \Gamma(t)} \left| \Gamma\left(\frac{t + i(x - y)}{2}\right) \right|^2 \tag{5.11}$$

$$= c_2 p_t^M(0) e^{-\delta_{M,t}^2(x,y)}$$

where

$$\delta_{M,t}^2(x, y) = \sum_{j=0}^{\infty} \ln\left(1 + \frac{|x - y|^2}{\left(\frac{1}{t} + 2j\right)^2}\right). \tag{5.12}$$

Here p_t^C is the transition density corresponding to the Cauchy process with $\psi_C(\xi) = |\xi|$, and p_t^M corresponds to the one-dimensional symmetric Meixner process associated with $\psi_M(\xi) = \ln \cosh \xi$.

So far we do not have general results to ensure (5.8) to hold but in [44] some classes of examples were discussed.

In [14], see also [13], an observation already made at the end of [44] was investigated in more detail. Consider the convolution semigroup $(\mu_t)_{t \geq 0}$ associated with ψ satisfying our standard assumptions. In particular we have

$$\hat{\mu}_t(\xi) = (2\pi)^{-n/2} e^{-t\psi(\xi)} \tag{5.13}$$

and for the density we find

$$p_t(x) = (2\pi)^{-n} \int_{\mathbb{R}^n} e^{ix \cdot \xi} e^{-t\psi(\xi)} \, d\xi . \tag{5.14}$$

The first observation is that

$$\varrho_t := \frac{e^{-t\psi(.)}}{(2\pi)^n p_t(0)} \lambda^{(n)} \tag{5.15}$$

is for $t > 0$ a symmetric probability measure on $\mathbb{R}^n$ and for

$$\nu_t := \varrho_{1/t} = \frac{e^{-\frac{1}{t}\psi(.)}}{(2\pi)^n p_{1/t}(0)} \lambda^{(n)} = \pi_{t,0}(.)\lambda^{(n)} \tag{5.16}$$

we find of course $\nu_t(\mathbb{R}^n) = 1$ and in addition

$$\lim_{n \to 0} \nu_t = \varepsilon_0 \quad \text{(weakly)} , \tag{5.17}$$

compare [14] or [45]. With

$$\sigma_t(\xi) = \frac{p_{1/t}(\xi)}{p_{1/t}(0)} \tag{5.18}$$

we define $\left(\text{first on } S\left(\mathbb{R}^n\right)\right)$ the operators

$$S_t u(x) = (2\pi)^{-n/2} \int_{\mathbb{R}^n} e^{ix \cdot \xi} \sigma_t(\xi) \hat{u}(\xi) \, d\xi \tag{5.19}$$

which are symmetric, positivity preserving L^2- and C_∞-contractions and satisfy

$$\lim_{t \to 0} \|S_t u - u\|_{L^2} = \lim_{t \to 0} \|S_t u - u\|_\infty = 0 \,, \tag{5.20}$$

i.e. they are strongly continuous at 0 in $L^2\left(\mathbb{R}^n\right)$ and $C_\infty\left(\mathbb{R}^n\right)$, respectively. With

$$q(t, \xi) = -\frac{\partial}{\partial t} \ln \frac{p_{1/t}(\xi)}{p_{1/t}(0)} \tag{5.21}$$

it follows that $(S_t u)_{t>0}$ solves

$$\frac{\partial}{\partial t} S_t u(x) + q(t, D) S_t u(x) = 0 \quad \text{and} \quad \lim_{t \to 0} S_t u = u \tag{5.22}$$

either in $L^2\left(\mathbb{R}^n\right)$ or in $C_\infty\left(\mathbb{R}^n\right)$. Here $q(t, D)$ is the pseudo-differential operator

$$q(t, D)w(t, x) = (2\pi)^{-n/2} \int_{\mathbb{R}^n} e^{ix \cdot \xi} q(t, \xi) \hat{w}(t, \xi) \, d\xi \,. \tag{5.23}$$

The Eq. (5.22) we can handle in the following way: Define

$$\gamma_{t,s} = \pi_{t,s}(.)\lambda^{(n)} \tag{5.24}$$

with

$$\pi_{t,s}(x) = (2\pi)^{-n/2} \int_{\mathbb{R}^n} e^{ix \cdot \xi} e^{-\int_s^t q(\tau, \xi) \, d\tau} \, d\xi \,, \tag{5.25}$$

i.e. we have

$$\hat{\gamma}_{t,s}(\xi) = (2\pi)^{-n/2} e^{-\int_s^t q(\tau, \xi) \, d\tau} \tag{5.26}$$

and with

$$Q_{t,s}(\xi) = \int_s^t q(\tau, \xi) \, d\tau \tag{5.27}$$

we have

$$\hat{\gamma}_{t,s}(\xi) = (2\pi)^{-n/2} e^{-Q_{t,s}(\xi)} . \tag{5.28}$$

The operators

$$V(t,s)u(x) = \int_{\mathbb{R}^n} u(x-y)\,\gamma_{t,s}\,(\mathrm{d}y) \tag{5.29}$$

$$= \int_{\mathbb{R}^n} \pi_{t,s}(x-y)u(y)\,\mathrm{d}y$$

$$= (2\pi)^{-n/2} \int_{\mathbb{R}^n} e^{\mathrm{i}x\cdot\xi} e^{-Q_{t,s}(\xi)} \hat{u}(\xi)\,\mathrm{d}\xi$$

give a fundamental solution to

$$\frac{\partial v}{\partial t} + q(t,D)v = 0, \quad \lim_{t\to s} v(t,s) = v_0 \tag{5.30}$$

in the sense that

$$V(s,s)u = u ;$$

$$V(t,r) \circ V(r,s)u = V(t,s)u , \quad s < r < t ;$$

$$V(t,s)u \to u \quad \text{as} \quad s \to t , \quad s < t ;$$

$$V(t,s)u \to u \quad \text{as} \quad t \to s , \quad s < t .$$

Moreover, if in addition we assume that $\xi \mapsto q(t,\xi)$ is a continuous negative definite function then we have, see [13], [14], and compare with [56],

Theorem 5.2 *We can associate with* $q(t,\xi)$, $t > 0$, $\xi \in \mathbb{R}^n$, *a canonical additive process in law* $(X_t)_{t\geq 0}$ *by the relations*

$$P_{X_t - X_s} = \gamma_{t,s} , \quad t > s > 0 . \tag{5.31}$$

The assumption that $\xi \mapsto q(t,\xi)$ is a continuous negative definite function is crucial, recall that this assumption means that the function

$$\xi \mapsto -\frac{\partial}{\partial t} \ln \frac{p_{1/t}(\xi)}{p_{1/t}(0)}$$

is a continuous negative definite function. We will now see how this assumption leads to a geometric interpretation of transition densities. First we note that this assumption implies, see [13] or [14] that

$$d_{Q_{t,s}}(\xi,\eta) := Q_{t,s}^{1/2}(\xi - \eta) , \quad 0 < s < t , \tag{5.32}$$

is a metric on $\mathbb{R}^n$ and we find by calculations already known to us

$$\pi_{t,s}(0) = (2\pi)^{-n} \int_0^\infty \lambda^{(n)} \left(\{ \xi \in \mathbb{R}^n \mid Q_{t,s}(\xi) \le r \} \right) e^{-r} \, dr \ . \tag{5.33}$$

If we add further the assumption that the inequalities

$$\beta_0 q(t_0, \xi) \le q(t, \xi) \le \beta_1 q(t_0, \xi) \ , \quad \beta_0 > 0 \ , \tag{5.34}$$

holds for all $\xi \in \mathbb{R}^n$ and all $t > \tau_0$, $t_0 > \tau_0$, then an easy modification of the calculation in the proof of Theorem 5.1 in [14] yields that if $\left(\mathbb{R}^n, d_{q(t_0, .)}, \lambda^{(n)} \right)$ has the volume doubling property then we have

$$\pi_{t,s}(0) \asymp \lambda^{(n)} \left(B^{d_{Q_{t,s}}} \left(0, \sqrt{\frac{\beta_1}{\beta_0}} \right) \right) \tag{5.35}$$

for all $t > s > \tau_0$.

Returning to p_t we observe that

$$p_t(x - y) = p_t(0) \frac{p_t(x - y)}{p_t(0)} = p_t(0) e^{-\ln q(1/t, x - y)} \ , \tag{5.36}$$

and under the assumption that $\xi \mapsto q(t, \xi)$ is a continuous negative definite function we find with

$$\delta_{\psi,t}^2(x, y) = \ln q \left(1/t, x - y \right) \tag{5.37}$$

that

$$p_t(x - y) = t \mathcal{L} \left(\tilde{V}_\psi \right)(t) e^{-\delta_{\psi,t}^2(x,y)} \ . \tag{5.38}$$

In the case that the metric measure space $\left(\mathbb{R}^n, d_\psi, \lambda^{(n)} \right)$ has the volume doubling property it follows further

$$p_t(x - y) \asymp \lambda^{(n)} \left(B^{d_\psi} \left(0, 1/\sqrt{t} \right) \right) e^{-\delta_{\psi,t}^2(x,y)} \tag{5.39}$$

and when replacing d_ψ by $d_{\psi,t}(x, y) = \sqrt{t} \psi(x - y)$ we arrive at

$$p_t(x - y) \asymp \lambda^{(n)} \left(B^{d_{\psi,t}}(0, 1) \right) e^{-\delta_{\psi,t}^2(x,y)} \ . \tag{5.40}$$

On the other hand, under the assumptions made above, we find for $\pi_{t,s}$

$$\pi_{t,s}(x - y) \asymp \lambda^{(n)} \left(B^{d_{Q_{t,s}}} \left(0, \sqrt{\frac{\beta_1}{\beta_0}} \right) \right) e^{-\int_s^t q(\tau, x-y) \, d\tau} \ . \tag{5.41}$$

Since

$$\int_s^t q(\tau, \xi)\, d\tau = -\ln \frac{p_{1/t}(\xi)}{p_{1/t}(0)} + \ln \frac{p_{1/s}(\xi)}{p_{1/s}(0)} \,,$$

and since by Theorem 5.7 in [45]

$$\lim_{s \to 0} \frac{p_{1/s}(\xi)}{p_{1/s}(0)} = 1$$

we find in the limit $s \to 0$

$$\pi_{t,0}(x - y) \asymp \lambda^{(n)} \left(B^{\delta_{\psi,1/t}} \left(0, \sqrt{\frac{\beta_1}{\beta_0}} \right) \right) e^{-d^2_{\psi,1/t}(x,y)} \,. \tag{5.42}$$

Comparing (5.40) with (5.42) we are reminded at the considerations starting with Lewis [50] on "adjoint" or pairs of probability densities, see also [49] or [29].

6 Pseudo-Differential Operators with Negative Definite Symbols: Ideas and Challenges

In Sects. 3–5 we have discussed translation invariant operators and related objects such as translation invariant Dirichlet forms associated with convolution semigroups $(\mu_t)_{t\geq 0}$ of (sub-)probability measures. As translation invariant operators they admit a representation as convolution operator, for our purposes it is however more important to represent the operators as (translation invariant) pseudo-differential operators, i.e. to look at T_t^ψ, A^ψ and R_λ^ψ as being given by

$$T_t^\psi u(x) = (2\pi)^{-n/2} \int_{\mathbb{R}^n} e^{ix\cdot\xi} e^{-t\psi(\xi)} \hat{u}(\xi)\, d\xi \,, \quad t \geq 0 \,, \tag{6.1}$$

$$A^\psi u(x) = (2\pi)^{-n/2} \int_{\mathbb{R}^n} e^{ix\cdot\xi} (-\psi(\xi)) \hat{u}(\xi)\, d\xi \tag{6.2}$$

and

$$R_\lambda u(x) = (2\pi)^{-n/2} \int_{\mathbb{R}^n} e^{ix\cdot\xi} \frac{1}{\lambda + \psi(\xi)} \hat{u}(\xi)\, d\xi \,, \quad \lambda > 0 \,. \tag{6.3}$$

So far only for investigating the density p_t^ψ of T_t^ψ, i.e. the function

$$p_t^\psi(x - y) = (2\pi)^{-n} \int_{\mathbb{R}^n} e^{i(x-y)\cdot\xi} e^{-t\psi(\xi)}\, d\xi \,, \tag{6.4}$$

the symbol of the generator A^ψ was used. Properties of $\left(T_t^\psi\right)_{t\geq 0}$ can be derived from $(\mu_t)_{t\geq 0}$ and/or the Lévy–Khinchine triple (or quadruple), see Sect. 2.

It turns out that introducing "variable coefficients" is best achieved by starting with the generator given in the form (6.2). A nice interpretation was given in [3]: Let $\psi = \psi_{a,b,c,\ldots}$ be a continuous negative definite function depending on certain parameters. Suppose now that these parameters become x-dependent (state space dependent), i.e. we are dealing with functions $x \mapsto a(x)$, $x \mapsto b(x)$, $x \mapsto c(x)$, ... having the property that for each $x_0 \in \mathbb{R}^n$ fixed $\xi \mapsto \psi_{a(x_0),b(x_0),c(x_0)\ldots}(\xi)$ is a continuous negative definite function and hence determines (for x_0 fixed) a convolution semigroup with corresponding operator semigroup, generator and resolvent. Formally we may consider

$$\tilde{T}_t u(x) = (2\pi)^{-n/2} \int_{\mathbb{R}^n} e^{ix\cdot\xi} e^{-t\psi_{a(x),b(x),c(x),\ldots}(\xi)}\hat{u}(\xi)\,\mathrm{d}\xi \tag{6.5}$$

and

$$\tilde{R}_\lambda u(x) = (2\pi)^{-n/2} \int_{\mathbb{R}^n} e^{ix\cdot\xi} \frac{1}{\lambda + \psi_{a(x),b(x),c(x),\ldots}(\xi)}\hat{u}(\xi)\,\mathrm{d}\xi\ , \tag{6.6}$$

but neither we can expect $\left(\tilde{T}_t\right)_{t\geq 0}$ to be a semigroup nor $\left(\tilde{R}_\lambda\right)_{\lambda>0}$ to be a resolvent on $L^2(\mathbb{R}^n)$ or $C_\infty(\mathbb{R}^n)$. However, we may expect that

$$Au(x) = -(2\pi)^{-n/2} \int_{\mathbb{R}^n} e^{ix\cdot\xi} \psi_{a(x),b(x),c(x),\ldots}(\xi)\hat{u}(\xi)\,\mathrm{d}\xi \tag{6.7}$$

extends from $S(\mathbb{R}^n)$ to a generator of a strongly continuous, positivity preserving contraction semigroup on $C_\infty(\mathbb{R}^n)$ with generator $(A, D(A))$.

Indeed, let $(T_t)_{t\geq 0}$ be a Feller semigroup, i.e. a strongly continuous, positivity preserving contraction semigroup on $C_\infty(\mathbb{R}^n)$ with generator $(A, D(A))$. A theorem due to Courrège [17] states that if $C_0^\infty(\mathbb{R}^n) \subset D(A)$ then A is of type

$$Au(x) = -q(x, D)u(x) = (2\pi)^{-n/2} \int_{\mathbb{R}^n} e^{ix\cdot\xi}(-q(x,\xi))\hat{u}(\xi)\,\mathrm{d}\xi \tag{6.8}$$

where $q : \mathbb{R}^n \times \mathbb{R}^n \longrightarrow \mathbb{C}$ is a continuous function such that $\xi \mapsto q(x,\xi)$ is for every $x \in \mathbb{R}^n$ negative definite. We refer to [38] and in particular to [12] for more details. We call a pseudo-differential operator $q(x, D)$ a **pseudo-differential operator with negative definite symbol** if its symbol $q(x, \xi)$ is for every $x \in \mathbb{R}^n$ with respect to the co-variable ξ a continuous negative definite function.

Pseudo-differential operators with negative definite symbols can be obtained as in (6.7). A further possibility is to look at variable order subordination, see [24, 33, 42] or [23], i.e. to look at operators of type

$$Au(x) = -(2\pi)^{-n/2} \int_{\mathbb{R}^n} e^{ix\cdot\xi} f(x, \psi(\xi)) \hat{u}(\xi) \, d\xi \tag{6.9}$$

where for $x \in \mathbb{R}^n$ fixed the function $s \mapsto f(x, s)$ is a Bernstein function and ψ is a continuous negative definite function. It is a highly non-trivial question whether operators of type (6.7), (6.9) or (6.8) have an extension to a generator of a Feller semigroup or an L^2-sub-Markovian semigroup.

Let us agree to call any operator $q(x, D)$ defined at least on $\mathcal{S}(\mathbb{R}^n)$ a **pseudo-differential operator with symbol** $q(x, \xi)$ if it is of the form

$$q(x, D)u(x) := (2\pi)^{-n/2} \int_{\mathbb{R}^n} e^{ix\cdot\xi} q(x, \xi) \hat{u}(\xi) \, d\xi \,, \tag{6.10}$$

where $q : \mathbb{R}^n \times \mathbb{R}^n \longrightarrow \mathbb{C}$ is a continuous function such that $\xi \mapsto q(x, \xi)$ is of at most polynomial growth. If A is a pseudo-differential operator we denote its symbol by $\sigma(A)(x, \xi)$. For x-independent symbols, i.e. translation invariant operators we have the simple rule

$$\sigma(q_1(D) \circ q_2(D))(x, \xi) = q_1(\xi)q_2(\xi) \,, \tag{6.11}$$

and in the case that $q_1(D)^{-1}$ exists and is a pseudo-differential operator we find

$$\sigma\left(q_1(D)^{-1}\right)(x, \xi) = \frac{1}{q_1(\xi)} \,. \tag{6.12}$$

(Note that it is sometimes possible that $\hat{u}(.)/q_1(.)$ is not integrable, i.e. (6.10) is not defined, for all $u \in \mathcal{S}(\mathbb{R}^n)$, however, one may find a subclass of functions in $\mathcal{S}(\mathbb{R}^n)$ for which $F^{-1}\left(\hat{u}/q_1\right)$ is defined.)

For pseudo-differential operators with symbols depending on x we cannot expect the analogue to (6.11) or (6.12) to hold, i.e. we have in general

$$\sigma(q_1(x, D) \circ q_2(x, D))(x, \xi) \neq q_1(x, \xi)q_2(x, \xi) \tag{6.13}$$

or

$$\sigma\left(q_1(x, D)^{-1}\right)(x, \xi) \neq \frac{1}{q_1(x, \xi)} \,. \tag{6.14}$$

The basic idea is to develop a **symbolic calculus** which allows us to consider

$$q_1(x, D) \circ q_2(x, D) - (q_1 \cdot q_2)(x, D) \tag{6.15}$$

or

$$q_1^{-1}(x, D) - \left(\frac{1}{q_1}\right)(x, D) \tag{6.16}$$

as a "lower order" perturbation of $q_1(x, D) \circ q_2(x, D)$ and $q_1^{-1}(x, D)$, respectively. Here we used the notations

$$(q_1 \cdot q_2)(x, D)u(x) = (2\pi)^{-n/2} \int_{\mathbb{R}^n} e^{ix\cdot\xi} q_1(x, \xi) q_2(x, \xi) \hat{u}(\xi) \, d\xi \tag{6.17}$$

and

$$\left(\frac{1}{q_1}\right)(x, D)u(x) = (2\pi)^{-n/2} \int_{\mathbb{R}^n} e^{ix\cdot\xi} \frac{1}{q_1(x, \xi)} \hat{u}(\xi) \, d\xi \ . \tag{6.18}$$

Perturbation should be understood in terms of suitable norm estimates which we want to link to growth conditions of symbols with respect to the co-variable ξ. For example, one might require for

$$r(x, \xi) := \frac{\sigma\left(q_1(x, D) \circ q_2(x, D)\right)(x, \xi) - q_1(x, \xi)q_2(x, \xi)}{q_1(x, \xi)q_2(x, \xi)} \tag{6.19}$$

that uniformly for x in compact sets we have

$$\lim_{|\xi|\to\infty} r(x, \xi) = 0 \ . \tag{6.20}$$

The classical Kohn–Nirenberg and Hörmander calculus uses a grading of symbols with respect to decreasing homogeneity or power growth, i.e. symbols are assumed to admit an expansion

$$\sigma(A)(x, \xi) \sim \sum_{k=-m}^{\infty} q_k(x, \xi) \tag{6.21}$$

where $q_k(x, \xi)$ is either with respect to ξ homogeneous of degree $m - k$, or $q_k(x, \xi)$ satisfies $|q_k(x, \xi)| \leq c_k \left(1 + |\xi|^2\right)^{m_k/2}$ uniformly for $x \in K \subset \mathbb{R}^n$ compact and $m_k > m_{k-1}$, $\lim_{k\to\infty} m_k = -\infty$. For partial differential operators $L(x, D) = \sum_{|\alpha|\leq m} a_\alpha(x) D^\alpha$ we have of course

$$\sigma(L(x, D))(x, \xi) = \sum_{|\alpha|\leq m} a_\alpha(x)\xi^\alpha = \sum_{k=0}^{m} \sum_{|\alpha|=k} a_\alpha(x)\xi^k \ , \tag{6.22}$$

i.e. an expansion as (6.21). For general symbols, obtaining (6.21) is non-trivial, in fact not always possible, moreover, once such an expansion is given, it is non-trivial to associate with it a symbol, i.e. an operator. We want to achieve in addition a deeper relation of an operator functional calculus and a calculus for symbols. For example, for $-q(x, D)$ being a generator of a Feller semigroup $(T_t)_{t \geq 0}$ with corresponding resolvent $(R_\lambda)_{\lambda > 0}$ we want to have

$$\sigma(T_t)(x, \xi) = e^{-tq(x,\xi)} (1 + \text{perturbation}) \qquad (6.23)$$

or

$$\sigma(R_\lambda)(x, \xi) = \frac{1}{\lambda + q(x, \xi)} (1 + \text{perturbation}). \qquad (6.24)$$

Typically, expressions such as (6.21), (6.23) or (6.24) will need smoothness of symbols in x and in particular in ξ, often Taylor series or at least the Taylor formula are the key tool. However, continuous negative definite functions are in general not differentiable and do not admit decomposition into homogeneous terms. Therefore classical calculi such as those based on symbol classes S^m or $S^m_{\varrho,\delta}$ are in general not suitable. In Hoh [32] a quite successful symbolic calculus for pseudo-differential operators with negative definite symbols was developed which could be used by Böttcher in [9], see also [10] and [11], to construct a parametrix for certain evolution equations, and by Evans [23] to study variable order subordination. But still it is too restrictive for some of our problems, as is the Weyl-calculus approach taken up in [2]. Nevertheless, once such a calculus is adapted as frame many results and phenomena discussed in the following sections will have a more satisfactory form. We will eventually avoid a symbolic calculus and use ad hoc calculations to outline ideas, see also Sect. 8.

In addition to providing a symbolic calculus the theory of pseudo-differential operators considered as micro-local analysis develops tools for an analysis of operators based on an analysis of symbols, i.e. functions (sections) on the co-tangent bundle. Typically this is put into the frame of global analysis and requires symbols to admit an invariantly defined principal symbol. Spectral analysis or the propagation of singularities of solutions of equations are key objects of this theory. The Hamiltonian dynamics associated with the (principal) symbol is a tool of great importance within these considerations. We will take ideas and some results as guideline for our investigations, however for pseudo-differential operators with negative definite symbols many key assumptions needed in the "classical" theory do in general not hold: we can in general not define a principal symbol, we do not work with smooth or invariantly defined symbols, etc. Still we can consider the full symbol $q(x, \xi)$ of a generator of a Feller semigroup as a function on the co-tangent bundle to the manifold $\mathbb{R}^n$ and we can look at the corresponding Hamiltonian dynamics. Such an approach allows us, for example, to give an answer to problems when looking at Feynman–Kac formulae and semi-classical asymptotics: we can define the corresponding "classical" objects.

To sum up: generators of Feller semigroups or L^2-sub-Markovian semigroups should be considered as pseudo-differential operators with negative definite symbols. However these symbols do in general not allow us to apply the general, classical theory (micro-local analysis), hence there is a need to modify the classical approach, even in the case of operators with constant coefficients, maybe with a potential perturbation. Getting first results for non-trivial examples is our goal and we do not investigate the indeed promising situation where a negative definite symbol also belongs to a classical symbol class, for more recent results in this direction we refer to [25] and [26].

7 Towards a Hamiltonian Dynamics Associated with Continuous Negative Definite Symbols

As indicated in the Introduction the study of pseudo-differential operators often makes use of the study of the Hamiltonian system which we can associate with the (principal) symbol. In this section we describe some first investigations on some Hamiltonian systems associated with negative definite symbols.

As a model we want to study pseudo-differential operators $A(x, D)$ on $\mathbb{R}^n$ with symbol $A(x, \xi) = \psi(\xi) + V(x)$ which lead to a Hamilton function

$$H(q, p) = \psi(p) + V(q) \tag{7.1}$$

where $\psi : \mathbb{R}^n \longrightarrow \mathbb{R}$ is a continuous negative definite function and $V : \mathbb{R}^n \longrightarrow \mathbb{R}$ is a potential. Note that for potentials satisfying a so-called Kato–Feller condition with respect to $\psi(D)$ the semigroup associated with $A(x, D)$ admits a Feynman–Kac representation

$$T_t u(x) = \mathbb{E}^x \left(u \left(X_t^\psi \right) e^{-\int_0^t V \left(X_s^\psi \right) \, ds} \right), \tag{7.2}$$

where $\left(X_t^\psi \right)_{t \geq 0}$ is the Lévy process associated with ψ.

Let us first have a look at the **free Hamilton function** $H(p) = \psi(p)$. In order to satisfy some minimal conditions needed to apply results from the classical theory we have to assume that ψ satisfies the following conditions:

Assumption 7.1 The function $\psi : \mathbb{R}^n \longrightarrow \mathbb{R}$ is a continuous negative definite function of class C^1 which is convex and coercive.

Recall that a function $\psi : \mathbb{R}^n \longrightarrow \mathbb{R}$ is **coercive** if

$$\lim_{\|p\| \to \infty} \frac{\psi(p)}{\|p\|} = +\infty . \tag{7.3}$$

The **conjugate convex function** ψ^* is the **Legendre transform** of ψ, i.e. we have

$$\psi^*(\eta) := \sup_{p \in \mathbb{R}^n} (\langle \eta, p \rangle - \psi(p)) . \tag{7.4}$$

The following result is proved in [68], see also [41]:

Theorem 7.2 *If $\psi : \mathbb{R}^n \longrightarrow \mathbb{R}$ is a coercive convex function, then its conjugate convex function exists and is a convex and coercive function. Moreover we have $(\psi^*)^* = \psi$.*

Example 7.3

A. For $1 < \alpha \leq 2$ the continuous negative definite function $\psi_\alpha(\xi) := \frac{1}{\alpha}|\xi|^\alpha$ is of class C^1, convex and coercive. Its conjugate function is given by

$$\psi_\alpha^*(\xi) = \frac{1}{\alpha^*}|\xi|^{\alpha^*} , \tag{7.5}$$

where $\frac{1}{\alpha} + \frac{1}{\alpha^*} = 1$, i.e. $\alpha^* = \frac{\alpha}{\alpha-1}$.

B. The continuous negative definite function ψ_{ER} from Example 3.18 is convex and coercive. The latter statement follows easily from (3.47), for a proof of the convexity of ψ_{ER} we refer to [55]. Note that ψ_{ER} is C^1 but not C^2, i.e. we cannot apply the standard criterion to check its convexity. The proof provided in [55] uses an approximation of ψ_{ER} by C^2-functions which are convex and coercive. We do not know ψ_{ER}^* explicitly, however from (3.47) and (3.49) we can conclude that

$$\psi_{ER}^*(\xi, \eta) \leq C \left(|\xi|^2 + |\eta|^2 \right)^{(\alpha_1\gamma_1)^*/2} . \tag{7.6}$$

Moreover, using ideas from [22], Theorem 3 on page 87, we can deduce that with $\zeta = (\xi, \eta) \in \mathbb{R}^2$ it follows that

$$\psi_{ER}^*(\zeta) \geq \inf_{\zeta_1+\zeta_2=\zeta} \left(c_0 \|\zeta_1\|^{\frac{(\gamma_1\alpha_1)^*}{2}} + c_1 \|\zeta_2\|^{\frac{(\gamma_2\alpha_2)^*}{2}} \right)$$

for some suitable constants c_0 and c_1, which implies

$$0 \leq \psi_{ER}^*(\xi, \eta) \leq C \left(|\xi|^2 + |\eta|^2 \right)^{\frac{(\alpha_1\gamma_1)^*}{2}} . \tag{7.7}$$

Given a Hamilton function $H(q, p) = \psi(p) + V(q)$ where ψ satisfies Assumption 7.1 and $V(q)$ is a potential which we assume to be of class C^1, the corresponding **Hamilton system** is given by

$$\dot{q}_i = \frac{\partial H}{\partial p_i}(q, p) \quad \text{and} \quad \dot{p}_i = -\frac{\partial H}{\partial q_i}(q, p) \tag{7.8}$$

or

$$\dot{q}_i = \frac{\partial \psi}{\partial p_i}(p) \quad \text{and} \quad \dot{p}_i = -\frac{\partial V}{\partial q_i}(q) . \tag{7.9}$$

Under Assumption 7.1 the partial Legendre transform of $H(q, p)$ with repsect to p exists and leads to the **Lagrange function** $L(q, \eta)$ associated with $H(q, p)$:

$$L(q, \eta) := \sup_{p \in \mathbb{R}^n} (\langle \eta, p \rangle - H(q, p)) \tag{7.10}$$

$$\eta_i = H_{p_i}(q, p) . \tag{7.11}$$

Note that $\eta \mapsto L(q, \eta)$ is a convex and coercive function, however in general we cannot make a statement on its smoothness. Indeed, Example 7.3 extends to $\xi \mapsto \frac{1}{\kappa}|\xi|^\kappa$, $\kappa > 1$ with conjugate function $\xi \mapsto \frac{1}{\kappa^*}|\xi|^{\kappa^*}$, which shows that for $1 < \kappa < 2$ the conjugate function of $\xi \mapsto \frac{1}{\kappa}|\xi|^\kappa$ is more smooth, at least of class C^2, while for $\kappa > 2$ the conjugate function is less smooth, not even of class C^2, an observation which will become important later.

In the case that the Lagrange function is a C^2-function, the Hamilton system (7.8) (or (7.9)) is equivalent to the **Euler–Lagrange equations**

$$\frac{d}{dt} \frac{\partial L}{\partial \eta_i} - \frac{\partial L}{\partial q_i} = 0 , \tag{7.12}$$

and they are indeed the necessary conditions a smooth extremal of

$$\inf \int_{t_1}^{t_2} L(q(t), \dot{q}(t)) \, dt , \quad q(t_1) = a, \; q(t_2) = b , \tag{7.13}$$

has to satisfy.

Taking the structure (7.1) into account we find

$$L(q, \eta) = \psi^*(\eta) - V(q) \tag{7.14}$$

and in order to form (7.12) we need $\psi^* \in C^2(\mathbb{R}^n)$ and $V \in C^1(\mathbb{R}^n)$. Then we obtain with $\eta_j = \dot{q}_j$

$$\sum_{j=1}^n \left(\frac{\partial^2 \psi^*}{\partial \eta_j \partial \eta_i} \right)(\dot{q})\ddot{q}_j + \frac{\partial V}{\partial q_i} = 0 \quad \text{for} \quad i = 1, \ldots, n . \tag{7.15}$$

Furthermore, since $H(q, p)$ is independent of t, i.e. the Hamilton system is autonomous, we have "conservation of energy" along trajectories solving (7.8), i.e. for solutions of the Hamilton system we find

$$\frac{dH(q(t), p(t))}{dt} = 0.$$ (7.16)

Given initial data $q_0 = q(t_0)$ and $p_0 = p(t_0)$ this observation allows us to introduce the "energy" of the system at t_0 as $E_0(t_0) = H(q_0, p_0)$ and thus we can find integration constants for (7.8), see below.

Before continuing our considerations we would like to remind the reader of our main problem: in general the functions ψ we are interested in are not C^2-functions (nor in general convex) and the standard textbook theory is not (in general) applicable to our case and we have to take some care to transfer (parts of) the standard theory.

Next we want to enlarge our reservoir of examples. The following result is taken from [55]:

Theorem 7.4 *Let f be a Bernstein function such that $\xi \mapsto \psi(\xi) = f(\|\xi\|^2)$ is convex and coercive. In this case ψ^* is given by*

$$\psi^*(\xi) = 2f'\left(\zeta_f^{-1}\left(\|\xi\|^2\right)\right)\zeta_f^{-1}\left(\|\xi\|^2\right) - f\left(\zeta_f^{-1}\left(\|\xi\|^2\right)\right)$$ (7.17)

where $\zeta_f(s) = 4\left(f'(s)\right)^2 s$ and we assume ζ_f^{-1} to exist.

Note that $\psi(\xi) = f(\|\xi\|^2)$ is coercive if and only if $\lim_{r\to\infty}\frac{f(r)}{\sqrt{r}} = \infty$, the convexity of ψ is non-trivial to check, recall that f is a Bernstein function, hence $f'' \leq 0$, i.e. f is concave.

Example 7.5 For $f(s) = s^\beta, 0 < \beta < 1$, we find $\zeta_f(s) = 4\beta^2 s^{2\beta-1}$ which yields

$$\zeta_f^{-1}(t) = \left(\frac{1}{4\beta^2}\right)^{\frac{1}{2\beta-1}} t^{\frac{1}{2\beta-1}}.$$

In the classical case, i.e. in classical dynamics, we have $\psi(\xi) = \frac{\|\xi\|^2}{2}$ and $H(q, p) = \frac{\|p\|^2}{2} + V(q)$. Therefore it follows that $L(q, \eta) = \frac{\|\eta\|^2}{2} - V(q)$. Thus with the kinetic energy $E_{kin} = \|p\|^2/2$ and the potential energy $E_{pot} = V(q)$ we have $H = E_{kin} + E_{pot}$ and $L = E_{kin} - E_{pot}$. For example, the harmonic oscillator is modelled by $H(q, p) = \frac{\|p\|^2}{2} + \frac{\|q\|^2}{2}$ and $L(q, \eta) = \frac{\|\eta\|^2}{2} - \frac{\|q\|^2}{2}$. For our case we suggest as generalizations of the harmonic oscillator the Hamilton function

$$H(q, p) = \psi(p) + \psi^*(q)$$ (7.18)

and as Coulomb potential $V_C(q) = \pm F^{-1}(1/\psi)(q)$ where F^{-1} is the inverse Fourier transform in $S'(\mathbb{R}^n)$.

The following considerations are taken from [55]. For $n = 1$ we consider the Hamilton function

$$H(q, p) = \frac{1}{\beta}|p|^\beta + \frac{1}{\beta^*}|q|^{\beta^*}, \quad 1 < \beta < 2, \tag{7.19}$$

which gives the Hamiltonian system

$$\dot{q} = \text{sgn}(p)|p|^{\beta-1} \tag{7.20}$$

and

$$\dot{p} = -\text{sgn}(q)|q|^{\beta^*-1}. \tag{7.21}$$

We want to discuss this system under the initial conditions $q(t_0) = q_0$ and $p(t_0) = p_0$ which leads to an initial energy $E_0 = H(q_0, p_0) > 0$ which must be conserved, i.e. for a solution to (7.20) and (7.21) we expect for all $t > t_0$

$$\frac{1}{\beta}|p(t)|^\beta + \frac{1}{\beta^*}|q(t)|^{\beta^*} = E_0 \tag{7.22}$$

to hold. In the following we choose $t_0 = 0$. Moreover, by requiring the solution to be symmetric to the q-axis and the p-axis we may assume for the following that $q(t) > 0$ and $p(t) > 0$, and (7.20), (7.21) reduces to the two non-coupled equations

$$\dot{q} = \beta^{\frac{1}{\beta^*}} \left(E_0 - \frac{1}{\beta^*}q^{\beta^*} \right)^{\frac{1}{\beta^*}} \tag{7.23}$$

and

$$\dot{p} = -(\beta^*)^{\frac{1}{\beta}} \left(E_0 - \frac{1}{\beta}p^\beta \right)^{\frac{1}{\beta}}. \tag{7.24}$$

The solutions of these equations are obtained with the help of the Gaussian hypergeometric function $_2F_1$:

$$(E_0\beta)^{-\frac{1}{\beta^*}} q \, _2F_1 \left(\frac{\beta-1}{\beta}, \frac{\beta-1}{\beta}, \frac{2\beta-1}{\beta}, \frac{q^{\beta^*}}{E_0\beta^*} \right) = t + A_1 \tag{7.25}$$

where

$$A_1 = (E_0\beta)^{-\frac{1}{\beta^*}} q_0 \, _2F_1 \left(\frac{\beta-1}{\beta}, \frac{\beta-1}{\beta}, \frac{2\beta-1}{\beta}, \frac{q_0^{\beta^*}}{E_0\beta^*} \right), \tag{7.26}$$

and

$$- (E_0\beta^*)^{-\frac{1}{\beta}} p \, {}_2F_1 \left(\frac{1}{\beta}, \frac{1}{\beta}, \frac{\beta+1}{\beta}, \frac{p^\beta}{E_0\beta} \right) = t + A_2 \tag{7.27}$$

where

$$A_2 = -(E_0\beta^*)^{-\frac{1}{\beta}} p_0 \, {}_2F_1 \left(\frac{1}{\beta}, \frac{1}{\beta}, \frac{\beta+1}{\beta}, \frac{p_0^\beta}{E_0\beta} \right) . \tag{7.28}$$

Note that the convergence of the hypergeometric series is in each case granted by (7.22), i.e. by the conditions (recall we are assuming in the moment $q > 0$ and $p > 0$)

$$\frac{q^{\beta^*}}{E_0\beta^*} < 1 \quad \text{and} \quad \frac{p^\beta}{E_0\beta} < 1 . \tag{7.29}$$

Replacing in (7.25) and (7.27) now q and p in the argument of ${}_2F_1$ by $|q|$ and $|p|$, we obtain the solution for (7.20) and (7.21).

As pointed out in [55] for $\beta \to 2$ the solution converges to the solution of the Hamiltonian system corresponding to the classical harmonic oscillator with Hamilton function $H(q, p) = \frac{1}{2}|p|^2 + \frac{1}{2}|q|^2$.

From the considerations made in this section it is clear that we can under certain additional assumptions develop the Hamiltonian dynamics corresponding to a pseudo-differential operator with negative definite symbols. Of course we do not expect that every Hamiltonian system admits a solution which we can represent by special functions, but at least we can consider the "classical" analogue to the "Schrödinger operator" $\psi(D) + V(q)$. Using the standard symplectic structure on $\mathbb{R}^n \times \mathbb{R}^n$ (the phase space corresponding to $H(q, p)$) we can develop the dynamics further by introducing the Hamiltonian vector field

$$X_H = \sum_{j=1}^n \left(\frac{\partial H}{\partial p_j} \frac{\partial}{\partial q_j} - \frac{\partial H}{\partial q_j} \frac{\partial}{\partial p_j} \right) , \tag{7.30}$$

by discussing flows or by studying the associated Hamilton–Jacobi theory, we refer to first ideas in [55].

However, so far we lack a possibility to define on $\mathbb{R}^n$ looked at as state space (manifold) an intrinsic metric by employing the length of curves connecting points. In the case where H is with respect to $p \in \mathbb{R}^n$ a positive semi-definite quadratic form this is possible with the help of sub-unit trajectories leading to sub-Riemannian geometry and this allows us to study heat kernels in a metric or geometric context. An attempt to develop such tools for certain pseudo-differential operators is due to C. Fefferman and A. Parmeggiani, see [53], but so far this approach has not led to full success, see also the remarks in Fefferman [27]. In this context it might be worth to observe that for the off-diagonal estimates discussed in Sect. 5 we have used two families of balls, i.e. we worked with two different metrics.

8   Some Perturbation Techniques and Results

As explained in Sect. 6, symbolic calculi to handle pseudo-differential operators with negative definite symbols are only under stronger regularity assumptions available. The most far reaching of these calculi is due to Hoh [32], but it is still for certain considerations too restrictive. In this section we want to sketch a direct perturbation theory for symbols with constant coefficients or parameters which is of zero order, i.e. the perturbation is of the same (growth) order as the original term but small with respect to some norm estimates. In other words, we look at a decomposition of a symbol by freezing the coefficients or parameters, i.e.

$$h(x, \xi) = h(x_0, \xi) + (h(x, \xi) - h(x_0, \xi)) = h_1(\xi) + h_2(x, \xi) \, . \tag{8.1}$$

We follow essentially the ideas of [36], see also [39]. All considerations in this section are made under the following assumptions:

A continuous negative definite symbol $h : \mathbb{R}^n \times \mathbb{R}^n \longrightarrow \mathbb{C}$ is given with decomposition $h(x, \xi) = h_1(\xi) + h_2(x, \xi)$. Further $\psi : \mathbb{R}^n \longrightarrow \mathbb{R}$ is a fixed continuous negative definite reference function.

A.1. The function h_1 is assumed to be itself a continuous negative definite function and to satisfy

$$\gamma_0 \psi(\xi) \leq \operatorname{Re} h_1(\xi) \leq \gamma_1 \psi(\xi) \quad \text{for all} \quad |\xi| \geq 1 \, , \tag{8.2}$$

and

$$|\operatorname{Im} h_1(\xi)| \leq \gamma_2 \operatorname{Re} h_1(\xi) \quad \text{for all} \quad \xi \in \mathbb{R}^n \, . \tag{8.3}$$

A.2.m For $m \in \mathbb{N}_0$ the function $x \mapsto h_2(x, \xi)$ belongs to $C^m (\mathbb{R}^n)$ and we have the estimates

$$|\partial_x^\alpha h_2(x, \xi)| \leq \varphi_\alpha(x)(1 + \psi(\xi)) \tag{8.4}$$

for all $\alpha \in \mathbb{N}_0^n$, $|\alpha| \leq m$, with $\varphi_\alpha \in L^1 (\mathbb{R}^n)$.

In addition we assume that

$$\psi(\xi) \geq c_0 |\xi|^{r_0} \tag{8.5}$$

for some $c_0 > 0$, $r_0 > 0$ and all $\xi \in \mathbb{R}^n$ with $|\xi| \geq R$.

Note that (8.5) implies that $H^{\psi,s}(\mathbb{R}^n)$ is continuously embedded into $C_\infty(\mathbb{R}^n)$ provided $s > n/r_0$, and we have the estimate

$$\|u\|_\infty \leq c_{s,r_0,n}\|u\|_{\psi,s} . \tag{8.6}$$

We associate with $h(x,\xi) = h_1(\xi) + h_2(x,\xi)$ the corresponding pseudo-differential operators $h(x,D)$, $h_1(D)$ and $h_2(x,D)$, respectively, which are at least on $C_0^\infty(\mathbb{R}^n)$ (or $S(\mathbb{R}^n)$) well defined. The properties of h_1 allows us to derive the following estimates

$$\|h_1(D)u\|_{\psi,s-2} \leq \tau_1\|u\|_{\psi,s} \tag{8.7}$$

and

$$\|h_1(D)u\|_{\psi,s-2} \geq \gamma_{0,s}\|u\|_{\psi,s} - \lambda_{0,s}\|u\|_{\psi,s-2} , \tag{8.8}$$

which entails for every $\varepsilon > 0$ the estimate

$$\|h_1(D)u\|_{\psi,s-2} \geq (\gamma_{0,s} - \varepsilon)\|u\|_{\psi,s} - \lambda_{\varepsilon,s}\|u\|_0 . \tag{8.9}$$

Estimates for $h_2(x,D)$ are more difficult to derive since for calculating the norm $\|h_2(x,D)u\|_{\psi,s}$ we need to control the Fourier transform $(h_2(x,D)u)^\wedge$ which of course depends on x and the co-variable ξ. Under Assumption **A.2.m** we find for $m > n + 2s$ that in the case $s \geq 1/2$ it follows that

$$\|[(1 + \psi(D))^s, h_2(x,D)]u\|_0 \leq \kappa_{n,m,s,\psi} \sum_{|\alpha| \leq m} \|\varphi_\alpha\|_{L^1}\|u\|_{\psi,2s+1} , \tag{8.10}$$

which extends to

$$\|[(1 + \psi(D))^s, h_2(x,D)]u\|_{\psi,2t} \leq \tilde{c} \sum_{|\alpha| \leq m} \|\varphi_\alpha\|_{L^1}\|u\|_{\psi,2s+2t+1} \tag{8.11}$$

for $s \geq 0$, $t \geq \frac{1}{2}$, $m > n + 2s + 2t$ and $\tilde{c} = \tilde{c}_{n,m,s,t,\psi}$, see Theorem 2.3.9 and Corollary 2.3.10 in [39].

From (8.11) we can now deduce taking (8.8) and (8.9) into account that

$$\|h_2(x,D)u\|_{\psi,t} \leq c_{n,m,t,\psi}\|u\|_{\psi,t+2} \tag{8.12}$$

with

$$c_{n,m,t,\psi} = \tilde{c}_{n,m,t,\psi} \sum_{|\alpha| \leq m} \|\varphi_\alpha\|_{L^1} , \tag{8.13}$$

and

$$\|h(x, D)u\|_{\psi,t} \le c\|u\|_{\psi,t+2} \,, \quad t \ge 1 \,, \tag{8.14}$$

as well as

$$\|h(x, D)u\|_{\psi,t} \ge \delta_0\|u\|_{\psi,t+2} - \gamma_{\eta,t}\|u\|_0 \,, \tag{8.15}$$

where $\eta \in (0, 1)$ and

$$\delta_0 = \eta\gamma_0 - \tilde{c}_{n,m,s,\psi} \sum_{|\alpha|\le m} \|\varphi_\alpha\|_{L^1} > 0 \tag{8.16}$$

and $m > n + [s] + 1$, $s \ge 1$, where the claim that $\delta_0 > 0$ is achieved by assuming for $\sum_{|\alpha|\le m} \|\varphi_\alpha\|_{L^1}$ an appropriate smallness condition (the perturbation of $h_1(D)$ shall be small).

We can now introduce the sesquilinear form

$$B(u, v) = (h(x, D)u, v)_0 = (h_1(D)u, v)_0 + (h_2(x, D)u, v)_0 \tag{8.17}$$

and for B we can derive, see again [39], the estimate

$$|B(u, v)| \le c\|u\|_{\psi,1}\|v\|_{\psi,1} \tag{8.18}$$

and more difficult to obtain is the Gårding inequality

$$\mathrm{Re}\, B(u, u) \ge \delta_1\|u\|_{\psi,1}^2 - \tilde{\lambda}_0\|u\|_0^2 \,, \tag{8.19}$$

provided **A.2.m** holds for $m \ge n + 2$ and we assume that

$$\delta_1 := \gamma_0 - \kappa_3 \sum_{|\alpha|\le n+2} \|\varphi_\alpha\|_{L^1} > 0 \,, \tag{8.20}$$

where κ_3 is a constant coming from the estimate

$$|(h_2(x, D)u, v)_0| \le \kappa_3 \sum_{|\alpha|\le n+2} \|\varphi_\alpha\|_{L^1}\|u\|_{\psi,1}\|v\|_{\psi,1} \,. \tag{8.21}$$

The estimates (8.14), (8.15), (8.18), and (8.19) allow us now to find for $\lambda \ge \lambda_0$ variational solutions in $H^{\psi,1}(\mathbb{R}^n)$ to the equation $h(x, D)u + \lambda u = f$ provided $f \in L^2(\mathbb{R}^n)$. Here we call $u \in H^{\psi,1}(\mathbb{R}^n)$ a variational solution to the equation $h(x, D)u + \lambda u = f$ if for all $v \in H^{\psi,1}(\mathbb{R}^n)$ we have

$$B(u, v) + \lambda(u, v)_0 = (f, v)_0. \tag{8.22}$$

In combination with some Friedrichs mollifier techniques we can conclude further that for $f \in H^{\psi,s}(\mathbb{R}^n)$ it follows that the variational solutions must belong to $H^{\psi,s+2}(\mathbb{R}^n)$. Using the Hille–Yosida–Ray theorem as well as (8.5), i.e. (8.6), we can prove, see [39] Theorem 2.6.4 and Theorem 2.6.6, that $-h(x, D)$ extends to the generator of a Feller semigroup and for λ sufficiently large, $-h_\lambda(x, D) := -h(x, D) - \lambda \mathrm{id}$ extends to the generator of a sub-Markovian semigroup.

In this section our aim was so far to give the reader some ideas of the type of estimates needed, not to provide all details. In particular we want to emphasize the smallness conditions (8.16) and (8.20) which means essentially that $h(x, \xi) - h_2(x, \xi)$ and certain derivatives with respect to x must be under control and small. Nonetheless, non-trivial examples can be constructed, and in [43] also some bounds for corresponding heat kernels were obtained.

In light of Sect. 7 we can now turn to the Hamilton function

$$H(q, p) = h(q, p) = h_1(p) + h_2(q, p) , \tag{8.23}$$

and try, for example, to study the corresponding Hamilton system

$$\dot{q}_i = \frac{\partial H}{\partial p_i}(q, p) = \frac{\partial h_1}{\partial p_i}(p) + \frac{\partial h_2}{\partial p_i}(q, p) , \tag{8.24}$$

$$\dot{p}_i = -\frac{\partial H}{\partial q_i}(q, p) = -\frac{\partial h_2}{\partial q_i}(q, p) . \tag{8.25}$$

We may try to relate its solutions to those of the free or unperturbed Hamilton system corresponding to $h_1(p)$. In fact we may even add a potential $V(q)$. Under suitable assumptions on $h(x, D)$ and $V(q)$ we may, by using the estimates from above, establish the existence of a self-adjoint extension of $h(x, D) + V(x)$ and now we may turn to the corresponding spectral problems. For example, we can start to study the operator $-\psi(D) + h_2(x, D) + \psi^*(x)$ for a suitable continuous negative definite function ψ, a "small" perturbation $h_2(x, D)$ and the potential $\psi^*(q)$.

9   The Symbol of a Feller Process

In the previous sections we have discussed pseudo-differential operators generating sub-Markovian or Feller semigroups and we have indicated to which extent the analysis of the symbol helps to understand the generator and the semigroup. The specific structure of the symbol, i.e. being a continuous negative definite function with respect to the co-variable ξ, does in general not allow us to use "standard" results or methods developed for "classical" pseudo-differential operators, and a larger part of our discussion is devoted to the "gap" of what we want to do in analogy to the classical theory and what so far we can do. As mentioned in the introduction, we do not pay here attention to an analysis of the generator (or the semigroup) based on the Lévy-triple associated with the symbol. Both approaches are complementary and not antagonistic.

In this section we want to investigate how we can use the symbol to study the process. The key observation is that given a "nice" Feller process we can define the symbol by pure probabilistic means. For Lévy processes this is of course nothing but a re-interpretation of the characteristic exponent. For other processes, to the best of our knowledge, it was first suggested in [37] and then much extended by Schilling in [58] and [59] how to define the symbol of a Feller process.

Let $(X_t)_{t\geq 0}$ be a Feller process with state space $\mathbb{R}^n$ and generator $(A, D(A))$, $D(A) \subset C_\infty(\mathbb{R}^n)$, and corresponding Feller semigroup $(T_t)_{t\geq 0}$. Each of the operators T_t we can extend to $C_b(\mathbb{R}^n)$ and therefore we can define pointwisely

$$\lambda_t(x, \xi) := \mathbb{E}^x\left(e^{i\xi\cdot(X_t-x)}\right) = \left(e_{-\xi}T_t e_\xi\right)(x) \tag{9.1}$$

where $e_\xi(x) = e^{ix\cdot\xi}$. For $u \in C_0^\infty(\mathbb{R}^n)$ it follows now that

$$T_t u(x) = \lambda_t(x, D)u(x) = (2\pi)^{-n/2}\int_{\mathbb{R}^n} e^{ix\cdot\xi}\lambda_t(x, \xi)\hat{u}(\xi)\,d\xi\ . \tag{9.2}$$

A formal calculation yields now that with

$$-q(x, \xi) := \left.\frac{d}{dt}\lambda_t(x, \xi)\right|_{t=0} \tag{9.3}$$

on $C_0^\infty(\mathbb{R}^n)$ the generator A should be

$$Au = -q(x, D)u\ , \tag{9.4}$$

i.e. the symbol of the generator is obtained at least formally in pure probabilistic terms by

$$-q(x, \xi) = \lim_{t\to 0}\frac{\mathbb{E}^x\left(e^{i\xi\cdot(X_t-x)}\right) - 1}{t}\ . \tag{9.5}$$

The following result is taken from [12] and gives a precise formulation of the statement when the symbol of A is given by (9.5). Recall, see the introduction, that if the generator A of a Feller semigroup maps $C_0^\infty(\mathbb{R}^n)$ into the continuous functions then it is on $C_0^\infty(\mathbb{R}^n)$ a pseudo-differential operator $-q(x, D)$ and $\xi \mapsto q(x, \xi)$ is a continuous negative definite function. Note further, see [12], Theorem 2.30, that if $x \mapsto q(x, 0)$ is continuous, then $x \mapsto q(x, \xi)$ is for all $\xi \in \mathbb{R}^n$ continuous.

Theorem 9.1 ([12], Corollary 2.39) *Let $(X_t)_{t\geq 0}$ be a Feller process with state space $\mathbb{R}^n$. Assume that its generator maps $C_0^\infty(\mathbb{R}^n)$ into $C_\infty(\mathbb{R}^n)$ and denote the corresponding Feller semigroup by $(T_t)_{t\geq 0}$ and the symbol by $q(x, \xi)$. Assume that*

$q(.,.)$ *is continuous in* $\mathbb{R}^n \times \mathbb{R}^n$. *Let*

$$\tau := \tau_r^x := \inf\{s > 0 \mid |X_s - x| > r\} \tag{9.6}$$

be the first exit time of the ball $B_r(x)$ *when* $(X_t)_{t \geq 0}$ *starts at* x, *i.e.* $X_0 = x$. *Then we have*

$$-q(x, \xi) = \lim_{t \to 0} \frac{\mathbb{E}^x \left(e^{i\xi \cdot (X_{t \wedge \tau} - x)}\right) - 1}{t} . \tag{9.7}$$

The proof of Theorem 9.1 relies on first ideas given in [37], but mainly on a combination of the results in [59] with those in [65]. From our point of view the paper [65] has also the interesting feature that it allows us to find the symbol of a process constructed by probabilistic techniques, i.e. solutions of stochastic differential equations.

With Theorem 9.1 in mind we may ask which (probabilistic) results for $(X_t)_{t \geq 0}$ we can derive from our knowledge of the symbol $q(x, \xi)$. For symbols with variable coefficients it was Schilling [57] who came up with first results and many further results have been obtained since then. Here are a few. After earlier results in [37] and [34], in [58] the following theorem was proved.

Theorem 9.2 *Suppose that* $C_0^\infty(\mathbb{R}^n) \subset D(q(x, D))$ *and that* $-q(x, D)$ *generates a Feller semigroup. If for all* $x \in \mathbb{R}^n$ *we have* $q(x, 0) = 0$ *and*

$$\limsup_{r \to \infty} \sup_{|x-y| \leq 2r} \sup_{|\eta| \leq 1/r} |q(y, \eta)| < \infty \tag{9.8}$$

then the corresponding Feller semigroup $(T_t)_{t \geq 0}$ *is conservative, i.e.* $T_t 1 = 1$ *for all* $t > 0$, *and the associated Feller process has infinite life time.*

Chapter 5 in [12] gives a lot of beautiful results relating properties of the symbol of a Feller process to properties of its paths. To get a feeling of the type of results we quote a result on the Hausdorff dimension of paths. Such results for Lévy processes are due to Blumenthal and Getoor [8], Pruitt [54] and Millar [52]. Let $q(x, \xi)$ be a negative definite symbol and $K \subset \mathbb{R}^n$ be a compact set. We define

$$\beta_\infty^K := \inf \left\{ \lambda > 0 \;\middle|\; \lim_{|\xi| \to \infty} \frac{\sup_{x \in K} \sup_{|\eta| \leq |\xi|} \sup_{|z-x| \leq 1/|\xi|} |q(z, \eta)|}{|\xi|^\lambda} = 0 \right\} . \tag{9.9}$$

The following result is essentially due to R. Schilling, see [60] and [61], and quoted from [12].

Theorem 9.3 *Let $(X_t)_{t \geq 0}$ be a Feller process with state space $\mathbb{R}^n$. Assume that the generator is a pseudo-differential operator $-q(x, D)$ with the test functions in its domain. For every bounded analytic time set $E \subset [0, \infty)$ we have*

$$\dim_H X(E) \leq \min\left(n, \left(\sup_K \beta^K_\infty\right) \dim_H E\right) , \tag{9.10}$$

where $\dim_H F$ denotes the Hausdorff dimension of F and the supremum is taken overall compact subsets $K \subset \mathbb{R}^n$.

One of the most striking results, again due to Schilling [61] and [63], is the fact that paths of Feller processes generated by pseudo-differential operators belong to weighted Besov spaces. We refer to [12], Section 8.5, for details.

We finish our short overview by a result on passage times. The interesting point is here for us the simple re-interpretation of the result when considering the metric induced by a continuous negative definite function.

Let us introduce the first passage time

$$\sigma_R := \sigma_R^x := \inf\{t \geq 0 \mid \|X_t - x\| > R\} , \tag{9.11}$$

but we do not require now that $(X_t)_{t \geq 0}$ starts at x. In [61] the following estimates for $\mathbb{E}^x(\sigma_R)$ were proved

$$\frac{C_1}{\substack{\sup \\ \|x-y\| \leq 2R \\ \|\xi\| \leq 1}} \mathrm{Re}\, q(y, \xi/R) \leq \mathbb{E}^x(\sigma_R) \leq \frac{C_2}{\substack{\inf \\ \|x-y\| \leq 2r}} \substack{\sup \\ \|\xi\| \leq 1} \mathrm{Re}\, q(y, \xi/4\kappa R) \tag{9.12}$$

where we assume that $q(x, \xi)$ satisfies the sector condition

$$|\mathrm{Im}\, q(y, \xi)| \leq C_0 \mathrm{Re}\, q(y, \xi) , \tag{9.13}$$

κ is a suitable constant as are C_1 and C_2. In the case where $(X_t)_{t \geq 0}$ is a symmetric Lévy process with characteristic exponent ψ we deduce from (9.12) the estimates

$$\frac{C_1}{\substack{\sup \\ \|\xi\| \leq 1}\, \psi(\xi/R)} \leq \mathbb{E}^x(\sigma_R) \leq \frac{C_2}{\substack{\sup \\ \|\xi\| \leq 1}\, \psi\left(\dfrac{\pi}{8}\dfrac{\xi}{R}\right)} . \tag{9.14}$$

If we now add our standard assumptions on ψ we can rewrite (9.14) as

$$\frac{C_1}{\substack{\sup \\ \|\xi\| \leq 1}\, d^2_\psi(0, \xi/R)} \leq \mathbb{E}^x(\sigma_R) \leq \frac{C_2}{\substack{\sup \\ \|\xi\| \leq 1}\, d^2_\psi\left(0, \dfrac{\pi}{8}\dfrac{\xi}{R}\right)} \tag{9.15}$$

and we find that the estimates (9.14) have a metric or geometric interpretation: we can look at d_ψ $(0, \xi/R)$ as a natural (intrinsic) distance of ξ/R to 0 and the estimates of $\mathbb{E}^x$ (σ_R) is determined by this distance when ξ runs through the Euclidean unit ball.

Appendix: On the Metric Balls $B^{d_\psi}(0, r)$

Continuous negative definite functions are in general not smooth, in fact their smoothness is determined by the moments of their Lévy measure. Moreover for $n \geq 2$ they can have rather anisotropic behaviour, i.e. in the case that $\psi^{\frac{1}{2}}$ gives rise to a metric, the metric balls can be rather anisotropic. Both facts must be taken into account in the analysis of the operator $\psi(D)$ and the corresponding operator semigroups $\left(T_t^\psi\right)_{t \geq 0}$.

The lack of smoothness has the effect that some "nice" looking estimates are not suitable for our analysis. The following example is taken from [44]: In $\mathbb{R}^2$ the two functions $\psi_1(\xi, \eta) = |\xi| + |\eta|$ and $\psi_2(\xi, \eta) = \sqrt{\psi^2 + \eta^2}$ are continuous negative definite functions and the estimates

$$\frac{1}{\sqrt{2}} (|\xi| + |\eta|) \leq \sqrt{|\xi|^2 + |\eta|^2} \leq |\xi| + |\eta| \tag{A.1}$$

hold. The corresponding densities of $\left(T_t^{\psi_j}\right)_{t \geq 0}$ are given by

$$p_t^{\psi_1}(x, y) = \frac{1}{\pi^2} \frac{t^2}{(x^2 + t^2)(y^2 + t^2)} \tag{A.2}$$

and

$$p_t^{\psi_2}(x, y) = \frac{1}{2\pi} \frac{t}{\left((x^2 + y^2) + t^2\right)^{\frac{3}{2}}}. \tag{A.3}$$

If we choose $x = 0$ and consider the limit $|y| \to \infty$ we find for $t = 1$

$$p_1^{\psi_1}(0, y) \asymp |y|^{-2}$$

and

$$p_1^{\psi_2}(0, y) \asymp |y|^{-3}.$$

Thus, although we have symbols which are comparable, the decay of the corresponding semigroups is not. The different degrees of smoothness leads to a different decay of the Fourier transforms of $e^{-t\psi_j}$, $j = 1, 2$, as we do expect. However the diagonal terms can be compared once we have an estimate such as (A.1). Indeed, $\psi_1 \asymp \psi_2$ leads to similar lower bounds of the corresponding Dirichlet forms:

$$\mathcal{E}^{\psi_1}(u, u) = \int_{\mathbb{R}^n} \psi_1(\xi) |\hat{u}(\xi)|^2 \, d\xi$$

$$\leq c_1 \int_{\mathbb{R}^n} \psi_2(\xi) |\hat{u}(\xi)|^2 \, d\xi$$

$$= c_1 \mathcal{E}^{\psi_2}(u, u),$$

and similarly we find $\mathcal{E}^{\psi_2}(u, u) \leq c_2 \mathcal{E}^{\psi_1}(u, u)$. Therefore, for example in the transient case, we obtain with the same $q > 2$ the estimates

$$\|u\|_{L^q}^2 \leq \tilde{c}_1 \mathcal{E}^{\psi_1}(u, u) \quad \text{and} \quad \|u\|_{L^q}^2 \leq \tilde{c}_2 \mathcal{E}^{\psi_2}(u, u).$$

This observation implies also that the diagonal behaviour of $p_t(\cdot)$ alone cannot determine the off-diagonal behaviour.

There are three classes of examples of continuous negative definite functions which we often use and each is requiring some different considerations when investigating the corresponding operator semigroups:

(i) On $\mathbb{R}^n$ we may look at the sum $\psi = \psi_1 + \psi_2$ of two continuous negative definite functions ψ_1 and ψ_2. Our running example ψ_{ER} is of this type. The convolution theorem yields $p_t^\psi = p_t^{\psi_1} * p_t^{\psi_2}$ and from this we derive using Young's inequality that

$$p_t^\psi(0) = \left\| p_t^\psi(0) \right\|_\infty \leq \left\| p_t^{\psi_1} \right\|_\infty \wedge \left\| p_t^{\psi_2} \right\|_\infty = p_t^{\psi_1}(0) \wedge p_t^{\psi_2}(0). \qquad \text{(A.4)}$$

Now special properties of ψ_1 and ψ_2 are needed to get further results. In some cases one can determine explicitly a time $T > 0$ such that for $t < T$ we have $p_t^{\psi_1}(0) \wedge p_t^{\psi_2}(0) = p_t^{\psi_1}(0)$ and for $t > T$ it follows that $p_t^{\psi_1}(0) \wedge p_t^{\psi_2}(0) = p_t^{\psi_2}(0)$. This is, for example, possible for $\psi(\xi) = \|\xi\|^\alpha + \|\xi\|^\beta$. This observation shows now that making use of the full symbol and not only the principal symbol gives more detailed information.

(ii) We may have a decomposition of $\mathbb{R}^n$, $\mathbb{R}^n = \mathbb{R}^{n_1} \times \mathbb{R}^{n_2}$, and $\psi(\xi, \eta) = \psi_1(\xi) + \psi_2(\eta)$. In this case we have of course $p_t^\psi(x, y) = p_t^{\psi_1}(x) p_t^{\psi_2}(y)$ and we can reduce the study of p_t^ψ directly to investigations on $p_t^{\psi_j}$. In this case we should work with $B^{d_{\psi_1}}(0, r) \times B^{d_{\psi_2}}(0, r)$ rather than with $B^{d_\psi}(0, r)$.

(iii) The final class of examples is obtained by subordination, i.e. by considering $f \circ \psi$ where f is a Bernstein function and $\psi : \mathbb{R}^n \to \mathbb{R}$ is a given continuous negative definite function. For some questions, compare with Theorem 3.15 and Corollary 3.16, a type of (operator) functional calculus is available. However, in general, subordination may destroy some structural properties: It may happen that for the symbol $\psi(\xi) = \psi_1(\xi) + \psi_2(\xi)$ we can consider $\psi_1(\xi)$ as a type of principal symbol, however $f(\psi_1(\xi) + \psi_2(\xi))$ need not allow a decomposition into a principal symbol and a "lower order" term with both being continuous negative definite functions.

The first geometric question we want to discuss is that of the convexity of metric balls. Note that two notions of convexity are possible, we may consider convexity in the vector space $\mathbb{R}^n$, and this notion is the important one for us, but we remind the reader on

Definition A.1 A subset G of a metric space (X, d) is called **metrically convex** if for every pair $p, q \in G$, $p \neq q$, there exists a point $r \in G$ such that $d(p, q) = d(p, r) + d(r, q)$.

Using the Lévy-Khinchine representation of ψ, Harris and Rhind could prove, see [30], that in general the metric space $(\mathbb{R}^n, d_\psi)$ is not metrically convex.

The following result is natural and we refer to [48] for a proof:

Proposition A.2 *Let* $\psi : \mathbb{R}^n \to \mathbb{R}$ *be a continuous negative definite function generating a metric on* $\mathbb{R}^n$ *and let* f *be a Bernstein function such that* $f \circ \psi$ *also generates a metric on* $\mathbb{R}^n$. *Then the balls* $B^{d_\psi}(0, r)$ *are convex if and only if the balls* $B^{d_{f \circ \psi}}(0, r)$ *are convex. In particular metric balls related to subordinate Brownian motion are for appropriate Bernstein functions convex.*

In general the metric balls $B^{d_\psi}(0, r)$ will not be convex, examples are easily constructed with the help of

$$\psi(\xi, \eta) = \left(\|\xi\|^\alpha + \|\eta\|^\beta \right)^{\frac{1}{2}}, \quad 0 < \alpha < 1 \quad \text{or} \quad 0 < \beta < 1,$$

where $\xi \in \mathbb{R}^{n_1}$ and $\eta \in \mathbb{R}^{n_2}$, see [48], or by looking at

$$\psi(\xi_1, \xi_2, \xi_3) = \text{arcosh}\left(|\xi_1|^2 + 1\right) + \text{arsinh}\left(|\xi_2|^2\right) + |\xi_3|^\alpha, \quad 0 < \alpha < 2,$$

see [30].

The following two examples do not only illustrate the failure of convexity of metric balls $B^{d_\psi}(0, r)$, they also illustrate the anisotropic behaviour of the balls. These examples are taken from [30].

The first example is the continuous negative definite function defined on $\mathbb{R}^3$ by

$$\psi_\alpha(\xi) = |\xi_1|^{\frac{3}{4}} + \text{arcosh}\left(|\xi_2|^2 + 1\right) + |\xi_3|^{2\alpha}.$$

The following graphic shows the corresponding metric balls $B^{d_{\psi_\alpha}}(0, r)$ for $r = 1$ and $\alpha = 0.35$, $\alpha = 0.5$, $\alpha = 0.75$ and $\alpha = 0.9$, respectively.

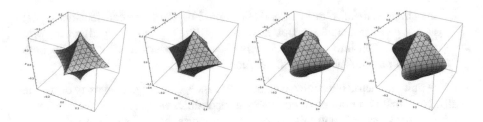

The second example is the continuous negative definite function

$$\psi(\xi) = \operatorname{arcosh}\left(|\xi_1|^2 + 1\right) + \operatorname{arsinh}\left(|\xi_2|^2\right) + |\xi_3|^{\frac{3}{5}},$$

and we consider the different radii $r = 0.5$, $r = 1$, $r = 1.5$ and $r = 2$.

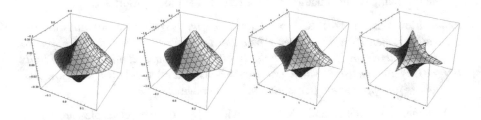

In Sect. 4 we have introduced the doubling property for the metric balls $B^{d_\psi}(0, r)$ and in the case where $\left(\mathbb{R}^n, d_\psi, \lambda^{(n)}\right)$ is a metric measure space in which the doubling property holds we could derive better estimates for $p_t^\psi(0)$ and therefore we want to study $\left(\mathbb{R}^n, d_\psi, \lambda^{(n)}\right)$ in relation to the doubling property.

First we note that the metric $d_\psi = \psi^{\frac{1}{2}}(\xi - \eta)$ is translation invariant which allows us to reduce all studies to metric balls with centre $0 \in \mathbb{R}^n$. Thus the conditions

$$\lambda^{(n)}\left(B^{d_\psi}(0, 2r)\right) \le c_0 \lambda^{(n)}\left(B^{d_\psi}(0, r)\right) \qquad \text{for all } r > 0, \tag{A.5}$$

and

$$\lambda^{(n)}\left(B^{d_\psi}(x, 2r)\right) \le c_0 \lambda^{(n)}\left(B^{d_\psi}(x, r)\right) \qquad \text{for all } r > 0, x \in \mathbb{R}^n, \tag{A.6}$$

with c_0 independent of r and x are equivalent. The doubling property implies power growth for $R \mapsto \lambda^{(n)}\left(B^{d_\psi}(0, R)\right)$, i.e. we have

$$\lambda^{(n)}\left(B^{d_\psi}(x, R)\right) \le \kappa R^{\ln c_0}, \qquad \kappa = \lambda^{(n)}\left(B^{d_\psi}(0, 1)\right). \tag{A.7}$$

We say that $(\mathbb{R}^n, d_\psi, \lambda^{(n)})$ has the **local volume doubling property** if (A.6) holds for all r, $0 < r < r_0$. We often say d_ψ has the volume doubling property when we mean that $(\mathbb{R}^n, d_\psi, \lambda^{(n)})$ has this property.

Example A.3

A. The metric measure space $(\mathbb{R}^n, d_{\psi_\alpha}, \lambda^{(n)})$ with $\psi_\alpha(\xi) = |\xi|^\alpha$, $0 < \alpha \leq 2$, has
the volume doubling property since $\lambda^{(n)} \left(B^{d_{\psi_\alpha}}(0, r) \right) = c_{n,\alpha} r^{\frac{2n}{\alpha}}$.
B. The metric measure space $(\mathbb{R}^n, d_\psi, \lambda^{(n)})$ with

$$\psi(\xi) = 1 - e^{-\gamma |\xi|^2}, \quad \gamma > 0, \tag{A.8}$$

has the local volume doubling property for $0 < r < 1$, but not the volume doubling property.

From this observation we deduce that in general, if d_ψ has the volume doubling property, $d_{f \circ \psi}$, where f is a Bernstein function, need not have the doubling property. In [44] some conditions on f and ψ are discussed for $d_{f \circ \psi}$ having the doubling property, we just quote as one result Corollary 3.11 from [44].

Corollary A.4 *If d_ψ has the volume doubling property and f is a Bernstein function such that for some $C > 1$*

$$\liminf_{r \to 0} \frac{f(Cr)}{f(r)} > 1 \quad and \quad \liminf_{r \to \infty} \frac{f(Cr)}{f(r)} > 1,$$

then $d_{f \circ \psi}$ has the volume doubling property too.

Many of our examples are of type $\psi(\xi) = \sum_{j=1}^N \psi_j(\xi_j)$ with $\xi = (\xi_1, \ldots, \xi_N)$, $\xi_j \in \mathbb{R}^{n_j}$. Suppose that each of the continuous negative definite functions ψ_j generates a metric d_{ψ_j} on $\mathbb{R}^{n_j}$. On $\mathbb{R}^n$, $n = n_1 + \ldots + n_N$, the natural choice of a metric is $d_\psi^{(1)} = \sum_{j=1}^N d_{\psi_j}$ and the question arises whether $d_\psi^{(1)}$ has the volume doubling property (with respect to $\lambda^{(n)}$) if each d_{ψ_j} has the doubling property with respect to $\lambda^{(n_j)}$? The metric balls with respect to $d_\psi^{(1)}$ are not as easy to treat as the metric balls with respect to $d_\psi^{(\infty)} = \max_{1 \leq j \leq N} d_{\psi_j}$. It is helpful to note the following result from [30].

Proposition A.5 *Let $d_\psi^{(p)} = \left(\sum_{j=1}^N d_{\psi_j}^p \right)^{\frac{1}{p}}$, $1 \leq p < \infty$, and let $d_\psi^{(\infty)}$ be defined as above. If one of the metrics $d_\psi^{(p)}$, $1 \leq p \leq \infty$ has the volume doubling property, then they all have the volume doubling property.*

Corollary A.6 *Suppose that each metric measure space* $\left(\mathbb{R}^{n_j}, d_{\psi_j}, \lambda^{(n_j)}\right)$, $1 \leq j \leq N$, *has the volume doubling property. Then the metric measure spaces* $\left(\mathbb{R}^n, d_\psi^{(p)}, \lambda^{(n)}\right)$, $\psi = \sum_{j=1}^N d_{\psi_j}$, $n = \sum_{j=1}^N n_j$, *has the volume doubling property for all* $1 \leq p \leq \infty$.

Proof We prove the doubling property for $p = \infty$ and Proposition A.5 will imply the result. Now we observe

$$\lambda^{(n)}\left(B^{d_\psi^{(\infty)}}(0, 2r)\right) = \prod_{j=1}^N \lambda^{(n_j)}\left(B^{d_{\psi_j}}(0, 2r)\right)$$

$$\leq \prod_{j=1}^N c_j \lambda^{(n_j)}\left(B^{d_{\psi_j}}(0, r)\right) \leq c\lambda^{(n)}\left(B^{d_\psi^{(\infty)}}(0, r)\right).$$

$\square$

Now let $q : \mathbb{R}^n \times \mathbb{R}^n \to \mathbb{R}$ be a continuous negative definite symbol, i.e. for all $x \in \mathbb{R}^n$ the function $q(x, \cdot) : \mathbb{R}^n \to \mathbb{R}$ is negative definite and q as a function on $\mathbb{R}^n \times \mathbb{R}^n$ is continuous. Assume that for a fixed continuous negative definite function ψ we have the estimates

$$\kappa_0 \psi(\xi) \leq q(x, \xi) \leq \kappa_1 \psi(\xi) \tag{A.9}$$

for all $x \in \mathbb{R}^n$, $\xi \in \mathbb{R}^n$, and $0 < \kappa_0 \leq \kappa_1$ are independent of x and ξ. Suppose that ψ satisfies our standard assumptions and the corresponding metric d_ψ has the volume doubling property. The following result taken from [30] is a first step to enable us to use "freezing the coefficients techniques" to investigate the pseudo-differential operator $-q(x, D)$ and in the case it generates a sub-Markovian or Feller semigroup to study associated transition densities with the help of the metric $d_{q(x, \cdot)}$.

Proposition A.7 *Let* $\psi : \mathbb{R}^n \to \mathbb{R}$ *be a fixed continuous negative definite function satisfying our standard conditions and let* $q : \mathbb{R}^n \times \mathbb{R}^n \to \mathbb{R}$ *be a continuous negative definite symbol. Further assume that uniformly in x the estimates* (A.9) *hold. For every* $x \in \mathbb{R}^n$, *now fixed,* $\left(\mathbb{R}^n, (q(x, \cdot))^{\frac{1}{2}}, \lambda^{(n)}\right)$ *is a metric measure space and the metrics* $(q(x, \xi - \eta))^{\frac{1}{2}}$ *and* $d_\psi(\xi, \eta) = \psi^{\frac{1}{2}}(\xi - \eta)$ *are equivalent. Moreover, if for some* $\gamma > 0$ *we can find two constants* $0 < c_0 < c_1$ *such that*

$$c_0 r^\gamma \leq \lambda^{(n)}\left(B^{d_\psi}(0, r)\right) \leq c_1 r^\gamma$$

holds, then for every $x \in \mathbb{R}^n$ *the metric* $(q(x, \cdot))^{\frac{1}{2}}$ *has the volume doubling property.*

Example A.8 Choose $0 < \alpha, \beta < 2$ and $\psi(\xi_1, \xi_2) = \|\xi_1\|^\alpha + \|\xi_2\|^\beta$, $\xi_1 \in \mathbb{R}^{n_1}$, $\xi_2 \in \mathbb{R}^{n_2}$. Further let $q : \mathbb{R}^n \times \mathbb{R}^n \to \mathbb{R}$ be a continuous negative definite symbol satisfying (A.9) with ψ as defined above. Since $\lambda^{(n)}\left(B^{d_\psi}(0,r)\right) = cr^{2\left(\frac{n_1}{\alpha} + \frac{n_2}{\beta}\right)}$ we may apply Proposition A.7 to $q(x, D)$.

For further results on the metric measure space $\left(\mathbb{R}^n, d_\psi, \lambda^{(n)}\right)$, we refer to [30] and [48]. We would like to mention once more that the Appendix is co-authored by J. Harris.

References

1. D. Bakry, J. Gentil, M. Ledoux, *Analysis and Geometry of Markov Diffusion Operators*. Grundlehren der Mathematischen Wissenschaften, vol. 348 (Springer, Berlin, 2014)
2. F. Baldus, Application of the Weyl–Hörmander calculus to generators of Feller semi-groups. Math. Nachr. **252**, 3–23 (2003)
3. O.E. Barndorff-Nielsen, S.Z. Levendorikĭ, Feller processes of normal inverse Gaussian type. Quant. Finance **1**, 318–331 (2001)
4. A. Bendikov, P. Maheux, Nash type inequalities for fractional powers of non-negative self-adjoint operators. Trans. Am. Math. Soc. **359**, 3085–3098 (2007)
5. Y. Benyamini, J. Lindenstrauss, *Geometric Nonlinear Functional Analysis*, vol. 1. Colloquim Publications, vol. 48 (American Mathematical Society, Providence, 2000)
6. C. Berg, G. Forst, *Potential Theory on Locally Compact Abelian Groups*. Ergebnisse der Mathematik und ihrer Grenzgebiete (Ser.2), vol. 87 (Springer, Berlin 1975)
7. A. Beurling, J. Deny, Dirichlet spaces. Proc. Natl. Acad. Sci. U. S. A. **45**, 208–215 (1959)
8. R. Blumenthal, R. Getoor, Sample functions of stochastic processes with stationary independent increments. J. Math. Mech. **10**, 493–516 (1961)
9. B. Böttcher, Some Investigations on Feller Processes Generated by Pseudo-differential Operators, PhD thesis, University of Wales, Swansea, 2004
10. B. Böttcher, A parametrix construction for the fundamental solution of the evolution equation associated with a pseudo-differential operator generating a Markov process. Math. Nachr. **278**, 1235–1241 (2005)
11. B. Böttcher, Construction of time inhomogeneous Markov processes via evolution equations using pseudo-differential operators. J. Lond. Math. Soc. **78**, 605–621 (2008)
12. B. Böttcher, R.L. Schilling, J. Wang, *Lévy-Type Processes: Construction, Approximation and Sample Path Properties*. Lecture Notes in Mathematics, vol. 2099 (Springer, Berlin, 2013)
13. L.J. Bray, Investigations on Transition Densities of Certain Classes of Stochastic Processes, PhD thesis, Swansea University, Swansea, 2016
14. L.J. Bray, N. Jacob, Some considerations on the structure of transition densities of symmetric Lévy processes. Commun. Stoch. Anal. **10**, 405–420 (2016)
15. E. Carlen, S. Kusuoka, D.W. Stroock, Upper bounds for symmetric Markov transition functions. Ann. Henri Poincaré Probabilités et Statistiques, Sup, au n° **23**(2), 245–287 (1987)
16. R. Coifman, G. Weiss, *Analyse harmonique non-commutative sur certains espaces homogènes*. Étude de certaines intégrales singulières. Lecture Notes in Mathematics, vol. 242 (Springer, Berlin, 1971)
17. P. Courrège, Sur la forme intégro-differentielle des opérateurs de C_K^∞ dans C satisfaisant au principe du maximum, in *Sém. Théorie du Potential 1965/66*. Exposé 2, 38 pp
18. E.B. Davies, Explicit constants for Gaussian upper bounds on heat kernels. Am. J. Math. **109**, 319–334 (1987)
19. E.B. Davies, Heat kernel bounds for second order elliptic operators on Riemannian manifolds. Am. J. Math. **109**, 545–570 (1987)
20. E.B. Davies, *Heat Kernels and Spectral Theory*. Cambridge Tracts in Mathematics, vol. 92 (Cambridge University Press, Cambridge 1989)

21. J. Deny, Méthodes Hilbertiemmes et théorie du potential, in *Potential Theory*, ed. by M. Brelot (Edizione Cremonese, Roma, 1970), pp. 123–201
22. J. Ekeland, *Convexity Methods in Hamiltonian Mechanics*. Ergebnisse der Mathematik und ihrer Grenzgebiete (Ser.2), vol. 19 (Springer, Berlin, 1990)
23. K.P. Evans, Subordination in the Sense of Bochner of Variable Order, PhD thesis, Swansea University, Swansea, 2008
24. K.P. Evans, N. Jacob, Feller semigroups obtained by variable order subordination. Rev. Mat. Complut. **20**, 293–307 (2007)
25. M.A. Fahrenwaldt, Heat trace asymptotics of subordinated Brownian motion on Euclidean space. Potential Anal. **44**, 331–354 (2016)
26. M.A. Fahrenwaldt, Off-diagonal heat kernel asymptotics of pseudo differential operators on closed manifolds and subordinate Brownian motion. Integr. Equ. Oper. Theory **87**, 327–347 (2017)
27. C.L. Fefferman, Symplectic subunit balls and algebraic functions, in *Harmonic Analysis and Partial Differential Equations. Essays in Honor of Alberto P. Calderon*, ed. by M. Christ, C.E. Kenig, C. Sadovsky (University of Chicago Press, Chicago, 1999), pp. 199–205
28. M. Fukushima, Y. Oshima, M. Takeda, *Dirichlet Forms and Symmetric Markov Processes*. De Gruyter Studies in Mathematics, vol. 19, 2nd edn. (Walter de Gruyter, Berlin, 2011)
29. B.W. Gnedenko, *Einführung in die Wahrscheinlichkeitstheorie*. Mathematische Lehrbücher, Bd.39 (Akademie, Berlin, 1991)
30. J. Harris, Investigations on Metric Spaces Associated with Continuous Negative Definite Functions and Bounds for Transition Densities of Certain Lévy Processes, PhD thesis, Swansea University, Swansea, 2016
31. J. Harris, N. Jacob, Some Thoughts and Investigations on Densities of One-Parameter Operator Semi-groups, in *Stochastic Partial Differential Equations and Related Fields. In Honor of Michael Röckner*, ed. by A. Eberle et al., Springer Series in Mathematics of Statistics, vol. 229 (Springer Verlag, Berlin, 2018), pp. 451–460
32. W. Hoh, A symbolic calculus for pseudo differential operators generating Feller semigroups. Osaka J. Math. **35**, 758–820 (1998)
33. W. Hoh, Pseudo differential operators with negative definite symbols of variable order. Rev. Mat. Iberoamericana **16**, 219–241 (2000)
34. W. Hoh, N. Jacob, On the Dirichlet problem for pseudo differential operators generating Feller semigroups. J. Funct. Anal. **137**, 19–48 (1996)
35. N. Jacob, Dirichlet forms and pseudo differential operators. Expo. Math. **6**, 313–351 (1988)
36. N. Jacob, A class of Feller semigroups generated by pseudo differential operators. Math. Z. **215**, 151–166 (1994)
37. N. Jacob, Characteristic functions and symbols in the theory of Feller processes. Potential Anal. **8**, 61–68 (1998)
38. N. Jacob, *Pseudo-Differential Operators and Markov Processes. Vol. 1: Fourier Analysis and Semigroups* (Imperial College Press, London, 2001)
39. N. Jacob, *Pseudo-Differential Operators and Markov Processes. Vol. 2: Generators and Their Potential Theory* (Imperial College Press, London, 2002)
40. N. Jacob, *Pseudo-Differential Operators and Markov Processes. Vol. 3: Markov Processes and Applications* (Imperial College Press, London, 2005)
41. N. Jacob, K.P. Evans, *A Course in Analysis. Vol. 2: Differentiation and Integration of Functions of Several Variables, Vector Calculus* (World Scientific, Singapore, 2016)
42. N. Jacob, H.-G. Leopold, Pseudo-differential operators with variable order of differentiation generating Feller semigroups. Integr. Equ. Oper. Theory **17**, 544–553 (1993)
43. N. Jacob, R.L. Schilling, Estimates for Feller semigroups generated by pseudo differential operators, in *Function Spaces, Differential Operators and Nonlinear Analysis*, ed. by J. Rakošnik (Prometheus Publishing House, Praha, 1996), pp. 27–49
44. N. Jacob, V. Knopova, S. Landwehr, R.L. Schilling, A geometric interpretation of the transition density of a symmetric Lévy process. Science China Ser. A Math. **55**, 1099–1126 (2012)

45. V. Knopova, R.L. Schilling, A note on the existence of transition probability densities for Lévy processes. Forum Math. **25**, 125–149 (2013)
46. A.N. Kochubei, Parabolic pseudodifferential equations, hypersingular integrals and Markov processes. Math. USSR Izvestija **33**, 233–259 (1989)
47. T. Komatsu, Pseudo-differential operators and Markov processes. J. Math. Soc. Jpn. **36**, 387–418 (1984)
48. S. Landwehr, On the Geometry Related to Jumps Processes, PhD thesis, Swansea University, Swansea, 2010
49. G. Laue, M. Riedel, H.-J. Roßberg, *Unimodale und positiv definite Dichten* (B.G. Teunrer Verlag, Stuttgart, 1999)
50. T. Lewis, Probability functions which are proportional to characteristic functions and the infinite divisibility of the von Mises distribution, in *Perspectives in Probability and Statistics* (Academic, New York, 1976, pp. 19–28)
51. P.-A. Meyer, Démonstration probabiliste de certaines inégalités de Littlewood-Paley. Exposé 2: L'opérateur carré du champ. Séminaire de Probabilités, vol. 10. Lecture Notes in Mathematics, vol. 511 (Springer, Berlin, 1976), pp. 142–163
52. P. Millar, Path behaviour of processes with stationary independent increments. Z. Wahrscheinlichkeitstheor. verw. Geb. **17**, 53–73 (1971)
53. A. Parmeggiani, Subunit balls for symbols of pseudo differential operators. Adv. Math. **131**, 357–452 (1997)
54. W.E. Pruitt, The Hausdorff dimension of the range of a process with stationary independent increments. Indiana J. Math. **19**, 371–378 (1969)
55. E.O.T. Rhind, PhD thesis, Swansea University, Swansea, 2018
56. K. Sato, *Lévy Processes and Infinitely Divisible Distributions*. Cambridge Studies in Advanced Mathematics, vol. 68 (Cambridge University Press, Cambridge, 1999)
57. R.L. Schilling, Zum Pfadverhalten von Markovschen Prozessen, die mit Lévy-Prozessen vergleichbar sind. Dissertation, Universität Erlangen-Nürnberg, Erlangen (1994)
58. R.L. Schilling, Conservativeness and extensions of Feller semigroups. Positivity **2**, 239–256 (1998)
59. R.L. Schilling, Conservativeness of semigroups generated by pseudo differential operators. Potential Anal. **9**, 91–104 (1998)
60. R.L. Schilling, Feller processes generated by pseudo-differential operators: on the Hausdorff dimension of their sample paths. J. Theor. Probab. **11**, 303–330 (1998)
61. R.L. Schilling, Growth and Hölder conditions for the sample paths of Feller processes. Probab. Theory Relat. Fields **112**, 565–611 (1998)
62. R.L. Schilling, Subordination in the sense of Bochner and a related functional calculus. J. Aust. Math. Soc. (Ser. A) **64**, 368–396 (1998)
63. R.L. Schilling, Function spaces as path spaces of Feller processes. Math. Nachr. **217**, 147–174 (2000)
64. R.L. Schilling, *Measures, Integrals and Martingales* (Cambridge University Press, Cambridge, 2005)
65. R.L. Schilling, A. Schnurr, The symbol associated with the solution of a stochastic differential equation. Electron. J. Probab. **15**, 1369–1393 (2010)
66. R.L. Schilling, J. Wang, Functional inequalities and subordination: stability of Nash and Poincaré inequalities. Math. Z. **272**, 921–936 (2012)
67. R.L. Schilling, R. Song, Z. Vondraček, *Bernstein Functions*. De Gruyter Studies in Mathematics, vol. 37, 2nd edn. (De Gruyter, Berlin, 2012)
68. B. Simon, *Convexity: An Analytic Viewpoint*. Cambridge Tracts in Mathematics, vol. 187 (Cambridge University Press, Cambridge, 2011)
69. E.M. Stein, *Singular Integrals and Differentiability Properties of Functions*. Princeton Mathematical Series, vol. 30 (Princeton University Press, Princeton, 1970)
70. M. Tomisaki, Comparison theorems on Dirichlet forms and their applications. Forum Math. **2**, 277–295 (1990)

71. N. Varopoulos, L. Saloff-Coste, T. Coulhon, *Analysis and Geometry on Groups*. Cambridge Tracts in Mathematics, vol. 100 (Cambridge University Press, Cambridge, 1992)
72. F.Y. Wang, *Functional Inequalities, Markov Semigroups and Spectral Theory*. Mathematical Monograph Series, vol. 4 (Science Press, Beijing, 2005)
73. F.Y. Wang, *Analysis for Diffusion Processes on Riemannian Manifolds*. Advanced Series on Statistical Science of Applied Probability, vol. 18 (World Scientific, Singapore, 2014)
74. K. Yosida, Abstract potential operators on Hilbert spaces. Publ. R.I.M.S. **8**, 201–205 (1972)
75. Y. Zhuang, Some Geometric Considerations Related to Transition Densities of Jump-Type Markov Processes, PhD thesis, Swansea University, Swansea, 2012

Lectures on Entropy. I: Information-Theoretic Notions

Vojkan Jakšić

Abstract These lecture notes concern information-theoretic notions of entropy. They are intended for, and have been successfully taught to, undergraduate students interested in research careers. Besides basic notions of analysis related to convergence that are typically taught in the first or second year of undergraduate studies, no other background is needed to read the notes. The notes might be also of interest to any mathematically inclined reader who wishes to learn basic facts about notions of entropy in an elementary setting.

1 Introduction

As the title indicates, this is the first in a planned series of four lecture notes. Part II concerns notions of entropy in the study of statistical mechanics, and III/IV are the quantum information theory/quantum statistical mechanics counterparts of I/II. All four parts target a similar audience and are on a similar technical level. Eventually, Parts I–IV together are intended to be an introductory chapter to a comprehensive volume dealing with the topic of entropy from a certain point of view on which I will elaborate below.

The research program that leads to these lecture notes concerns the elusive notion of entropy in non-equilibrium statistical mechanics. It is for this pursuit that the notes are preparing a research-oriented reader, and it is the pursuit to which the later more advanced topics hope to contribute. Thus, it is important to emphasize that the choice of topics and their presentation have a specific motivation which may not be obvious until at least the Part II of the lecture notes is completed. Needless to say, the lecture notes can be read independently of its motivation, as they provide a concise, elementary, and mathematically rigorous introduction to the topics they cover.

V. Jakšić (✉)
Department of Mathematics and Statistics, McGill University, Montreal, QC, Canada
e-mail: vojkan.jaksic@mcgill.ca

© Springer Nature Switzerland AG 2019   141
D. Bahns et al. (eds.), *Open Quantum Systems*, Tutorials, Schools, and Workshops
in the Mathematical Sciences, https://doi.org/10.1007/978-3-030-13046-6_4

The theme of this Part I is the Boltzmann–Gibbs–Shannon (BGS) entropy of a finite probability distribution $(p_1, \cdots, p_n)$, and its various deformations such as the Rényi entropy, the relative entropy, and the relative Rényi entropy. The BGS entropy and the relative entropy have intuitive and beautiful axiomatic characterizations discussed in Sects. 3.4 and 5. The Rényi entropies also have axiomatic characterizations, but those are perhaps less natural, and we shall not discuss them in detail. Instead, we shall motivate the Rényi entropies by the so-called Large Deviation Principle (LDP) in probability theory. The link between the LDP and notions of entropy runs deep and will play a central role in this lecture notes. For this reason Cramér's theorem is proven right away in the introductory Sect. 2 (the more involved proof of Sanov's theorem is given in Sect. 5.4). It is precisely this emphasis on the LDP that makes this lecture notes somewhat unusual in comparison with other introductory presentations of the information-theoretic entropy.

The Fisher entropy and a related topic of parameter estimation are also an important part of this lecture notes. The historical background and most of applications of these topics are in the field of statistics. There is a hope that they may play an important role in study of entropy in non-equilibrium statistical mechanics, and that is the reason for including them in the lecture notes. Again, Sects. 6 and 7 can be read independently of this motivation by anyone interested in an elementary introduction to the Fisher entropy and parameter estimation.

These notes are work in progress, and additional topics may be added in the future.

The notes benefited from the comments of numerous McGill undergraduate students who attended the seminars and courses in which I have taught the presented material. I am grateful for their help and for their enthusiasm which to a large extent motivated my decision to prepare the notes for publication. In particular, I am grateful to Sherry Chu, Wissam Ghantous, and Jane Panangaden whose McGill's undergraduate summer research projects were linked to the topics of the lecture notes and whose research reports helped me in writing parts of the notes. I am also grateful to Laurent Bruneau, Noé Cuneo, Tomas Langsetmo, Renaud Raquépas, and Armen Shirikyan for comments and suggestions. I wish to thank Jacques Hurtubise and David Stephens who, as the chairmans of the McGill Department of Mathematics and Statistics, enabled me to teach the material of the notes in a course format. Finally, I am grateful to Marisa Rossi for her exceptional hospitality and support during the period when Sect. 7 was written.

A part of the material of these lecture notes was presented in Göttingen's Second Summer/Winter School on Dynamical Approaches in Spectral Geometry titled "Dynamical Methods in Open Quantum Systems". I am grateful to Dorothea Bahns, Anke Pohl, and Ingo Witt for their invitation to lecture at this school and for their hospitality.

This research that has led to this lecture notes was partly funded by NSERC, *Agence Nationale de la Recherche* through the grant NONSTOPS (ANR-17-CE40-0006-01, ANR-17-CE40-0006-02, ANR-17-CE40-0006-03), the CNRS collaboration grant *Fluctuation theorems in stochastic systems*, and the *Initiative d'excellence Paris-Seine*.

1.1 Notes and References

Shannon's seminal 1948 paper [44], reprinted in [45], remains a must-read for anyone interested in notions of entropy. Khintchine's reworking of the mathematical foundations of Shannon's theory in the early 1950s, summarized in the monograph [32], provides a perspective on the early mathematically rigorous developments of the subject. For further historical perspective, we refer the reader to [52] and the detailed list of references provided there. There are many books dealing with entropy and information theory. The textbook [9] is an excellent introduction to the subject, [4, 22, 46] are recommended to mathematically more advanced reader. Another instructive reference is [11], where a substantial part of the material covered in this lecture notes is left as an exercise for the reader!

Discussions of a link between information and statistical mechanics preceded Shannon's work. Although Weaver's remark[1] on page 3 of [45] appears to be historically inaccurate, the discussions of the role of information in foundations of statistical mechanics goes back at least to the work of Szillard [49] in 1929, see also https://plato.stanford.edu/entries/information-entropy/, and remains to this day a hotly disputed subject; see [21] for a recent discussion. An early discussion can be found in [27, 28]. The textbook [39] gives an additional perspective on this topic.

In contrast to equilibrium statistical mechanics whose mathematically rigorous foundations, based on the nineteenth century works of Boltzmann and Gibbs, were laid in the 1960s and 1970s, the physical and mathematical theory of non-equilibrium statistical mechanics remains in its infancy. The introduction of non-equilibrium steady states and the discovery of the fluctuation relations in the context of chaotic dynamical systems in the early 1990s (see [26] for references) revolutionized our understanding of some important corners of the field, and have generated an enormous amount of theoretical, experimental, and numerical works with applications extending to chemistry and biology. The research program of Claude-Alain Pillet and myself mentioned in the introduction is rooted in these developments.[2] In this program, the search for a notion of entropy for systems out of equilibrium plays a central role. The planned four parts lecture notes are meant as an introduction to this search, with this Part I focusing on the information-theoretic notions of entropy.

---

[1] "Dr. Shannon's work roots back, as von Neumann has pointed out, to Boltzmann's observation, in some of his work on statistical physics (1894), that entropy is related to "missing information," inasmuch as it is related to the number of alternatives which remain possible to a physical system after all the macroscopically observable information concerning it has been recorded."

[2] The references to results of this program are not relevant for this Part I of the lectures and they will be listed in the latter installments.

2 Elements of Probability

2.1 Prologue: Integration on Finite Sets

Let Ω be a finite set. Generic element of Ω is denoted by ω. When needed, we will enumerate elements of Ω as $\Omega = \{\omega_1, \cdots, \omega_L\}$, where $|\Omega| = L$.

A measure on Ω is a map

$$\mu : \Omega \to \mathbb{R}_+ = [0, \infty[.$$

The pair (Ω, μ) is called measurable space. The measure of $S \subset \Omega$ is

$$\mu(S) = \sum_{\omega \in S} \mu(\omega).$$

By definition, $\mu(\emptyset) = 0$.

Let $f : \Omega \to \mathbb{C}$ be a function. The integral of f over $S \subset \Omega$ is defined by

$$\int_S f \mathrm{d}\mu = \sum_{\omega \in S} f(\omega)\mu(\omega).$$

Let Ω and $\mathcal{E}$ be two finite sets and $T : \Omega \to \mathcal{E}$ a map. Let μ be a measure on Ω. For $\zeta \in \mathcal{E}$ set

$$\mu_T(\zeta) = \mu(T^{-1}(\zeta)) = \sum_{\omega : T(\omega) = \zeta} \mu(\omega).$$

μ_T is a measure on $\mathcal{E}$ induced by (μ, T). If $f : \mathcal{E} \to \mathbb{C}$, then

$$\int_{\mathcal{E}} f \mathrm{d}\mu_T = \int_{\Omega} f \circ T \mathrm{d}\mu.$$

If $f : \Omega \to \mathbb{C}$, we denote by μ_f the measure on the set of values $\mathcal{E} = \{f(\omega) \mid \omega \in \Omega\}$ induced by (Ω, f). μ_f is called the distribution measure of the function f.

We denote by

$$\Omega^N = \{\omega = (\omega_1, \cdots, \omega_N) \mid \omega_k \in \Omega\},$$

$$\mu_N(\omega = (\omega_1, \cdots, \omega_N)) = \mu(\omega_1) \cdots \mu(\omega_N),$$

the N-fold product set and measure of the pair (Ω, μ).

Let $\Omega_{l/r}$ be two finite sets and μ a measure on $\Omega_l \times \Omega_r$. The marginals of μ are measures $\mu_{l/r}$ on $\Omega_{l/r}$ defined by

$$\mu_l(\omega) = \sum_{\omega' \in \Omega_r} \mu(\omega, \omega'), \qquad \omega \in \Omega_l,$$

$$\mu_r(\omega) = \sum_{\omega' \in \Omega_l} \mu(\omega', \omega), \qquad \omega \in \Omega_r.$$

If $\mu_{l/r}$ are measures on $\Omega_{l/r}$. we denote by $\mu_l \otimes \mu_r$ the product measure defined by

$$\mu_l \otimes \mu_r(\omega, \omega') = \mu_l(\omega)\mu_r(\omega').$$

The support of the measure μ is the set

$$\operatorname{supp} \mu = \{\omega \mid \mu(\omega) \neq 0\}.$$

Two measures μ_1 and μ_2 are mutually singular, denoted $\mu_1 \perp \mu_2$, iff $\operatorname{supp}\mu_1 \cap \operatorname{supp}\mu_2 = \emptyset$. A measure μ_1 is absolutely continuous w.r.t. another measure μ_2, denoted $\mu_1 \ll \mu_2$, iff $\operatorname{supp}\mu_1 \subset \operatorname{supp}\mu_2$, that is, iff $\mu_2(\omega) = 0 \Rightarrow \mu_1(\omega) = 0$. If $\mu_1 \ll \mu_2$, the Radon-Nikodym derivative of μ_1 w.r.t. μ_2 is defined by

$$\Delta_{\mu_1 | \mu_2}(\omega) = \begin{cases} \frac{\mu_1(\omega)}{\mu_2(\omega)} & \text{if } \omega \in \operatorname{supp} \mu_1 \\ 0 & \text{if } \omega \notin \operatorname{supp} \mu_1. \end{cases}$$

Note that

$$\int_\Omega f \Delta_{\mu_1 | \mu_2} \mathrm{d}\mu_2 = \int_\Omega f \mathrm{d}\mu_1.$$

Two measures μ_1 and μ_2 are called equivalent iff $\operatorname{supp} \mu_1 = \operatorname{supp} \mu_2$.

Let μ, ρ be two measures on Ω. Then there exists a unique decomposition (called the Lebesgue decomposition) $\mu = \mu_1 + \mu_2$, where $\mu_1 \ll \rho$ and $\mu_2 \perp \rho$. Obviously,

$$\mu_1(\omega) = \begin{cases} \mu(\omega) & \text{if } \omega \in \operatorname{supp} \rho \\ 0 & \text{if } \omega \notin \operatorname{supp} \rho, \end{cases} \qquad \mu_2(\omega) = \begin{cases} 0 & \text{if } \omega \in \operatorname{supp} \rho \\ \mu(\omega) & \text{if } \omega \notin \operatorname{supp} \rho. \end{cases}$$

A measure μ is called faithful if $\mu(\omega) > 0$ for all $\omega \in \Omega$.

Proposition 2.1 *Let* $f : \Omega \to \mathbb{R}_+$, $a > 0$, *and* $S_a = \{\omega \mid f(\omega) \geq a\}$. *Then*

$$\mu(S_a) \leq \frac{1}{a} \int_\Omega f \mathrm{d}\mu.$$

Proof The statement is obvious is $S_a = \emptyset$. If S_a is non-empty,

$$\mu(S_a) = \sum_{\omega \in S_a} \mu(\omega) \leq \frac{1}{a} \sum_{\omega \in S_a} f(\omega)\mu(\omega) \leq \frac{1}{a} \int_\Omega f \, d\mu.$$

□

We recall the Minkowski inequality

$$\left(\int_\Omega |f + g|^p d\mu \right)^{1/p} \leq \left(\int_\Omega |f|^p d\mu \right)^{1/p} + \left(\int_\Omega |g|^p d\mu \right)^{1/p},$$

where $p \geq 1$, and the Hölder inequality

$$\int_\Omega fg \, d\mu \leq \left(\int_\Omega |f|^p d\mu \right)^{1/p} \left(\int_\Omega |g|^q d\mu \right)^{1/q},$$

where $p, q \geq 1$, $p^{-1} + q^{-1} = 1$. For $p = q = 2$ the Hölder inequality reduces to the Cauchy-Schwarz inequality.

If $f : \Omega \to]-\infty, \infty]$ or $[-\infty, \infty[$, we again set $\int_\Omega f \, d\mu = \sum_\omega f(\omega)\mu(\omega)$ with the convention that $0 \cdot (\pm\infty) = 0$.

2.2 Probability on Finite Sets

We start with a change of vocabulary adapted to the probabilistic interpretation of measure theory.

A measure P on a finite set Ω is called a probability measure if $P(\Omega) = \sum_{\omega \in \Omega} P(\omega) = 1$. The pair (Ω, P) is called probability space. A set $S \subset \Omega$ is called an event and $P(S)$ is the probability of the event S. Points $\omega \in \Omega$ are sometimes called elementary events.

A perhaps most basic example of a probabilistic setting is a fair coin experiment, where a coin is tossed N times and the outcomes are recorded as Head $= 1$ and Tail $= -1$. The set of outcomes is

$$\Omega = \{\omega = (\omega_1, \cdots, \omega_N) \mid \omega_k = \pm 1\},$$

and

$$P(\omega = (\omega_1, \cdots, \omega_N)) = \frac{1}{2^N}.$$

Let S be the event that k Heads and $N - k$ Tails are observed. The binomial formula gives

$$P(S) = \binom{N}{k} \frac{1}{2^N}.$$

As another example, let

$$S_j = \left\{ \omega = (\omega_1, \cdots, \omega_N) \mid \sum_k \omega_k = j \right\},$$

where $-N \le j \le N$. $P(S_j) = 0$ if $N + j$ is odd. If $N + j$ is even, then

$$P(S_j) = \binom{N}{\frac{N+j}{2}} \frac{1}{2^N}.$$

A function $X : \Omega \to \mathbb{R}$ is called random variable.

The measure P_X induced by (P, X) is called the probability distribution of X. The expectation of X is

$$E(X) = \int_\Omega X \mathrm{d}P.$$

The moments of X are

$$M_k = E(X^k), \qquad k = 1, 2 \cdots,$$

and the moment generating function is

$$M(\alpha) = E(e^{\alpha X}) = \sum_{\omega \in \Omega} e^{\alpha X(\omega)} P(\omega),$$

where $\alpha \in \mathbb{R}$. Obviously,

$$M_k = \frac{\mathrm{d}^k}{\mathrm{d}\alpha^k} M(\alpha)\big|_{\alpha=0}.$$

The cumulant generating function of X is

$$C(\alpha) = \log E(e^{\alpha X}) = \log \left(\sum_{\omega \in \Omega} e^{\alpha X(\omega)} P(\omega) \right).$$

The cumulants of X are

$$C_k = \frac{\mathrm{d}^k}{\mathrm{d}\alpha^k} C(\alpha)\big|_{\alpha=0}, \qquad k = 1, 2, \cdots.$$

$C_1 = M_1 = E(X)$ and

$$C_2 = E(X^2) - E(X)^2 = E((X - E(X))^2).$$

C_2 is called the variance of X and is denoted by $\text{Var}(X)$. Note that $\text{Var}(X) = 0$ iff X is constant on supp P. When we wish to indicate the dependence of the expectation and variance on the underlying measure P, we shall write $E_P(X)$, $\text{Var}_P(X)$, etc.

Exercise 2.1 The sequences $\{M_k\}$ and $\{C_k\}$ determine each other, i.e., there are functions F_k and G_k such that

$$C_k = F_k(M_1, \cdots, M_k), \qquad M_k = G_k(C_1, \cdots, C_k).$$

Describe recursive relations that determine F_k and G_k.

In probabilistic setup Proposition 2.1 takes the form

$$P(\{\omega \in \Omega \mid |X(\omega)| \geq a\}) \leq \frac{1}{a} E(|X|), \tag{2.1}$$

and is often called Markov or Chebyshev inequality. We shall often use a shorthand and abbreviate the l.h.s in (2.1) as $P\{|X(\omega)| \geq a\}$, etc.

2.3 Law of Large Numbers

Let (Ω, P) be a probability space and $X : \Omega \to \mathbb{R}$ a random variable. On the product probability space (Ω^N, P_N) we define

$$S_N(\omega = (\omega_1, \cdots, \omega_N)) = \sum_{k=1}^{N} X(\omega_k).$$

We shall refer to the following results as the *Law of large numbers (LLN)*.

Proposition 2.2 *For any $\epsilon > 0$,*

$$\lim_{N \to \infty} P_N \left\{ \left| \frac{S_N(\omega)}{N} - E(X) \right| \geq \epsilon \right\} = 0.$$

Remark 2.3 An equivalent formulation of the LLN is that for any $\epsilon > 0$,

$$\lim_{N \to \infty} P_N \left\{ \left| \frac{S_N(\omega)}{N} - E(X) \right| \leq \epsilon \right\} = 1.$$

Proof Denote by E_N the expectation w.r.t. P_N. Define $X_k(\omega) = X(\omega_k)$ and note that $E_N(X_k) = E(X)$, $E_N(X_k^2) = E(X^2)$, $E_N(X_k X_j) = E(X)^2$ for $k \neq j$. Then

$$
P_N \left\{ \left| \frac{S_N(\omega)}{N} - E(X) \right| \geq \epsilon \right\} = P_N \left\{ \left(\frac{S_N(\omega)}{N} - E(X) \right)^2 \geq \epsilon^2 \right\}
$$

$$
\leq \frac{1}{\epsilon^2} E_N \left(\left(\frac{S_N(\omega)}{N} - E(X) \right)^2 \right)
$$

$$
= \frac{1}{N^2 \epsilon^2} E_N \left(\sum_{k,j} (X_k - E(X_k))(X_j - E(X_j)) \right)
$$

$$
= \frac{1}{N \epsilon^2} \mathrm{Var}(X),
$$

and the statement follows. $\qquad \square$

2.4 Cumulant Generating Function

Let (Ω, P) be a probability space and $X : \Omega \to \mathbb{R}$ a random variable. In this section we shall study in some detail the properties of the cumulant generating function

$$
C(\alpha) = \log E(e^{\alpha X}).
$$

To avoid discussion of trivialities, until the end of this chapter we shall assume that X is not constant on supp P, i.e. that X assumes at least two distinct values on supp P. Obviously, the function $C(\alpha)$ is infinitely differentiable and

$$
\begin{aligned}
\lim_{\alpha \to \infty} C'(\alpha) &= \max_{\omega} X(\omega), \\
\lim_{\alpha \to -\infty} C'(\alpha) &= \min_{\omega} X(\omega).
\end{aligned}
\tag{2.2}
$$

Proposition 2.4 $C''(\alpha) > 0$ *for all* α. *In particular, the function* C *is strictly convex.*

Remark 2.5 By strictly convex we mean that C' is strictly increasing, i.e., that the graph of C does not have a flat piece.

Proof Set

$$
Q_\alpha(\omega) = \frac{e^{\alpha X(\omega)} P(\omega)}{\sum_{\omega} e^{\alpha X(\omega)} P(\omega)},
\tag{2.3}
$$

and note that Q_α is a probability measure on Ω equivalent to P.

One easily verifies that

$$C'(\alpha) = E_{Q_\alpha}(X), \qquad C''(\alpha) = \mathrm{Var}_{Q_\alpha}(X).$$

The second identity yields the statement.   □

Proposition 2.6 *C extends to an analytic function in the strip*

$$|\mathrm{Im}\,\alpha| < \frac{\pi}{2}\frac{1}{\max_\omega |X(\omega)|}. \tag{2.4}$$

Proof Obviously, the function $\alpha \mapsto E(e^{\alpha X})$ is entire analytic. If $\alpha = a + ib$, then

$$E(e^{\alpha X}) = \sum_{\omega \in \Omega} e^{aX(\omega)}\cos(bX(\omega))P(\omega) + i\sum_{\omega \in \Omega} e^{aX(\omega)}\sin(bX(\omega))P(\omega).$$

If $|bX(\omega)| < \pi/2$ for all ω, then the real part of $E(e^{\alpha X})$ is strictly positive. It follows that the function

$$\mathrm{Log}\,E(e^{\alpha X}),$$

where Log is the principal branch of complex logarithm, is analytic in the strip (2.4) and the statement follows.   □

Remark 2.7 Let $\Omega = \{-1, 1\}$, $P(-1) = P(1) = 1/2$, $X(1) = 1$, $X(-1) = -1$. Then

$$C(\alpha) = \log\cosh\alpha.$$

Since $\cosh(\pi i/2) = 0$, we see that Proposition 2.6 is an optimal result.

2.5   Rate Function

We continue with the framework of the previous section. The *rate function* of the random variable X is defined by

$$I(\theta) = \sup_{\alpha \in \mathbb{R}} (\alpha\theta - C(\alpha)), \qquad \theta \in \mathbb{R}.$$

In the language of convex analysis, I is the Fenchel-Legendre transform of the cumulant generating function C. Obviously, $I(\theta) \geq 0$ for all θ. Set

$$m = \min_\omega X(\omega), \qquad M = \max_\omega X(\omega),$$

and recall the relations (2.2). By the intermediate value theorem, for any θ in $]m, M[$ there exists unique $\alpha(\theta) \in \mathbb{R}$ such that

$$\theta = C'(\alpha(\theta)).$$

The function

$$\alpha(\theta) = (C')^{-1}(\theta)$$

is infinitely differentiable on $]m, M[$, strictly increasing on $]m, M[$, $\alpha(\theta) \downarrow -\infty$ iff $\theta \downarrow m$, and $\alpha(\theta) \uparrow \infty$ iff $\theta \uparrow M$.

Exercise 2.2 Prove that the function $]m, M[\ni \theta \mapsto \alpha(\theta)$ is real-analytic. Hint: Apply the analytic implicit function theorem.

Proposition 2.8

(1) For $\theta \in]m, M[$,

$$I(\theta) = \alpha(\theta)\theta - C(\alpha(\theta)).$$

(2) The function I is infinitely differentiable on $]m, M[$.
(3) $I'(\theta) = \alpha(\theta)$. In particular, I' is strictly increasing on $]m, M[$ and

$$\lim_{\theta \downarrow m} I'(\theta) = -\infty, \qquad \lim_{\theta \uparrow M} I'(\theta) = \infty.$$

(4) $I''(\theta) = 1/C''(\alpha(\theta))$.
(5) $I(\theta) = 0$ iff $\theta = E(X)$.

Proof To prove (1), note that for $\theta \in]m, M[$ the function

$$\frac{\mathrm{d}}{\mathrm{d}\alpha}(\alpha\theta - C(\alpha)) = \theta - C'(\alpha)$$

vanishes at $\alpha(\theta)$, is positive for $\alpha < \alpha(\theta)$, and is negative for $\alpha > \alpha(\theta)$. Hence, the function $\alpha \mapsto \alpha\theta - C(\alpha)$ has the global maximum at $\alpha = \alpha(\theta)$ and Part (1) follows. Parts (2), (3), and (4) are obvious. To prove (5), note that if $I(\theta) = 0$ for some $\theta \in]m, M[$, then, since I is non-negative, we also have $0 = I'(\theta) = \alpha(\theta)$, and the relation $\theta = C'(\alpha(\theta)) = C'(0) = E(X)$ follows. On the other hand, if $\theta = E(X) = C'(0)$, then $\alpha(\theta) = 0$, and $I(\theta) = -C(0) = 0$.   □

Exercise 2.3 Prove that the function I is real-analytic in $]m, M[$.

Let

$$S_m = \{\omega \in \Omega \mid X(\omega) = m\}, \qquad S_M = \{\omega \in \Omega \mid X(\omega) = M\}.$$

Proposition 2.9

(1) $I(\theta) = \infty$ *for* $\theta \notin [m, M]$.
(2)

$$I(m) = \lim_{\theta \downarrow m} I(\theta) = -\log P(S_m),$$

$$I(M) = \lim_{\theta \uparrow M} I(\theta) = -\log P(S_M).$$

Proof

(1) Suppose that $\theta > M$. Then

$$\frac{d}{d\alpha}(\alpha\theta - C(\alpha)) = \theta - C'(\alpha) > \theta - M.$$

Integrating this inequality over $[0, \alpha]$ we derive

$$\alpha\theta - C(\alpha) > (\theta - M)\alpha,$$

and so

$$I(\theta) = \sup_{\alpha \in \mathbb{R}}(\alpha\theta - C(\alpha)) = \infty.$$

The case $\theta < m$ is similar.

(2) We shall prove only the second formula, the proof of the first is similar. Since the function $\alpha M - C(\alpha)$ is increasing,

$$I(M) = \lim_{\alpha \to \infty}(\alpha M - C(\alpha)).$$

Since

$$C(\alpha) = \alpha M + \log P(S_M) + \log(1 + A(\alpha)), \tag{2.5}$$

where

$$A(\alpha) = \frac{1}{P(S_M)} \sum_{\omega \notin S_M} e^{\alpha(X(\omega) - M)} P(\omega),$$

we derive that $I(M) = -\log P(S_M)$.

Since $C'(\alpha(\theta)) = \theta$, Part (1) of Proposition 2.8 gives that

$$\lim_{\theta \uparrow M} I(\theta) = \lim_{\alpha \to \infty}(\alpha C'(\alpha) - C(\alpha)).$$

Write

$$C'(\alpha) = M\frac{1 + B(\alpha)}{1 + A(\alpha)}, \tag{2.6}$$

where

$$B(\alpha) = \frac{1}{MP(S_M)} \sum_{\omega \notin S_M} X(\omega)e^{\alpha(X(\omega)-M)}P(\omega).$$

The formulas (2.5) and (2.6) yield

$$\alpha C'(\alpha) - C(\alpha) = \alpha M\frac{B(\alpha) - A(\alpha)}{1 + A(\alpha)} - \log P(S_M) - \log(1 + A(\alpha)).$$

Since $A(\alpha)$ and $B(\alpha)$ converge to 0 as $\alpha \to \infty$,

$$\lim_{\theta \uparrow M} I(\theta) = \lim_{\alpha \to \infty} (\alpha C'(\alpha) - C(\alpha)) = -\log P(S_M).$$

$\square$

Proposition 2.10

$$C(\alpha) = \sup_{\theta \in \mathbb{R}} (\theta\alpha - I(\theta)). \tag{2.7}$$

Proof To avoid confusion, fix $\alpha = \alpha_0$. Below, $\alpha(\theta) = (C')^{-1}(\theta)$ is as in Proposition 2.8.

The supremum in (2.7) is achieved at θ_0 satisfying

$$\alpha_0 = I'(\theta_0).$$

Since $I'(\theta_0) = \alpha(\theta_0)$, we have $\alpha_0 = \alpha(\theta_0)$, and

$$I(\theta_0) = \theta_0\alpha(\theta_0) - C(\alpha(\theta_0)) = \theta_0\alpha_0 - C(\alpha_0).$$

Hence

$$\sup_{\theta \in \mathbb{R}} (\theta\alpha_0 - I(\theta)) = \alpha_0\theta_0 - I(\theta_0) = C(\alpha_0).$$

$\square$

Returning to the example of Remark 2.7, $m = -1$, $M = 1$, $C(\alpha) = \log\cosh\alpha$, and $C'(\alpha) = \tanh\alpha$. Hence, for $\theta \in]-1, 1[$,

$$\alpha(\theta) = \tanh^{-1}(\theta) = \frac{1}{2}\log\frac{1+\theta}{1-\theta}.$$

It follows that

$$I(\theta) = \theta\alpha(\theta) - C(\alpha(\theta)) = \frac{1}{2}(1+\theta)\log(1+\theta) + \frac{1}{2}(1-\theta)\log(1-\theta).$$

2.6 Cramér's Theorem

This section is devoted to the proof of Cramér's theorem:

Theorem 2.11 For any interval $[a, b]$,

$$\lim_{N\to\infty}\frac{1}{N}\log P_N\left\{\frac{S_N(\omega)}{N} \in [a, b]\right\} = -\inf_{\theta\in[a,b]} I(\theta).$$

Remark 2.12 To prove this result without loss of generality we may assume that $[a, b] \subset [m, M]$.

Remark 2.13 Note that

$$\inf_{\theta\in[a,b]} I(\theta) = \begin{cases} 0 & \text{if } \mathbb{E}(X) \in [a, b] \\ I(a) & \text{if } a > \mathbb{E}(X) \\ I(b) & \text{if } b < \mathbb{E}(X), \end{cases}$$

and that

$$\lim_{N\to\infty}\frac{1}{N}\log P_N\left\{\frac{S_N(\omega)}{N} = M\right\} = \log P(S_M) = -I(M),$$

$$\lim_{N\to\infty}\frac{1}{N}\log P_N\left\{\frac{S_N(\omega)}{N} = m\right\} = \log P(S_m) = -I(m).$$

We start the proof with

Proposition 2.14

(1) For $\theta \geq \mathbb{E}(X)$,

$$\limsup_{N\to\infty}\frac{1}{N}\log P_N\left\{\frac{S_N(\omega)}{N} \geq \theta\right\} \leq -I(\theta).$$

(2) For $\theta \leq \mathbb{E}(X)$,

$$\limsup_{N \to \infty} \frac{1}{N} \log P_N \left\{ \frac{S_N(\omega)}{N} \leq \theta \right\} \leq -I(\theta).$$

Remark 2.15 Note that if $\theta < \mathbb{E}(X)$, then by the LLN

$$\lim_{N \to \infty} \frac{1}{N} \log P_N \left\{ \frac{S_N(\omega)}{N} \geq \theta \right\} = 0.$$

Similarly, if $\theta > \mathbb{E}(X)$,

$$\lim_{N \to \infty} \frac{1}{N} \log P_N \left\{ \frac{S_N(\omega)}{N} \leq \theta \right\} = 0.$$

Proof For $\alpha > 0$,

$$P_N \left\{ S_N(\omega) \geq N\theta \right\} = P_N \left\{ e^{\alpha S_N(\omega)} \geq e^{\alpha N\theta} \right\}$$

$$\leq e^{-\alpha N\theta} \mathbb{E}_N \left(e^{\alpha S_N(\omega)} \right)$$

$$= e^{-\alpha N\theta} \mathbb{E} \left(e^{\alpha X} \right)^N$$

$$= e^{N(C(\alpha) - \alpha\theta)}.$$

It follows that

$$\limsup_{N \to \infty} \frac{1}{N} \log P_N \left\{ \frac{S_N(\omega)}{N} \geq \theta \right\} \leq \inf_{\alpha > 0} (C(\alpha) - \alpha\theta) = -\sup_{\alpha > 0} (\alpha\theta - C(\alpha)).$$

If $\theta \geq \mathbb{E}(X)$, then $\alpha\theta - C(\alpha) \leq 0$ for $\alpha \leq 0$ and

$$\sup_{\alpha > 0} (\alpha\theta - C(\alpha)) = \sup_{\alpha \in \mathbb{R}} (\alpha\theta - C(\alpha)) = I(\theta).$$

This yields Part (1). Part (2) follows by applying Part (1) to the random variable $-X$. □

Exercise 2.4 Using Proposition 2.14 prove that for any $\epsilon > 0$ there exist $\gamma_\epsilon > 0$ and N_ϵ such that for $N \geq N_\epsilon$,

$$P_N \left\{ \left| \frac{S_N(\omega)}{N} - E(X) \right| \geq \epsilon \right\} \leq e^{-\gamma_\epsilon N}.$$

Proposition 2.16

(1) For $\theta \geq \mathbb{E}(X)$,

$$\liminf_{N \to \infty} \frac{1}{N} \log P_N \left\{ \frac{S_N(\omega)}{N} \geq \theta \right\} \geq -I(\theta).$$

(2) For $\theta \leq \mathbb{E}(X)$,

$$\liminf_{N \to \infty} \frac{1}{N} \log P_N \left\{ \frac{S_N(\omega)}{N} \leq \theta \right\} \geq -I(\theta).$$

Remark 2.17 Note that Part (1) trivially holds if $\theta < \mathbb{E}(X)$. Similarly, Part (2) trivially holds if $\theta > \mathbb{E}(X)$.

Proof We again need to prove only Part (1) (Part (2) follows by applying Part (1) to the random variable $-X$). If $\theta \geq M$, the statement is obvious and so without loss of generality we may assume that $\theta \in [\mathbb{E}(X), M[$. Fix such θ and choose s and $\epsilon > 0$ such that $\theta < s - \epsilon < s + \epsilon < M$.

Let Q_α be the probability measure introduced in the proof of Proposition 2.4, and let $Q_{\alpha,N}$ be the induced product probability measure on Ω^N. The measures P_N and $Q_{\alpha,N}$ are equivalent, and for $\omega \in \text{supp } P_N$

$$\Delta_{P_N | Q_{\alpha,N}}(\omega) = e^{-\alpha S_N(\omega) + N C(\alpha)}.$$

We now consider the measure $Q_{\alpha,N}$ for $\alpha = \alpha(s)$. Recall that

$$C'(\alpha(s)) = s = \mathbb{E}_{Q_{\alpha(s)}}(X).$$

Set

$$T_N = \left\{ \omega \in \Omega^N \,\Big|\, \frac{S_N(\omega)}{N} \in [s - \epsilon, s + \epsilon] \right\},$$

and note that the LLN implies

$$\lim_{N \to \infty} Q_{\alpha(s),N}(T_N) = 1. \tag{2.8}$$

The estimates

$$P_N \left\{ \frac{S_N(\omega)}{N} \geq \theta \right\} \geq P_N(T_N) = \int_{T_N} \Delta_{P_N | Q_{\alpha(s),N}} dQ_{\alpha(s),N}$$

$$= \int_{T_N} e^{-\alpha(s)S_N + NC(\alpha(s))} dQ_{\alpha(s),N}$$

$$\geq e^{N(C(\alpha(s)) - s\alpha(s) - \epsilon|\alpha(s)|)} Q_{\alpha(s),N}(T_N)$$

and (2.8) give

$$\liminf_{N\to\infty} \frac{1}{N} \log P_N \left\{ \frac{S_N(\omega)}{N} \geq \theta \right\} \geq C(\alpha(s)) - s\alpha(s) - \epsilon |\alpha(s)| = -I(s) - \epsilon |\alpha(s)|.$$

The statement now follows by taking first $\epsilon \downarrow 0$ and then $s \downarrow \theta$.   □

Combining Propositions 2.14 and 2.16 we derive

Corollary 2.18 *For* $\theta \geq \mathbb{E}(X)$,

$$\lim_{N\to\infty} \frac{1}{N} \log P_N \left\{ \frac{S_N(\omega)}{N} \geq \theta \right\} = -I(\theta).$$

For $\theta \leq \mathbb{E}(X)$,

$$\lim_{N\to\infty} \frac{1}{N} \log P_N \left\{ \frac{S_N(\omega)}{N} \leq \theta \right\} = -I(\theta).$$

We are now ready to complete

Proof of Theorem 2.11 If $\mathbb{E}(X) \in]a, b[$ the result follows from the LLN. Suppose that $M > a \geq \mathbb{E}(X)$. Then

$$P_N \left\{ \frac{S_N(\omega)}{N} \in [a, b] \right\} = P_N \left\{ \frac{S_N(\omega)}{N} \geq a \right\} - P_N \left\{ \frac{S_N(\omega)}{N} > b \right\}.$$

It follows from Corollary 2.18 that

$$\lim_{N\to\infty} \frac{1}{N} \log \left[1 - \frac{P_N \left\{ \frac{S_N(\omega)}{N} > b \right\}}{P_N \left\{ \frac{S_N(\omega)}{N} \geq a \right\}} \right] = 0, \qquad (2.9)$$

and so

$$\lim_{N\to\infty} \frac{1}{N} \log P_N \left\{ \frac{S_N(\omega)}{N} \in [a, b] \right\} = \lim_{N\to\infty} \frac{1}{N} \log P_N \left\{ \frac{S_N(\omega)}{N} \geq a \right\} = -I(a).$$

The case $m < b \leq \mathbb{E}(X)$ is similar.   □

Exercise 2.5 Write down the proof of (2.9) and of the case $m < b \leq \mathbb{E}(X)$.

Exercise 2.6 Consider the example introduced in Remark 2.7 and prove Cramér's theorem in this special case by using Stirling's formula and a direct combinatorial argument.
Hint: See Theorem 1.3.1 in [15].

2.7   Notes and References

Although it is assumed that the student reader had no previous exposure to probability theory, a reading of additional material could be helpful at this point. Recommended textbooks are [8, 42, 43].

For additional information and original references regarding Cramer's theorem, we refer the reader to Chapter 2 of [13]. Reader interested to learn more about theory of large deviations may consult classical references [13–15], and the lecture notes of S.R.S. Varadhan https://math.nyu.edu/~varadhan/LDP.html.

It is possible to give a combinatorial proof of Theorem 2.11, as indicated in the Exercise 2.6. The advantage of the argument presented in this chapter is that it naturally extends to a proof of much more general results (such as the Gärtner-Ellis theorem) which will be discussed in the Part II of the lecture notes.

3   Boltzmann–Gibbs–Shannon Entropy

3.1   Preliminaries

Let Ω be a finite set, $|\Omega| = L$, and let $\mathcal{P}(\Omega)$ be the collection of all probability measures on Ω. $\mathcal{P}(\Omega)$ is naturally identified with the set

$$\mathcal{P}_L = \left\{ (p_1, \cdots, p_L) \mid p_k \geq 0, \ \sum_{k=1}^{L} p_k = 1 \right\} \tag{3.1}$$

(the identification map is $P \mapsto (P(\omega_1), \cdots P(\omega_L))$. We shall often use this identification without further notice. A convenient metric on $\mathcal{P}(\Omega)$ is the variational distance

$$d_V(P, Q) = \sum_{\omega \in \Omega} |P(\omega) - Q(\omega)|. \tag{3.2}$$

We denote by $\mathcal{P}_f(\Omega)$ the set of all faithful probability measures on $\mathcal{P}(\Omega)$ (recall that $P \in \mathcal{P}_f(\Omega)$ iff $P(\omega) > 0$ for all $\omega \in \Omega$). $\mathcal{P}_f(\Omega)$ coincides with the interior of $\mathcal{P}(\Omega)$ and is identified with

$$\mathcal{P}_{L,f} = \left\{ (p_1, \cdots, p_L) \mid p_k > 0, \ \sum_{k=1}^{L} p_k = 1 \right\}.$$

Note that $\mathcal{P}(\Omega)$ and $\mathcal{P}_f(\Omega)$ are convex sets.

The probability measure P is called *pure* if $P(\omega) = 1$ for some $\omega \in \Omega$. The *chaotic* probability measure is $P_{ch}(\omega) = 1/L$, $\omega \in \Omega$.

We shall often make use of Jensen's inequality. This inequality states that if $f : [a, b] \to \mathbb{R}$ is concave, then for $x_k \in [a, b], k = 1, \cdots , n$, and $(p_1, \cdots , p_n) \in \mathcal{P}_{n,\mathrm{f}}$ we have

$$\sum_{k=1}^{n} p_k f(x_k) \leq f \left(\sum_{k=1}^{n} p_k x_k \right). \tag{3.3}$$

Moreover, if f is strictly concave the inequality is strict unless $x_1 = \cdots = x_n$. A similar statement holds for convex functions.

Exercise 3.1 Prove Jensen's inequality.

3.2 Definition and Basic Properties

The *entropy function* (sometimes called the *information function*) of $P \in \mathcal{P}(\Omega)$ is[3]

$$S_P(\omega) = -c \log P(\omega), \tag{3.4}$$

where $c > 0$ is a constant that does not depend on P or Ω, and $-\log 0 = \infty$. The function S_P takes values in $[0, \infty]$. The *Boltzmann–Gibbs–Shannon entropy* (in the sequel we will often call it just *entropy*) of P is

$$S(P) = \int_{\Omega} S_P \mathrm{d}P = -c \sum_{\omega \in \Omega} P(\omega) \log P(\omega). \tag{3.5}$$

The value of the constant c is linked to the choice of units (or equivalently, the base of logarithm). The natural choice in the information theory is $c = 1/\log 2$ (that is, the logarithm is taken in the base 2). The value of c plays no role in these lecture

---

[3]Regarding the choice of logarithm, in the introduction of [44] Shannon comments: "(1) It is practically more useful. Parameters of engineering importance such as time, bandwidth, number of relays, etc., tend to vary linearly with the logarithm of the number of possibilities. For example, adding one relay to a group doubles the number of possible states of the relays. It adds 1 to the base 2 logarithm of this number. Doubling the time roughly squares the number of possible messages, or doubles the logarithm, etc. (2) It is nearer to our intuitive feeling as to the proper measure. This is closely related to (1) since we intuitively measure entities by linear comparison with common standards. One feels, for example, that two punched cards should have twice the capacity of one for information storage, and two identical channels twice the capacity of one for transmitting information. (3) It is mathematically more suitable. Many of the limiting operations are simple in terms of the logarithm but would require clumsy restatement in terms of the number of possibilities."

notes, and from now on we set $c = 1$ and call

$$S(P) = - \sum_{\omega \in \Omega} P(\omega) \log P(\omega)$$

the Boltzmann–Gibbs–Shannon entropy of P. We note, however, that the constant c will reappear in the axiomatic characterizations of entropy given in Theorems 3.4 and 3.7.

The basic properties of entropy are:

Proposition 3.1

(1) $S(P) \geq 0$ and $S(P) = 0$ iff P is pure.
(2) $S(P) \leq \log L$ and $S(P) = \log L$ iff $P = P_{\mathrm{ch}}$.
(3) The map $\mathcal{P}(\Omega) \ni P \mapsto S(P)$ is continuous and concave, that is, if p_k's are as in (3.3) and $P_k \in \mathcal{P}(\Omega)$, then

$$p_1 S(P_1) + \cdots + p_n S(P_n) \leq S(p_1 P_1 + \cdots p_n P_n), \tag{3.6}$$

with equality iff $P_1 = \cdots = P_n$.
(4) The concavity inequality (3.6) has the following "almost convexity" counterpart:

$$S(p_1 P_1 + \cdots + p_n P_n) \leq p_1 S(P_1) + \cdots + p_n S(P_n) + S(p_1, \cdots, p_n),$$

with equality iff supp $P_k \cap$ supp $P_j = \emptyset$ for $k \neq j$.

Proof Parts (1) and (3) follow from the obvious fact that the function $[0, 1] \ni x \mapsto -x \log x$ is continuous, strictly concave, non-negative, and vanishing iff $x = 0$ or $x = 1$. Part (2) follows from Jensen's inequality. Part (4) follows from the monotonicity of $\log x$:

$$S(p_1 P_1 + \cdots + p_n P_n) = \sum_{\omega \in \Omega} \sum_{k=1}^{n} -p_k P_k(\omega) \log \left(\sum_{j=1}^{n} p_j P_j(\omega) \right)$$

$$\leq \sum_{\omega \in \Omega} \sum_{k=1}^{n} -p_k P_k(\omega) \log \left(p_k P_k(\omega) \right)$$

$$= \sum_{k=1}^{n} p_k \left(\sum_{\omega \in \Omega} -P_k(\omega) \log P_k(\omega) \right)$$

$$- \sum_{k=1}^{n} \left(\sum_{\omega \in \Omega} P_k(\omega) \right) p_k \log p_k$$

$$= \sum_{k=1}^{n} p_k S(P_k) + S(p_1, \cdots, p_n).$$

The equality holds if for all ω and $k \neq j$, $p_k P_k(\omega) > 0 \Rightarrow p_j P_j(\omega) = 0$, which is equivalent to supp $P_k \cap$ supp $P_j = \emptyset$ for all $k \neq j$. $\qquad\square$

Suppose that $\Omega = \Omega_l \times \Omega_r$ and let $P_{l/r}$ be the marginals of $P \in \mathcal{P}(\Omega)$. For a given $\omega \in$ supp P_l the conditional probability measure $P_{r|l}^\omega$ on Ω_r is defined by

$$P_{r|l}^\omega(\omega') = \frac{P(\omega, \omega')}{P_l(\omega)}.$$

Note that

$$\sum_{\omega \in \text{supp} P_l} P_l(\omega) P_{r|l}^\omega = P_r.$$

Proposition 3.2

(1)

$$S(P) = S(P_l) + \sum_{\omega \in \Omega_l} P_l(\omega) S(P_{r|l}^\omega).$$

(2) The entropy is strictly sub-additive:

$$S(P) \le S(P_l) + S(P_r),$$

with the equality iff $P = P_l \otimes P_r$.

Proof Part (1) and the identity $S(P_l \otimes P_r) = S(P_l) + S(P_r)$ follow by direct computation. To prove (2), note that Part (3) of Proposition 3.1 gives

$$\sum_{\omega \in \text{supp} P_l} P_l(\omega) S(P_{r|l}^\omega) \le S\left(\sum_{\omega \in \text{supp} P_l} P_l(\omega) P_{r|l}^\omega \right) = S(P_r),$$

and so it follows from Part (1) that $S(P) \le S(P_l) + S(P_r)$ with the equality iff all the probability measures $P_{r|l}^\omega$, $\omega \in$ supp P_l, are equal. Thus, if the equality holds, then for all $(\omega, \omega') \in \Omega_l \times \Omega_r$, $P(\omega, \omega') = C(\omega') P_l(\omega)$. Summing over ω's gives that $P = P_l \otimes P_r$. $\qquad\square$

Exercise 3.2 The Hartley entropy of $P \in \mathcal{P}(\Omega)$ is defined by

$$S_H(P) = \log |\{\omega \mid P(\omega) > 0\}|.$$

1. Prove that the Hartley entropy is also strictly sub-additive: $S_H(P) \le S_H(P_l) + S_H(P_r)$, with the equality iff $P = P_l \otimes P_r$.
2. Show that the map $P \mapsto S_H(P)$ is not continuous if $L \ge 2$.

3.3   Covering Exponents and Source Coding

To gain further insight into the concept of entropy, assume that P is faithful and consider the product probability space (Ω^N, P_N). For given $\epsilon > 0$ let

$$
T_{N,\epsilon} = \left\{ \omega = (\omega_1, \cdots, \omega_N) \in \Omega^N \,\middle|\, \left| \frac{S_P(\omega_1) + \cdots S_P(\omega_N)}{N} - S(P) \right| < \epsilon \right\}
$$

$$
= \left\{ \omega \in \Omega^N \,\middle|\, \left| -\frac{\log P_N(\omega)}{N} - S(P) \right| < \epsilon \right\}
$$

$$
= \left\{ \omega \in \Omega^N \,\middle|\, e^{-N(S(P)+\epsilon)} < P_N(\omega) < e^{-N(S(P)-\epsilon)} \right\}.
$$

The LLN gives

$$
\lim_{N \to \infty} P_N(T_{N,\epsilon}) = 1.
$$

We also have the following obvious bounds on the cardinality of $T_{N,\epsilon}$:

$$
P_N(T_{N,\epsilon}) e^{N(S(P)-\epsilon)} < |T_{N,\epsilon}| < e^{N(S(P)+\epsilon)}.
$$

It follows that

$$
S(P) - S(P_{\mathrm{ch}}) - \epsilon \le \liminf_{N \to \infty} \frac{1}{N} \log \frac{|T_{N,\epsilon}|}{|\Omega|^N}
$$

$$
\le \limsup_{N \to \infty} \frac{1}{N} \log \frac{|T_{N,\epsilon}|}{|\Omega|^N} \le S(P) - S(P_{\mathrm{ch}}) + \epsilon.
$$

This estimate implies that if $P \ne P_{\mathrm{ch}}$, then, as $N \to \infty$, the measure P_N is "concentrated" and "equipartitioned" on the set $T_{N,\epsilon}$ whose size is "exponentially small" with respect to the size of Ω^N.

We continue with the analysis of the above concepts. Let $\gamma \in]0, 1[$ be fixed. The (N, γ) covering exponent is defined by

$$
c_N(\gamma) = \min \left\{ |A| \,\middle|\, A \subset \Omega^N, \ P_N(A) \ge \gamma \right\}. \tag{3.7}
$$

One can find $c_N(\gamma)$ according to the following algorithm:

(a) List the events $\omega = (\omega_1, \cdots, \omega_N)$ in order of decreasing probabilities.
(b) Count the events until the first time the total probability is $\ge \gamma$.

Proposition 3.3 *For all* $\gamma \in]0, 1[$,

$$\lim_{N \to \infty} \frac{1}{N} \log c_N(\gamma) = S(P).$$

Proof Fix $\epsilon > 0$ and recall the definition of $T_{N,\epsilon}$. For N large enough, $P_N(T_{N,\epsilon}) \geq \gamma$,
and so for such N's,

$$c_N(\gamma) \leq |T_{N,\epsilon}| \leq e^{N(S(P)+\epsilon)}.$$

It follows that

$$\limsup_{N \to \infty} \frac{1}{N} \log c_N(\gamma) \leq S(P).$$

To prove the lower bound, let $A_{N,\gamma}$ be a set for which the minimum in (3.7) is achieved. Let $\epsilon > 0$. Note that

$$\liminf_{N \to \infty} P_N(T_{N,\epsilon} \cap A_{N,\gamma}) \geq \gamma. \tag{3.8}$$

Since for $P_N(\omega) \leq e^{-N(S(P)-\epsilon)}$ for $\omega \in T_{N,\epsilon}$,

$$P_N(T_{N,\epsilon} \cap A_{N,\gamma}) = \sum_{\omega \in T_{N,\epsilon} \cap A_{N,\gamma}} P_N(\omega) \leq e^{-N(S(P)-\epsilon)} |T_{N,\epsilon} \cap A_{N,\gamma}|.$$

Hence,

$$|A_{N,\gamma}| \geq e^{N(S(P)-\epsilon)} P_N(T_{N,\epsilon} \cap A_{N,\gamma}),$$

and it follows from (3.8) that

$$\liminf_{N \to \infty} \frac{1}{N} \log c_N(\gamma) \geq S(P) - \epsilon.$$

Since $\epsilon > 0$ is arbitrary,

$$\liminf_{N \to \infty} \frac{1}{N} \log c_N(\gamma) \geq S(P),$$

and the proposition is proven. $\qquad \square$

We finish this section with a discussion of Shannon's source coding theorem. Given a pair of positive integers N, M, the *encoder* is a map

$$F_N : \Omega^N \to \{0, 1\}^M.$$

The *decoder* is a map

$$G_N : \{0, 1\}^M \to \Omega^N.$$

The error probability of the coding pair (F_N, G_N) is

$$P_N \{G_N \circ F_N(\omega) \neq \omega\}.$$

If this probability is less than some prescribed $1 > \epsilon > 0$, we shall say that the coding pair is ϵ-good. Note that to any ϵ-good coding pair one can associate the set

$$A = \{\omega \,|\, G_N \circ F_N(\omega) = \omega\}$$

which satisfies

$$P_N(A) \geq 1 - \epsilon, \qquad |A| \leq 2^M. \tag{3.9}$$

On the other hand, if $A \subset \Omega^N$ satisfies (3.9), we can associate to it an ϵ-good pair (F_N, G_N) by setting F_N to be one-one on A (and arbitrary otherwise), and $G_N = F_N^{-1}$ on $F_N(A)$ (and arbitrary otherwise).

In the source coding we wish to find M that minimizes the compression coefficients M/N subject to an allowed ϵ-error probability. Clearly, the optimal M is

$$M_N = \left[\log_2 \min \left\{ |A| \,\big|\, A \subset \Omega^N \; P_N(A) \geq 1 - \epsilon \right\} \right],$$

where $[\,\cdot\,]$ denotes the greatest integer part. Shannon's source coding theorem now follows from Proposition 3.3: the limiting optimal compression coefficient is

$$\lim_{N \to \infty} \frac{M_N}{N} = \frac{1}{\log 2} S(P).$$

3.4 Why is the Entropy Natural?

Set $\mathcal{P} = \cup_\Omega \mathcal{P}(\Omega)$. In this section we shall consider functions $\mathfrak{S} : \mathcal{P} \to \mathbb{R}$ that satisfy properties that correspond intuitively to those of *entropy* as a measure of *randomness* of probability measures. The goal is to show that those intuitive natural demands uniquely specify $\mathfrak{S}$ up to a choice of units, that is, that for some $c > 0$ and all $P \in \mathcal{P}$, $\mathfrak{S}(P) = cS(P)$.

We describe first three basic properties that any candidate for $\mathfrak{S}$ should satisfy. The first is the positivity and non-triviality requirement: $\mathfrak{S}(P) \geq 0$ and this inequality is strict for at least one $P \in \mathcal{P}$. The second is that if $|\Omega_1| = |\Omega_2|$ and $\theta : \Omega_1 \to \Omega_2$ is a bijection, then for any $P \in \mathcal{P}(\Omega_1)$, $\mathfrak{S}(P) = \mathfrak{S}(P \circ \theta)$. In other words, the entropy of P should not depend on the labeling of the elementary events. This second requirement gives that $\mathfrak{S}$ is completely specified by its restriction $\mathfrak{S} : \cup_{L \geq 1} \mathcal{P}_L \to [0, \infty[$ which satisfies

$$\mathfrak{S}(p_1, \cdots, p_L) = \mathfrak{S}(p_{\pi(1)}, \cdots, p_{\pi(L)}) \tag{3.10}$$

for any $L \geq 1$ and any permutation π of $\{1, \cdots, L\}$. In the proof of Theorem 3.7 we shall also assume that

$$\mathfrak{S}(p_1, \cdots, p_L, 0) = \mathfrak{S}(p_1, \cdots, p_L) \tag{3.11}$$

for all $L \geq 1$ and $(p_1, \cdots, p_L) \in \mathcal{P}_L$. In the literature, the common sense assumption (3.11) is sometimes called *expansibility*.

Throughout this section we shall assume that the above three properties hold. We remark that the assumptions of Theorem 3.7 actually imply the positivity and non-triviality requirement.

Split Additivity Characterization

If Ω_1, Ω_2 are two disjoint sets, we denote by $\Omega_1 \oplus \Omega_2$ their union (the symbol $\oplus$ is used to emphasize the fact that the sets are disjoint). If μ_1 is a measure on Ω_1 and μ_2 is a measure on Ω_2, then $\mu = \mu_1 \oplus \mu_2$ is a measure on $\Omega_1 \oplus \Omega_2$ defined by $\mu(\omega) = \mu_1(\omega)$ if $\omega \in \Omega_1$ and $\mu(\omega) = \mu_2(\omega)$ if $\omega \in \Omega_2$. Two measurable spaces (Ω_1, μ_1), (Ω_2, μ_2) are called disjoint if the sets Ω_1, Ω_2, are disjoint.

The split additivity characterization has its roots in the identity

$$S(p_1 P_1 + \cdots + p_n P_n) = p_1 S(P_1) + \cdots + p_n S(P_n) + S(p_1, \cdots, p_n)$$

which holds if $\operatorname{supp} P_k \cap \operatorname{supp} P_j = \emptyset$ for $k \neq j$.

Theorem 3.4 Let $\mathfrak{S} : \mathcal{P} \to [0, \infty[$ be a function such that:

(a) $\mathfrak{S}$ is continuous on $\mathcal{P}_2$.
(b) For any finite collection of disjoint probability spaces (Ω_j, P_j), $j = 1, \cdots, n$, and any $(p_1, \cdots, p_n) \in \mathcal{P}_n$,

$$\mathfrak{S}\left(\bigoplus_{k=1}^{n} p_k P_k\right) = \sum_{k=1}^{n} p_k \mathfrak{S}(P_k) + \mathfrak{S}(p_1, \cdots, p_n). \tag{3.12}$$

Then there exists $c > 0$ such that for all $P \in \mathcal{P}$,

$$\mathfrak{S}(P) = cS(P). \tag{3.13}$$

Remark 3.5 If the positivity and non-triviality assumptions are dropped, then the proof gives that (3.13) holds for some $c \in \mathbb{R}$.

Remark 3.6 The split-additivity property (3.12) is sometimes called the chain rule for entropy. It can be verbalized as follows: if the initial choices $(1, \cdots , n)$, realized with probabilities $(p_1, \cdots , p_n)$, are split into sub-choices described by probability spaces $(\Omega_k, P_k), k = 1, \cdots , n$, then the new entropy is the sum of the initial entropy and the entropies of sub-choices weighted by their probabilities.

Proof In what follows, $\overline{P}_n \in \mathcal{P}_n$ denotes the chaotic probability measure

$$\overline{P}_n = \left(\frac{1}{n}, \cdots , \frac{1}{n} \right),$$

and

$$f(n) = \mathfrak{S}(\overline{P}_n) = \mathfrak{S}\left(\frac{1}{n}, \cdots , \frac{1}{n} \right).$$

We split the argument into six steps.

Step 1 $\mathfrak{S}(1) = \mathfrak{S}(0, 1) = 0$.

Suppose that $|\Omega| = 2$ and let $P = (q_1, q_2) \in \mathcal{P}_2$. Writing $\Omega = \Omega_1 \oplus \Omega_2$ where $|\Omega_1| = |\Omega_2| = 1$ and taking $P_1 = (1)$, $P_2 = (1)$, $p_1 = q_1$, $p_2 = q_2$, we get $\mathfrak{S}(q_1, q_2) = \mathfrak{S}(1) + \mathfrak{S}(q_1, q_2)$, and so $\mathfrak{S}(1) = 0$. Similarly, the relations

$$\mathfrak{S}(0, q_1, q_2) = q_1 \mathfrak{S}(0, 1) + q_2 \mathfrak{S}(1) + \mathfrak{S}(q_1, q_2),$$

$$\mathfrak{S}(0, q_1, q_2) = 0 \cdot \mathfrak{S}(1) + 1 \cdot \mathfrak{S}(q_1, q_2) + \mathfrak{S}(0, 1),$$

yield that $\mathfrak{S}(0, 1) = q_1 \mathfrak{S}(0, 1)$ for all q_1, and so $\mathfrak{S}(0, 1) = 0$.

Step 2 $f(nm) = f(n) + f(m)$.

Take $\Omega = \Omega_1 \oplus \cdots \oplus \Omega_m$ with $|\Omega_k| = n$ for all $1 \le k \le m$, and set $P_k = \overline{P}_n$, $p_k = 1/m$. It then follows from (3.12) that $f(nm) = m \cdot \frac{1}{m} f(n) + f(m) = f(n) + f(m)$.

Step 3 $\lim_{n \to \infty} (f(n) - f(n-1)) = 0$.

In the proof of this step we shall make use of the following elementary result regarding convergence of the Cesàro means: if $(a_n)_{n \ge 1}$ is a converging sequence of real numbers and $\lim_{n \to \infty} a_n = a$, then

$$\lim_{n \to \infty} \frac{1}{n} \sum_{k=1}^{n} a_k = a.$$

As an exercise, prove this result.

Set $d_n = f(n) - f(n-1)$, $\delta_n = \mathfrak{S}(\frac{1}{n}, 1 - \frac{1}{n})$. Since $f(1) = \mathfrak{S}(1) = 0$,

$$f(n) = d_n + \cdots + d_2.$$

The relation (3.12) gives

$$f(n) = \left(1 - \frac{1}{n}\right) f(n-1) + \delta_n,$$

and so

$$n\delta_n = nd_n + f(n-1).$$

It follows that

$$\sum_{k=2}^{n} k\delta_k = nf(n) = n(d_n + f(n-1)) = n(n\delta_n - (n-1)d_n),$$

which yields

$$d_n = \delta_n - \frac{1}{n(n-1)} \sum_{k=2}^{n-1} k\delta_k.$$

By Step 1, $\lim_{n\to\infty} \delta_n = 0$. Obviously,

$$0 \le \frac{1}{n(n-1)} \sum_{k=2}^{n-1} k\delta_k \le \frac{1}{n} \sum_{k=2}^{n-1} \delta_k,$$

and we derive

$$\lim_{n\to\infty} \frac{1}{n(n-1)} \sum_{k=2}^{n-1} k\delta_k = 0.$$

It follows that $\lim_{n\to\infty} d_n = 0$.

Step 4 There is a constant c such that $f(n) = c \log n$ for all n.

By Step 2, for any $k \ge 1$,

$$\frac{f(n^k)}{\log n^k} = \frac{n}{\log n}.$$

Hence, to prove the statement it suffices to show that the limit

$$c = \lim_{n \to \infty} \frac{f(n)}{\log n}$$

exists. To prove that, we will show that $g(n)$ defined by

$$g(n) = f(n) - \frac{f(2)}{\log 2} \log n \qquad (3.14)$$

satisfies

$$\lim_{n \to \infty} \frac{g(n)}{\log n} = 0.$$

The choice of integer 2 in (3.14) is irrelevant, and the argument works with 2 replaced by any integer $m \geq 2$.

Obviously, $g(nm) = g(n) + g(m)$ and $g(1) = g(2) = 0$. Set $\xi_m = g(m) - g(m-1)$ if n is odd, $\xi_m = 0$ if m is even. By Step 3, $\lim_{m \to \infty} \xi_m = 0$. Let $n > 1$ be given. Write $n = 2n_1 + r_1$, where $r_1 = 0$ or $r_1 = 1$. Then

$$g(n) = \zeta_n + g(2n_1) = \zeta_n + g(n_1),$$

where we used that $g(2) = 0$. If $n_1 > 1$, write again $n_1 = 2n_1 + r_2$, where $r_2 = 0$ or $r_2 = 1$, so that

$$g(n_1) = \zeta_{n_1} + g(n_2).$$

This procedure terminates after k_0 steps, that is, when we reach $n_{k_0} = 1$. Obviously,

$$k_0 \leq \frac{\log n}{\log 2}, \qquad g(n) = \sum_{k=0}^{k_0 - 1} \zeta_{n_k},$$

where we set $n_0 = n$. Let $\epsilon > 0$ and m_ϵ be such that for $m \geq m_\epsilon$ we have $|\xi_m| < \epsilon / \log 2$. Then

$$\frac{|g(n)|}{\log n} \leq \frac{1}{\log n} \left(\sum_{m \leq m_\epsilon} |\xi_m| \right) + \epsilon \frac{k_0 \log 2}{\log n} \leq \frac{1}{\log n} \left(\sum_{m \leq m_\epsilon} |\xi_m| \right) + \epsilon.$$

It follows that

$$\limsup_{n \to \infty} \frac{|g(n)|}{\log n} \leq \epsilon.$$

Since $\epsilon > 0$ is arbitrary, the proof is complete.

Step 5 If c is as in Step 4, then

$$\mathfrak{S}(q_1, q_2) = cS(q_1, q_2).$$

Let $\Omega = \Omega_1 \oplus \Omega_2$ with $|\Omega_1| = m$, $|\Omega_2| = m - n$. Applying (3.12) to $P_1 = \overline{P}_n$, $P_2 = \overline{P}_{n-m}$, $p_1 = \frac{n}{m}$, $p_2 = \frac{m-n}{m}$, we derive

$$f(m) = \frac{n}{m} f(n) + \frac{m-n}{m} f(m-n) + \mathfrak{S}\left(\frac{n}{m}, \frac{m-n}{m}\right).$$

Step 4 gives that

$$\mathfrak{S}\left(\frac{n}{m}, \frac{m-n}{m}\right) = cS\left(\frac{n}{m}, \frac{m-n}{m}\right).$$

Since this relation holds for any $m < n$, the continuity of $\mathfrak{S}$ and S on $\mathcal{P}_2$ yields the statement.

Step 6 We now complete the proof by induction on $|\Omega|$. Suppose that $\mathfrak{S}(P) = cS(P)$ holds for all $P \in \mathcal{P}(\Omega)$ with $|\Omega| = n - 1$, where c is as in Step 4. Let $P = (p_1, \cdots, p_n)$ be a probability measure on $\Omega = \Omega_{n-1} \oplus \Omega_1$, where $|\Omega_{n-1}| = n - 1$, $|\Omega_1| = 1$. Without loss of generality we may assume that $q_n < 1$. Applying (3.12) with

$$P_1 = \left(\frac{q_1}{1 - q_n}, \cdots, \frac{q_{n-1}}{1 - q_n}\right),$$

$P_2 = (1)$, $p_1 = 1 - q_n$, $p_2 = q_n$, we derive

$$\mathfrak{S}(P) = cS(P_1) + cS(p_1, p_2) = cS(P).$$

This completes the proof. The non-triviality assumption yields that $c > 0$.

$\square$

Sub-additivity Characterization

The sub-additivity of entropy described in Proposition 3.2 is certainly a very intuitive property. If the entropy quantifies randomness of a probability measure P, or equivalently, the amount of information gained by an outcome of a probabilistic experiment described by P, than the product of marginals $P_l \otimes P_r$ is certainly more random then $P \in \mathcal{P}(\Omega_l \times \Omega_r)$. The Boltzmann–Gibbs–Shannon entropy S and the Hartley entropy S_H introduced in Exercise 3.2 are strictly sub-additive, and so is any linear combination

$$\mathfrak{G} = cS + CS_H, \tag{3.15}$$

where $c \geq 0$, $C \geq 0$, and at least one of these constants is strictly positive. It is a remarkable fact that the strict sub-additivity requirement together with the obvious assumption (3.11) selects (3.15) as the only possible choices for entropy. We also note the strict sub-additivity assumption selects the sign of the constants in (3.15), and that here we can omit the assumption (a) of Theorem 3.4.

Theorem 3.7 Let $\mathfrak{G} : \mathcal{P} \to [0, \infty[$ be a strictly sub-additive map, namely if $\Omega = \Omega_l \times \Omega_r$ and $P \in \mathcal{P}(\Omega)$, then

$$\mathfrak{G}(P) \leq \mathfrak{G}(P_l) + \mathfrak{G}(P_r)$$

with equality iff $P = P_l \otimes P_r$. Then there are constants $c \geq 0$, $C \geq 0$, $c + C > 0$, such that for all $P \in \mathcal{P}$,

$$\mathfrak{G}(P) = cS(P) + CS_H(P). \tag{3.16}$$

If in addition $\mathfrak{G}$ is continuous on $\mathcal{P}_2$, then $C = 0$ and $\mathfrak{G} = cS$ for some $c > 0$.

Proof We denote by $\mathfrak{G}_n$ the restriction of $\mathfrak{G}$ to $\mathcal{P}_n$. Note that the sub-additivity implies that

$$\mathfrak{G}_{2n}(p_{11}, p_{12}, \cdots, p_{n1}, p_{n2}) \leq \mathfrak{G}_2(p_{11} + \cdots + p_{n1}, p_{12} + \cdots + p_{n2})$$
$$+ \mathfrak{G}_n(p_{11} + p_{12}, \cdots, p_{n1} + p_{n2}). \tag{3.17}$$

For $x \in [0, 1]$ we set $\overline{x} = 1 - x$. The function

$$F(x) = \mathfrak{G}_2(\overline{x}, x) \tag{3.18}$$

will play an important role in the proof. It follows from (3.10) that $F(x) = F(\overline{x})$. By taking $P_l = P_r = (1, 0)$, we see that

$$2F(0) = \mathfrak{G}(P_l) + \mathfrak{G}(P_r) = \mathfrak{G}(P_l \otimes P_r) = \mathfrak{G}(1, 0, 0, 0) = \mathfrak{G}(1, 0) = F(0),$$

and so $F(0) = 0$.

We split the proof into eight steps.

Step 1 For all $q, r \in [0, 1]$ and $(p, p_3, \cdots, p_n) \in \mathcal{P}_{n-1}, n \geq 3$, one has

$$\mathfrak{S}_2(\overline{q}, q) - \mathfrak{S}_2(\overline{p}\,\overline{q} + p\overline{r}, \overline{p}q + pr) \leq \mathfrak{S}_n(p\overline{q}, pq, p_3, \cdots, p_n)$$

$$-\mathfrak{S}_n(p\overline{r}, pr, p_3, \cdots, p_n)$$

$$\leq \mathfrak{S}_2(\overline{p}\,\overline{r} + p\overline{q}, \overline{p}r + rq)$$

$$-\mathfrak{S}_2(\overline{r}, r). \qquad (3.19)$$

By interchanging q and r, it suffices to prove the first inequality in (3.19). We have

$$\mathfrak{S}_2(\overline{q}, q) + \mathfrak{S}_n(p\overline{r}, pr, p_3, \cdots, p_n)$$

$$= \mathfrak{S}_{2n}(\overline{q}\,p\overline{r}, qp\overline{r}, \overline{q}\,pr, qpr, \overline{q}\,p_3, qp_3, \cdots, \overline{q}\,p_n, qp_n)$$

$$= \mathfrak{S}_{2n}(\overline{q}\,p\overline{r}, \overline{q}\,pr, qp\overline{r}, qpr, \overline{q}\,p_3, qp_3, \cdots, \overline{q}\,p_n, qp_n)$$

$$\leq \mathfrak{S}_2(\overline{q}\,p\overline{r} + qp\overline{r} + \overline{q}(p_3 + \cdots + p_n), \overline{q}\,pr + qpr + q(p_3 + \cdots + p_n))$$

$$+ \mathfrak{S}_n(\overline{q}\,p\overline{r} + \overline{q}\,pr, qp\overline{r} + qpr, \overline{q}\,p_3 + qp_3, \cdots, \overline{q}\,p_n + qp_n)$$

$$= \mathfrak{S}_2(\overline{p}\,\overline{q} + p\overline{r}, \overline{p}q + pr) + S_n(p\overline{r}, pr, p_3, \cdots, p_n).$$

The first equality follows from (3.10) and the first inequality from (3.17). The final equality is elementary (we used that $p + p_3 + \cdots p_n = 1$).

Step 2 The function F, defined by (3.18), is increasing on $[0, 1/2]$, decreasing on $[1/2, 1]$, and is continuous and concave on $]0, 1[$. Moreover, for $q \in]0, 1[$ the left and right derivatives

$$D^+ F(q) = \lim_{h \downarrow 0} \frac{F(q + h) - F(q)}{h}, \qquad D^- F(q) = \lim_{h \uparrow 0} \frac{F(q + h) - F(q)}{h}$$

exist, are finite, and $D^+ F(q) \geq D^- F(q)$.

We first establish the monotonicity statement. Note that the inequality of Step 1

$$\mathfrak{S}_2(\overline{q}, q) - \mathfrak{S}_2(\overline{p}\,\overline{q} + p\overline{r}, \overline{p}q + pr) \leq \mathfrak{S}_2(\overline{p}\,\overline{r} + p\overline{q}, \overline{p}r + rq) - \mathfrak{S}_2(\overline{r}, r) \qquad (3.20)$$

with $r = \overline{q}$ gives

$$2\mathfrak{S}_2(\overline{q}, q) \leq \mathfrak{S}_2((1 - p)(1 - q) + pq, (1 - p)q + p(1 - q))$$

$$+ \mathfrak{S}_2((1 - p)q + p(1 - q), (1 - p)(1 - q) + pq),$$

or equivalently, that

$$F(q) \leq F((1 - p)q + p(1 - q)). \qquad (3.21)$$

Fix $q \in [0, 1/2]$ and note that $[0, 1] \ni p \mapsto (1 - p)q + p(1 - q)$ is the parametrization of the interval $[q, 1 - q]$. Since $F(q) = F(1 - q)$, we derive that $F(q) \leq F(x)$ for $x \in [q, 1/2]$, and that $F(x) \geq F(1 - q)$ for $x \in [1/2, q]$. Thus, F is increasing on $[0, 1/2]$ and decreasing on $[1/2, 1]$. In particular, for all $x \in [0, 1]$,

$$F(1/2) \geq F(x) \geq 0, \tag{3.22}$$

where we used that $F(0) = F(1) = 0$.

We now turn to the continuity and concavity, starting with continuity first. The inequality (3.20) with $p = 1/2$ gives that for any $q, r \in [0, 1]$,

$$\frac{1}{2} F(q) + \frac{1}{2} F(r) \leq F\left(\frac{1}{2} q + \frac{1}{2} r\right). \tag{3.23}$$

Fix now $q \in]0, 1[$, set $\lambda_n = 2^{-n}$ and, starting with large enough n so that $q \pm \lambda_n \in [0, 1]$, define

$$\Delta_n^+(q) = \frac{F(q + \lambda_n) - F(q)}{\lambda_n}, \qquad \Delta_n^-(q) = \frac{F(q - \lambda_n) - F(-q)}{-\lambda_n}.$$

It follows from (3.23) that the sequence $\Delta_n^+(q)$ is increasing, that the sequence $\Delta_n^-(q)$ is decreasing, and that $\Delta_n^+(q) \leq \Delta_n^-(q)$ (write down the details!). Hence, the limits

$$\lim_{n \to \infty} \Delta_n^+(q), \qquad \lim_{n \to \infty} \Delta_n^-(q)$$

exist, are finite, and

$$\lim_{n \to \infty} F(q \pm \lambda_n) = F(q). \tag{3.24}$$

The established monotonicity properties of F yield that the limits $\lim_{h \downarrow 0} F(q + h)$ and $\lim_{h \uparrow 0} F(q + h)$ exist. Combining this observation with (3.24), we derive that

$$\lim_{h \to 0} F(q + h) = F(q),$$

and so F is continuous on $]0, 1[$. We now prove the concavity. Replacing r with $(q + r)/2$ in (3.23), we get that

$$\lambda F(q) + (1 - \lambda) F(r) \leq F(\lambda q + (1 - \lambda) r) \tag{3.25}$$

holds for $\lambda = 3/4$, while replacing q with $(q + r)/2$ shows that (3.25) holds for $\lambda = 1/4$. Continuing in this way shows that (3.25) holds for all dyadic fractions $\lambda = k/2^n$, $1 \leq k \leq 2^n$, $n = 1, 2, \cdots$. Since dyadic fractions are dense in $[0, 1]$, the continuity of F yields that (3.25) holds for $\lambda \in [0, 1]$ and $q, r \in]0, 1[$. Finally, to prove the statement about the derivatives, fix $q \in]0, 1[$ and for $h > 0$ small enough consider the functions

$$\Delta^+(h) = \frac{F(q + h) - F(q)}{h}, \qquad \Delta^-(h) = \frac{F(q - h) - F(q)}{-h}.$$

The concavity of F gives that the function $h \mapsto \Delta^+(h)$ is increasing, that $h \mapsto \Delta^-(h)$ is increasing, and that $\Delta^+(h) \leq \Delta^-(h)$. This establishes the last claim of Step 2 concerning left and right derivatives of F on $]0, 1[$.

Step 3 There exist functions $\mathcal{R}_n : \mathcal{P}_n \to \mathbb{R}, n \geq 2$, such that

$$\mathfrak{S}_n(p\overline{q}, pq, p_3, \cdots, p_n) = pF(q) + \mathcal{R}_{n-1}(p, p_3, \cdots, p_n) \tag{3.26}$$

for all $q \in]0, 1[, (p, p_3, \cdots, p_n) \in \mathcal{P}_{n-1}$ and $n \geq 2$.

To prove this, note that Step 1 and the relation $F(x) = F(\overline{x})$ give

$$\frac{F(\overline{p}q + pq) - F(\overline{p}q + pr)}{q - r} \leq \frac{S_n(p\overline{q}, pq, p_3, \cdots, p_n) - S_n(p\overline{r}, pr, p_3, \cdots, p_n)}{q - r}$$

$$\leq \frac{F(pq + \overline{p}r) - F(pr + \overline{p}r)}{q - r} \tag{3.27}$$

for $0 < r < q < 1$ and $(p, p_3, \cdots, p_n) \in \mathcal{P}_n$. Fix $(p, p_3, \cdots, p_n) \in \mathcal{P}_n$ and set

$$L(q) = \mathfrak{S}_n(p\overline{q}, pq, p_3, \cdots, p_n).$$

Taking $q \downarrow r$ in (3.27) we get

$$pD^- F(r) = D^- L(r),$$

while taking $r \uparrow q$ gives

$$pD^+ F(q) = D^+ L(q).$$

Since $D^\pm F(q)$ is finite by Step 2, we derive that the function $L(q) - pF(q)$ is differentiable on $]0, 1[$ with vanishing derivative. Hence, for $q \in]0, 1[$,

$$L(q) = pF(q) + \mathcal{R}_{n-1}(p, p_3, \cdots, p_n),$$

where the constant $\mathcal{R}_{n-1}$ depends on the values $(p, p_3, \cdots, p_n)$ we have fixed in the above argument.

Step 4 There exist constants $c \geq 0$ and C such that for all $q \in]0, 1[$,

$$F(q) = cS(1 - q, q) + C. \tag{3.28}$$

We start the proof by taking $(p_1, p_2, p_3) \in \mathcal{P}_{3,f}$. Setting

$$p = p_1 + p_2, \qquad q = \frac{p_2}{p_1 + p_2},$$

we write

$$\mathfrak{S}_3(p_1, p_2, p_3) = \mathfrak{S}_3(p\bar{q}, pq, p_3).$$

It then follows from Step 3 that

$$\mathfrak{S}_3(p_1, p_2, p_3) = (p_1 + p_2)\mathfrak{S}_2\left(\frac{p_1}{p_1 + p_2}, \frac{p_2}{p_1 + p_2}\right) + \mathcal{R}_2(p_1 + p_2, p_3).$$
$$(3.29)$$

By (3.10) we also have

$$\mathfrak{S}_3(p_1, p_2, p_3) = \mathfrak{S}_3(p_1, p_3, p_2) = (p_1 + p_3)\mathfrak{S}_2\left(\frac{p_1}{p_1 + p_3}, \frac{p_3}{p_1 + p_3}\right)$$
$$+ \mathcal{R}_2(p_1 + p_3, p_3). \qquad (3.30)$$

Setting $G(x) = \mathcal{R}_2(\bar{x}, x)$, $x = p_3$, $y = p_2$, we rewrite (3.29) = (3.30) as

$$(1 - x)F\left(\frac{y}{1 - x}\right) + G(x) = (1 - y)F\left(\frac{x}{1 - y}\right) + G(y), \qquad (3.31)$$

where $x, y \in]0, 1[$ and $x + y < 1$. The rest of the proof concerns analysis of the functional equation (3.31).

Since F is continuous on $]0, 1[$, fixing one variable one easily deduces from (3.31) that G is also continuous on $]0, 1[$. Let $0 < a < b < 1$ and fix $y \in]0, 1 - b[$. It follows that (verify this!)

$$\frac{x}{1 - y} \in \left]a, \frac{b}{1 - y}\right] \subset]0, 1[, \qquad \frac{y}{1 - x} \in \left]y, \frac{y}{1 - b}\right] \subset]0, 1[.$$

Integrating (3.31) with respect to x over $[a, b]$ we derive

$$(b - a)G(y) = \int_a^b G(y)\mathrm{d}x$$

$$= \int_a^b G(x)\mathrm{d}x + \int_a^b (1 - x)F\left(\frac{y}{1 - x}\right)\mathrm{d}x$$

$$- (1 - y)\int_a^b F\left(\frac{x}{1 - y}\right)\mathrm{d}x$$

$$= \int_a^b G(x)dx + y^2 \int_{y/(1-a)}^{y/(1-b)} s^{-3} F(s)ds$$

$$-(1-y)^2 \int_{a/(1-y)}^{b/(1-y)} F(t)dt, \qquad (3.32)$$

where we have used the change of variable

$$s = \frac{y}{1-x}, \qquad t = \frac{x}{1-y}. \qquad (3.33)$$

It follows that G is differentiable on $]0, b[$. Since $0 < b < 1$ is arbitrary, G is differentiable on $]0, 1[$.

The change of variable (3.33) maps bijectively $\{(x, y) \mid x, y > 0\}$ to $\{(s, t) \mid s, t \in]0, 1[\}$ (verify this!), and in this new variables the functional equation (3.31) reads

$$F(t) = \frac{1-t}{1-s} F(s) + \frac{1-st}{1-s} \left[G\left(\frac{t-st}{1-st}\right) - G\left(\frac{s-st}{1-st}\right) \right]. \qquad (3.34)$$

Fixing s, we see that the differentiability of G implies the differentiability of F on $]0, 1[$. Returning to (3.32), we get that G is twice differentiable on $]0, 1[$, and then (3.34) gives that F is also twice differentiable on $]0, 1[$. Continuing in this way we derive that both F and G are infinitely differentiable on $]0, 1[$. Differentiating (3.31) first with respect to x and then with respect to y gives

$$\frac{y}{(1-x)^2} F''\left(\frac{y}{1-x}\right) = \frac{x}{(1-y)^2} F''\left(\frac{x}{1-y}\right). \qquad (3.35)$$

The substitution (3.33) gives that for $s, t \in]0, 1[$,

$$s(1-s)F''(s) = t(1-t)F''(t).$$

It follows that for some $c \in \mathbb{R}$,

$$t(1-t)F''(t) = -c.$$

Integration gives

$$F(t) = cS(1-t, t) + Bt + C.$$

Since $F(t) = F(\bar{t})$, we have $B = 0$, and since F is increasing on $[0, 1/2]$, we have $c \geq 0$. This completes the proof of Step 4. Note that as a by-product of the proof we have derived that for some constant D,

$$G(x) = F(x) + D, \qquad x \in]0, 1[. \qquad (3.36)$$

To prove (3.36), note that (3.28) gives that F satisfies the functional equation

$$(1 - x)F\left(\frac{y}{1 - x}\right) + F(x) = (1 - y)F\left(\frac{x}{1 - y}\right) + F(y).$$

Combining this equation with (3.31) we derive that for $x, y > 0, 0 < x + y < 1$,

$$G(x) - F(x) = G(y) - F(y).$$

Hence, $G(x) - F(x) = D_y$ for $x \in]0, 1 - y[$. If $y_1 < y_2$, we must have $D_{y_1} = D_{y_2}$, and so $D = D_y$ does not depend on y, which gives (3.36).

Step 5 For any $n \geq 2$ there exists constant $C(n)$ such that for $(p_1, \cdots, p_n) \in \mathcal{P}_{n,\mathrm{f}}$,

$$\mathfrak{S}_n(p_1, \cdots, p_n) = cS(p_1, \cdots, p_n) + C(n), \tag{3.37}$$

where $c \geq 0$ is the constant from Step 4.

In Step 4 we established (3.37) for $n = 2$ (we set $C(2) = C$), and so we assume that $n \geq 3$. Set $p = p_1 + p_2, q = p_2/(p_1 + p_2)$. It then follows from Steps 3 and 4 that

$$\mathfrak{S}_n(p_1, \cdots, p_n) = (p_1 + p_2)\mathfrak{S}_2\left(\frac{p_1}{p_1 + p_2}, \frac{p_2}{p_1 + p_2}\right)$$

$$+ \mathcal{R}_{n-1}(p_1 + p_2, p_3, \cdots, p_n)$$

$$= (p_1 + p_2)cS\left(\frac{p_1}{p_1 + p_2}, \frac{p_2}{p_1 + p_2}\right)$$

$$+ \widehat{\mathcal{R}}_{n-1}(p_1 + p_2, p_3, \cdots, p_n), \tag{3.38}$$

where $\widehat{\mathcal{R}}_{n-1}(p, p_3, \cdots, p_n) = pC_2 + \mathcal{R}_{n-1}(p, p_3, \cdots, p_n)$. Note that since $\mathcal{R}_{n-1}$ is invariant under the permutations of the variables $(p_3, \cdots, p_n)$ (recall (3.26)), so is $\widehat{\mathcal{R}}_{n-1}$. The invariance of $\mathfrak{S}_n$ under the permutation of the variables gives

$$\mathfrak{S}_n(p_1, \cdots, p_n) = (p_1 + p_3)cS\left(\frac{p_1}{p_1 + p_3}, \frac{p_3}{p_1 + p_3}\right)$$

$$+ \widehat{\mathcal{R}}_{n-1}(p_1 + p_3, p_2, p_4 \cdots, p_n),$$

and so

$$(p_1 + p_2)cS\left(\frac{p_1}{p_1 + p_2}, \frac{p_2}{p_1 + p_2}\right) - (p_1 + p_3)cS\left(\frac{p_1}{p_1 + p_3}, \frac{p_3}{p_1 + p_3}\right)$$

$$= \widehat{\mathcal{R}}_{n-1}(p_1 + p_2, p_3, \cdots, p_n) - \widehat{\mathcal{R}}_{n-1}(p_1 + p_3, p_2, p_4 \cdots, p_n). \tag{3.39}$$

Until the end of the proof when we wish to indicate the number of variables in the Boltzmann–Gibbs–Shannon entropy we will write $S_n(p_1, \cdots, p_n)$. One easily verifies that

$$
S_n(p_1, \cdots, p_n) = (p_1 + p_2)S_2\left(\frac{p_1}{p_1 + p_2}, \frac{p_2}{p_1 + p_2}\right)
$$
$$
+ S_{n-1}(p_1 + p_2, p_3, \cdots, p_n)
$$
$$
= (p_1 + p_3)S_2\left(\frac{p_1}{p_1 + p_3}, \frac{p_3}{p_1 + p_3}\right)
$$
$$
+ S_{n-1}(p_1 + p_3, p_2, p_4, \cdots, p_n),
$$

and so

$$
(p_1 + p_2)S_2\left(\frac{p_1}{p_1 + p_2}, \frac{p_2}{p_1 + p_2}\right) - (p_1 + p_3)S_2\left(\frac{p_1}{p_1 + p_3}, \frac{p_3}{p_1 + p_3}\right)
$$
$$
= S_{n-1}(p_1 + p_2, p_3, \cdots, p_n) - S_{n-1}(p_1 + p_3, p_2, p_4 \cdots, p_n).
$$
$$(3.40)$$

Since in the formulas (3.39) and (3.40) $S = S_2$, we derive that the function

$$
T_{n-1}(p, q, p_4, \cdots, p_n) = \widehat{R}_{n-1}(p, q, p_4, \cdots, p_n) - cS_{n-1}(p, q, p_4, \cdots, p_n)
$$

satisfies

$$
T_{n-1}(p_1 + p_2, p_3, p_4, \cdots, p_n) = T_{n-1}(p_1 + p_3, p_2, p_4, \cdots, p_n) \qquad (3.41)
$$

for all $(p_1, \cdots p_n) \in \mathcal{P}_{n,f}$. Moreover, by construction, $T_{n-1}(p, q, p_4, \cdots, p_n)$ is invariant under the permutation of the variables $(q, p_4, \cdots, p_n)$. Set $s = p_1 + p_2 + p_3$. Then (3.41) reads as

$$
T_{n-1}(s - p_3, p_3, p_4, \cdots, p_n) = T_{n-1}(s - p_2, p_2, p - p_4, \cdots, p_n).
$$

Hence, the map

$$
]0, s[\ni p \mapsto T_{n-1}(s - p, p, p_4, \cdots, p_n)
$$

is constant. By the permutation invariance, the maps

$$
]0, s[\ni p \mapsto T_{n-1}(s - p, p_3, \cdots, p_{m-1}, p, p_{m+1}, \cdots)
$$

are also constant. Setting $s = p_1 + p_2 + p_3 + p_4$, we deduce that the map

$$
(p_3, p_4) \mapsto T_{n-1}(s - p_3 - p_4, p_3, p_4, \cdots, p_n)
$$

with domain $p_3 > 0$, $p_4 > 0$, $p_3 + p_4 < s$, is constant. Continuing inductively, we conclude that the map

$$(p_3, \cdots, p_n) \mapsto T_{n-1}(1 - (p_3 + \cdots + p_n), p_3, p_4, \cdots, p_n)$$

with domain $p_k > 0$, $\sum_{k=3}^{n} p_k < 1$ is constant. Hence, the map

$$\mathcal{P}_{n,f} \ni (p_1, \cdots, p_n) \mapsto T_{n-1}(p_1 + p_2, p_3, \cdots, p_n)$$

is constant, and we denote the value it assumes by $C(n)$. Returning now to (3.38), we conclude the proof of (3.37):

$$\mathfrak{S}_n(p_1, \cdots, p_n) = (p_1 + p_2)cS_2\left(\frac{p_1}{p_1 + p_2}, \frac{p_2}{p_1 + p_2}\right)$$
$$+ \widehat{\mathcal{R}}_{n-1}(p_1 + p_2, p_3, \cdots, p_n)$$
$$= (p_1 + p_2)cS_2\left(\frac{p_1}{p_1 + p_2}, \frac{p_2}{p_1 + p_2}\right)$$
$$+ cS_{n-1}(p_1 + p_2, p_3, \cdots, p_n) + C(n)$$
$$= cS_n(p_1, \cdots, p_n) + C(n). \tag{3.42}$$

Step 6 $C(n + m) = C(n)C(m)$ for $n, m \geq 2$, and

$$\liminf_{n \to \infty}(C(n + 1) - C(n)) = 0. \tag{3.43}$$

If $P_l \in \mathcal{P}_n$ and $P_r \in \mathcal{P}_m$, then the identity $\mathfrak{S}_{nm}(P_l \times P_r) = \mathfrak{S}_n(P_l) + \mathfrak{S}(P_r)$ and (3.37) give that $C(n + m) = C(n) + C(m)$. To prove (3.43), suppose that $n \geq 3$ and take in (3.19) $q = 1/2$, $r = 0$, $p = p_3 = \cdots = p_n = 1/(n - 1)$. Then, combining (3.19) with Step 5, we derive

$$F\left(\frac{1}{2}\right) - F\left(\frac{n - 2}{2(n - 1)}\right) \leq \mathfrak{S}_n\left(\frac{1}{2(n - 1)}, \frac{1}{2(n - 1)}, \frac{1}{n - 1}, \cdots, \frac{1}{n - 1}\right)$$
$$- \mathfrak{S}_n\left(\frac{1}{n - 1}, 0, \frac{1}{n - 1}, \cdots \frac{1}{n - 1}\right)$$
$$= cS_n\left(\frac{1}{2(n - 1)}, \frac{1}{2(n - 1)}, \frac{1}{n - 1}, \cdots, \frac{1}{n - 1}\right)$$
$$- cS_{n-1}\left(\frac{1}{n - 1}, \frac{1}{n - 1}, \cdots \frac{1}{n - 1}\right)$$
$$+ C(n) - C(n - 1)$$
$$= \frac{\log 2}{n - 1} + C(n) - C(n - 1).$$

The first inequality in (3.22) gives

$$0 \le \frac{\log 2}{n-1} + C(n) - C(n-1),$$

and the statement follows.

Step 7 There is a constant $C \ge 0$ such that for all $n \ge 2$, $C(n) = C \log n$.

Fix $\epsilon > 0$ and $n > 1$. Let $k \in \mathbb{N}$ be such that for all integers $p \ge n^k$, $C(p+1) - C(p) \ge -\epsilon$. It follows that for $p \ge p^k$ and $j \in \mathbb{N}$,

$$C(p+j) - C(p) = \sum_{i=1}^{j} (C(p+i) - C(p+i-1)) \ge -j\epsilon.$$

Fix now $p \ge n^k$ and let $m \in \mathbb{N}$ be such that $n^m \le p < n^{m+1}$. Obviously, $m \ge k$. Write

$$p = a_m n^m + a_{m-1} n^{m-1} + \cdots + a_1 p + a_0,$$

where a_k's are integers such that $1 \le a_m < n$ and $0 \le a_k < n$ for $k < m$. It follows that

$$C(p) > C(a_m n^m + \cdots + a_1 n) - n\epsilon = C(n) + C(a_m n^{m-1} + \cdots + a_2 n + a_1) - n\epsilon.$$

Continuing inductively, we derive that

$$C(p) > (m-k+1)C(n) + C(a_m n^{k-1} + a_{m-1} n^{k-2} + \cdots + a_{m-k+1}) - (m-k+1)\epsilon.$$

If $M = \max_{2 \le j \le n^{k+1}} |C(j)|$, then the last inequality gives

$$C(p) > (m-k+1)C(n) - M - (m-k+1)\epsilon.$$

By the choice of m, $\log p \le (m+1) \log n$, and so

$$\liminf_{p \to \infty} \frac{C(p)}{\log p} \ge \frac{C(n)}{\log n}.$$

Since

$$\liminf_{n \to \infty} \frac{C(p)}{\log p} \le \liminf_{j \to \infty} \frac{C(n^j)}{\log n^j} = \frac{C(n)}{n},$$

we derive that for all $n \ge 2$,

$$C(n) = C \log n,$$

where

$$C = \liminf_{p \to \infty} \frac{C(p)}{p}.$$

It remains to show that $C \geq 0$. Since

$$F(x) = cS_2(1 - x, x) + C \log 2,$$

we have $\lim_{x \downarrow 0} F(x) = C \log 2$, and (3.22) yields that $C \geq 0$.

Step 8 We now conclude the proof. Let $P = (p_1, \cdots, p_n) \in \mathcal{P}_n$. Write

$$P = (p_{j_1}, \cdots, p_{j_k}, 0, \cdots, 0),$$

where $p_{j_m} > 0$ for $m = 1, \cdots, k$. Then

$$\mathfrak{S}_n(P) = \mathfrak{S}_k(p_{j_1}, \cdots, p_{j_k}) = cS_k(p_{j_1}, \cdots, p_{j_k}) + C \log k = cS_n(P) + CS_H(P).$$

Since $\mathfrak{S}_n$ is strictly sub-additive, we must have $c + C > 0$. The final statement is a consequence of the fact that S_H is not continuous on $\mathcal{P}_n$ for $n \geq 2$.

$\square$

3.5 Rényi Entropy

Let Ω be a finite set and $P \in \mathcal{P}(\Omega)$. For $\alpha \in]0, 1[$ we set

$$S_\alpha(P) = \frac{1}{1 - \alpha} \log \left(\sum_{\omega \in \Omega} P(\omega)^\alpha \right).$$

$S_\alpha(P)$ is called the Rényi entropy of P.

Proposition 3.8

(1) $\lim_{\alpha \uparrow 1} S_\alpha(P) = S(P)$.
(2) $\lim_{\alpha \downarrow 0} S_\alpha(P) = S_H(P)$.
(3) $S_\alpha(P) \geq 0$ and $S_\alpha(P) = 0$ iff P is pure.
(4) $S_\alpha(P) \leq \log |\Omega|$ with equality iff $P = P_{\text{ch}}$.
(5) The map $]0, 1[\ni \alpha \mapsto S_\alpha(P)$ is decreasing and is strictly decreasing unless $P = P_{\text{ch}}$.
(6) The map $\mathcal{P}(\Omega) \ni P \mapsto S_\alpha(P)$ is continuous and concave.

(7) *If* $P = P_l \otimes P_r$ *is a product measure on* $\Omega = \Omega_l \times \Omega_r$, *then*
$S_\alpha(P) = S_\alpha(P_l) + S_\alpha(P_r)$.

(8) *The map* $\alpha \mapsto S_\alpha(P)$ *extends to a real analytic function on* $\mathbb{R}$ *by the formulas*
$S_1(P) = S(P)$ *and*

$$S_\alpha(P) = \frac{1}{1-\alpha} \log \left(\sum_{\omega \in \mathrm{supp} P} P(\omega)^\alpha \right), \qquad \alpha \neq 1.$$

Exercise 3.3 Prove Proposition 3.8.

Exercise 3.4 Describe properties of $S_\alpha(P)$ for $\alpha \notin]0, 1[$.

Exercise 3.5 Let $\Omega = \{-1, 1\} \times \{-1, 1\}, 0 < p, q < 1, p + q = 1, p \neq q$, and

$$P_\epsilon(-1, -1) = pq + \epsilon, \qquad P_\epsilon(-1, 1) = p(1 - q) - \epsilon,$$

$$P_\epsilon(1, -1) = (1 - p)q - \epsilon, \qquad P_\epsilon(1, 1) = (1 - p)(1 - q) + \epsilon.$$

Show that for $\alpha \neq 1$ and small non-zero ϵ,

$$S_\alpha(P_\epsilon) > S_\alpha(P_{\epsilon,l}) + S_\alpha(P_{\epsilon,r}).$$

Hence, Rényi entropy is not sub-additive (compare with Theorem 3.7).

3.6 Why is the Rényi Entropy Natural?

In introducing $S_\alpha(P)$ Rényi was motivated by a concept of generalized means. Let $w_k > 0, \sum_{k=1}^n w_k = 1$ be weights and $G :]0, \infty[\to]0, \infty[$ a continuous strictly increasing function. We shall call such G a *mean function*. The G-mean of strictly positive real numbers $x_1, \cdots, x_n$ is

$$S_G(x_1, \cdots, x_n) = G^{-1} \left(\sum_{k=1}^n w_k G(x_k) \right).$$

Set $\mathcal{P}_f = \cup_{n \geq 1} \mathcal{P}_{n,f}$.
   One then has:

Theorem 3.9 Let $\mathfrak{S} : \mathcal{P}_f \to [0, \infty[$ be a function with the following properties.

(a) If $P = P_l \otimes P_r$, then $\mathfrak{S}(P) = \mathfrak{S}(P_l) + \mathfrak{S}(P_r)$.
(b) There exists a mean function G such that for all $n \geq 1$ and $P = (p_1, \cdots, p_n) \in \mathcal{P}_{n,\mathrm{f}}$,

$$\mathfrak{S}(p_1, \cdots, p_n) = G^{-1}\left(\mathbb{E}_P(G(S_P))\right) = G^{-1}\left(\sum_{k=1}^{n} p_k G(-\log p_k)\right).$$

(c) $\mathfrak{S}(p, 1 - p) \to 0$ as $p \to 0$.

Then there exist $\alpha > 0$ and a constant $c \geq 0$ such that for all $P \in \mathcal{P}_{\mathrm{f}}$,

$$\mathfrak{S}(P) = c S_\alpha(P).$$

Remark 3.10 The assumption (c) excludes the possibility $\alpha \leq 0$.

Remark 3.11 If in addition one requires that the map $\mathcal{P}_{n,\mathrm{f}} \ni P \to \mathfrak{S}(P)$ is concave for all $n \geq 1$, then $\mathfrak{S}(P) = c S_\alpha(P)$ for some $\alpha \in]0, 1]$.

Although historically important, we find that Theorem 3.9 (and any other axiomatic characterization of the Rényi entropy) is less satisfactory than the powerful characterizations of the Boltzmann–Gibbs–Shannon entropy given in Sect. 3.4. Taking Boltzmann–Gibbs–Shannon entropy for granted, an alternative understanding of the Rényi entropy arises through Cramér's theorem for the entropy function S_P. For the purpose of this interpretation, without loss of generality we may assume that $P \in \mathcal{P}(\Omega)$ is faithful. Set

$$\widehat{S}_\alpha(P) = \log\left(\sum_{\omega \in \Omega} [P(\omega)]^{1-\alpha}\right), \qquad \alpha \in \mathbb{R}. \tag{3.44}$$

Obviously, for $\alpha \in \mathbb{R}$,

$$\widehat{S}_\alpha(P) = \alpha S_{1-\alpha}(P). \tag{3.45}$$

The naturalness of the choice (3.44) stems from the fact that the function $\alpha \mapsto \widehat{S}_\alpha(P)$ is the cumulant generating function of $S_P(\omega) = -\log P(\omega)$ with respect to P,

$$\widehat{S}_\alpha(P) = \log \mathbb{E}_P(e^{\alpha S_P}). \tag{3.46}$$

Passing to the products (Ω^N, P_N), the LLN gives that for any $\epsilon > 0$,

$$\lim_{N \to \infty} P_N\left\{\omega = (\omega_1, \cdots, \omega_N) \in \Omega^N \,\Big|\, \left|\frac{S_P(\omega_1) + \cdots S_P(\omega_N)}{N}\right.\right.$$
$$\left.\left. - S(P)\right| \geq \epsilon\right\} = 0. \tag{3.47}$$

It follows from Cramér's theorem that the rate function

$$I(\theta) = \sup_{\alpha \in \mathbb{R}} (\alpha\theta - \widehat{S}_\alpha(P)), \qquad \theta \in \mathbb{R}, \tag{3.48}$$

controls the fluctuations that accompany the limit (3.47):

$$\lim_{N \to \infty} \frac{1}{N} \log P_N \left\{ \omega = (\omega_1, \cdots, \omega_N) \in \Omega^N \mid \frac{S_P(\omega_1) + \cdots S_P(\omega_N)}{N} \in [a,b] \right\}$$
$$= - \inf_{\theta \in [a,b]} I(\theta). \tag{3.49}$$

We shall adopt a point of view that the relations (3.45), (3.48), and (3.49) constitute the foundational basis for introduction of the Rényi entropy. In accordance with this interpretation, the traditional definition of the Rényi entropy is somewhat redundant, and one may as well work with $\widehat{S}_\alpha(P)$ from the beginning and call it the *Rényi entropy* of P (or α-entropy of P when there is a danger of confusion).

The basic properties of the map $\alpha \mapsto \widehat{S}_\alpha(P)$ follow from (3.46) and results described in Sect. 2.4. Note that $S_0(P) = 0$ and $S_1(P) = \log|\Omega|$. The map $\mathcal{P}_f(\Omega) \ni P \mapsto \widehat{S}_\alpha(P)$ is convex for $\alpha \notin [0, 1]$ and concave for $\alpha \in]0, 1[$.

3.7 Notes and References

The celebrated expression (3.5) for entropy of a probability measure goes back to the 1870s and works of Boltzmann and Gibbs on the foundations of statistical mechanics. This will be discussed in more detail in Part II of the lecture notes. Shannon has rediscovered this expression in his work on foundations of mathematical information theory [44]. The results of Sects. 3.2 and 3.3 go back to this seminal work. Regarding Exercise 3.2, Hartley entropy was introduced in [23]. Hartley's work has partly motivated Shannon's [44].

Shannon was also first to give an axiomatization of entropy. The axioms in [44] are the continuity of $\mathfrak{S}$ on $\mathcal{P}_n$ for all n, the split-additivity (3.12), and the monotonicity $\mathfrak{S}(\overline{P}_{n+1}) < \mathfrak{S}(\overline{P}_n)$, where $\overline{P}_k \in \mathcal{P}_k$ is the chaotic probability measures. Shannon then proved that the only functions $\mathfrak{S}$ satisfying these properties are $cS, c > 0$. Theorem 3.4 is in spirit of Shannon's axiomatization, with the monotonicity axiom $\mathfrak{S}(\overline{P}_{n+1}) < \mathfrak{S}(\overline{P}_n)$ dropped and the continuity requirement relaxed; see Chapter 2 in [1] for additional information and Theorem 2.2.3 in [50] whose proof we roughly followed. We leave it as an exercise for the reader to simplify the proof of Theorem 3.4 under additional Shannon's axioms.

Shannon comments in [44] on the importance of his axiomatization as

This theorem, and the assumptions required for its proof, are in no way necessary for the present theory. It is given chiefly to lend a certain plausibility to some of our later definitions. The real justification of these definitions, however, will reside in their implications.

The others beg to differ on its importance, and axiomatizations of entropies became an independent research direction, starting with early works of Khintchine [32] and Faddeev [17]. Much of these efforts are summarized in the monograph [1], see also [10].

The magnificent Theorem 3.7 is due to Aczél, Forte, and Ng [2]. I was not able to simplify their arguments and the proof of Theorem 3.7 follows closely the original paper. Step 7 is due to [31]. The proof of Theorem 3.7 can be also found in [1, Section 4.4]. An interesting exercise that may elucidate a line of thought that has led to the proof of Theorem 3.7 is to simplify various steps of the proof by making additional regularity assumptions.

Rényi entropy has been introduced in [41]. Theorem 3.9 was proven in [12]; see Chapter 5 in [1] for additional information.

4 Relative Entropy

4.1 Definition and Basic Properties

Let Ω be a finite set and $P, Q \in \mathcal{P}(\Omega)$. If $P \ll Q$, the relative entropy function of the pair (P, Q) is defined for $\omega \in \operatorname{supp}P$ by

$$cS_{P|Q}(\omega) = cS_Q(\omega) - cS_P(\omega) = c \log P(\omega) - c \log Q(\omega) = c \log \Delta_{P|Q}(\omega),$$

where $c > 0$ is a constant that does not depend on Ω, P, Q. The relative entropy of P with respect to Q is

$$S(P|Q) = c \int_{\operatorname{supp}P} S_{P|Q}\mathrm{d}P = c \sum_{\omega \in \operatorname{supp}P} P(\omega) \log \frac{P(\omega)}{Q(\omega)}. \tag{4.1}$$

If P is not absolutely continuous with respect to Q (i.e., $Q(\omega) = 0$ and $P(\omega) > 0$ for some ω), we set

$$S(P|Q) = \infty.$$

The value of the constant c will play no role in the sequel, and we set $c = 1$. As in the case of entropy, the constant c will reappear in the axiomatic characterizations of relative entropy (see Theorems 5.1 and 5.3).

Note that

$$S(P|P_{ch}) = -S(P) + \log |\Omega|.$$

Proposition 4.1 $S(P|Q) \geq 0$ and $S(P|Q) = 0$ iff $P = Q$.

Proof We need to consider only the case $P \ll Q$. By Jensen's inequality,

$$\sum_{\omega \in \text{supp} P} P(\omega) \log \frac{Q(\omega)}{P(\omega)} \leq \log \left(\sum_{\omega \in \text{supp} P} Q(\omega) \right),$$

and so

$$\sum_{\omega \in \text{supp} P} P(\omega) \log \frac{Q(\omega)}{P(\omega)} \leq 0$$

with equality iff $P = Q$.   □

The next result refines the previous proposition. Recall that the variational distance $d_V(P, Q)$ is defined by (3.2).

Theorem 4.2

$$S(P|Q) \geq \frac{1}{2} d_V(P, Q)^2. \qquad (4.2)$$

The equality holds iff $P = Q$.

Proof We start with the elementary inequality

$$(1 + x) \log(1 + x) - x \geq \frac{1}{2} \frac{x^2}{1 + \frac{x}{3}}, \qquad x \geq -1. \qquad (4.3)$$

This inequality obviously holds for $x = -1$, so we may assume that $x > -1$. Denote the l.h.s by $F(x)$ and the r.h.s. by $G(x)$. One verifies that $F(0) = F'(0) = G(0) = G'(0) = 0$, and that

$$F''(x) = \frac{1}{1 + x}, \qquad G''(x) = \left(1 + \frac{x}{3}\right)^{-3}.$$

Obviously, $F''(x) > G''(x)$ for $x > -1$, $x \neq 0$. Integrating this inequality we derive that $F'(x) > G'(x)$ for $x > 0$ and $F'(x) < G'(x)$ for $x \in]-1, 0[$. Integrating these inequalities we get $F(x) \geq G(x)$ and that equality holds iff $x = 0$.

We now turn to the proof of the theorem. We need only to consider the case $P \ll Q$. Set

$$X(\omega) = \frac{P(\omega)}{Q(\omega)} - 1,$$

with the convention that $0/0 = 0$. Note that $\int_{\Omega} X dQ = 0$ and that

$$S(P|Q) = \int_{\Omega} ((X + 1) \log(X + 1) - X) \, dQ.$$

The inequality (4.3) implies

$$S(P|Q) \geq \frac{1}{2} \int_{\Omega} \frac{X^2}{1 + \frac{X}{3}} dQ, \tag{4.4}$$

with the equality iff $P = Q$. Note that

$$\int_{\Omega} \left(1 + \frac{X}{3}\right) dQ = 1,$$

and that Cauchy-Schwarz inequality gives

$$\int_{\Omega} \frac{X^2}{1 + \frac{X}{3}} dQ = \left(\int_{\Omega} \left(1 + \frac{X}{3}\right) dQ\right) \left(\int_{\Omega} \frac{X^2}{1 + \frac{X}{3}} dQ\right) \geq \left(\int_{\Omega} |X| dQ\right)^2$$

$$= d_V(P, Q)^2. \tag{4.5}$$

Combining (4.4) and (4.5) we derive the statement.   □

Exercise 4.1 Prove that the estimate (4.2) is the best possible in the sense that

$$\inf_{P \neq Q} \frac{S(P|Q)}{d_V(P, Q)^2} = \frac{1}{2}.$$

Set

$$\mathcal{A}(\Omega) = \{(P, Q) \mid P, Q \in \mathcal{P}(\Omega), P \ll Q\}. \tag{4.6}$$

One easily verifies that $\mathcal{A}(\Omega)$ is a convex subset of $\mathcal{P}(\Omega) \times \mathcal{P}(\Omega)$. Obviously,

$$\mathcal{A}(\Omega) = \{(P, Q) \mid S(P|Q) < \infty\}.$$

Note also that $\mathcal{P}(\Omega) \times \mathcal{P}_f(\Omega)$ is a dense subset of $\mathcal{A}(\Omega)$.

Proposition 4.3 *The map*

$$\mathcal{A}(\Omega) \ni (P, Q) \mapsto S(P|Q)$$

is continuous, and the map

$$\mathcal{P}(\Omega) \times \mathcal{P}(\Omega) \ni (P, Q) \mapsto S(P|Q) \tag{4.7}$$

is lower semicontinuous.

Exercise 4.2 Prove the above proposition. Show that if $|\Omega| > 1$ and Q is a boundary point of $\mathcal{P}(\Omega)$, then there is a sequence $P_n \to Q$ such that $\lim_{n\to\infty} S(P_n|Q) = \infty$. Hence, the map (4.7) is not continuous except in the trivial case $|\Omega| = 1$.

Proposition 4.4 *The relative entropy is jointly convex: for* $\lambda \in]0, 1[$ *and* $P_1, P_2, Q_1, Q_2 \in \mathcal{P}(\Omega)$,

$$S(\lambda P_1 + (1 - \lambda) P_2 | \lambda Q_1 + (1 - \lambda) Q_2) \le \lambda S(P_1|Q_1) + (1 - \lambda) S(P_2|Q_2). \tag{4.8}$$

Moreover, if the r.h.s. in (4.8) is finite, the equality holds iff for $\omega \in \operatorname{supp} Q_1 \cap \operatorname{supp} Q_2$ *we have* $P_1(\omega)/Q_1(\omega) = P_2(\omega)/Q_2(\omega)$.

Remark 4.5 In particular, if $Q_1 \perp Q_2$ and the r.h.s. in (4.8) is finite, then $P_1 \perp P_2$ and the equality holds in (4.8). On the other hand, if $Q_1 = Q_2 = Q$ and Q is faithful,

$$S(\lambda P_1 + (1 - \lambda) P_2 | Q) \le \lambda S(P_1|Q) + (1 - \lambda) S(P_2|Q).$$

with the equality iff $P_1 = P_2$. An analogous statement holds if $P_1 = P_2 = P$ and P is faithful.

Proof We recall the following basic fact: if $g :]0, \infty[\to \mathbb{R}$ is concave, then the function

$$G(x, y) = xg\left(\frac{y}{x}\right) \tag{4.9}$$

is jointly concave on $]0, \infty[\times]0, \infty[$. Indeed, for $\lambda \in]0, 1[$,

$$G(\lambda x_1 + (1 - \lambda)x_2, \lambda y_1 + (1 - \lambda)y_2)$$
$$= (\lambda x_1 + (1 - \lambda)x_2)g\left(\frac{\lambda x_1}{\lambda x_1 + (1 - \lambda)x_2}\frac{y_1}{x_1} + \frac{(1 - \lambda)x_2}{\lambda x_1 + (1 - \lambda)x_2}\frac{y_2}{x_2}\right)$$
$$\ge \lambda G(x_1, y_1) + (1 - \lambda)G(x_2, y_2),$$

$$\tag{4.10}$$

and if g is strictly concave, the inequality is strict unless $\frac{y_1}{x_1} = \frac{y_2}{x_2}$.

We now turn to the proof. Without loss of generality we may assume that $P_1 \ll Q_1$ and $P_2 \ll Q_2$. One easily shows that then also $\lambda P_1 + (1-\lambda) P_2 \ll \lambda Q_1 + (1-\lambda) Q_2$. For any $\omega \in \Omega$ we have that

$$(\lambda_1 P_1(\omega) + (1-\lambda) P_2(\omega)) \log \frac{\lambda_1 P_1(\omega) + (1-\lambda) P_2(\omega)}{\lambda_1 Q_1(\omega) + (1-\lambda) Q_2(\omega)}$$

$$\leq \lambda P_1(\omega) \log \frac{P_1(\omega)}{Q_1(\omega)} + (1-\lambda) P_2(\omega) \times \log \frac{P_2(\omega)}{Q_2(\omega)}.$$

$$(4.11)$$

To establish this relation, note that if $P_1(\omega) = P_2(\omega) = 0$, then (4.11) holds with the equality. If $P_1(\omega) = 0$ and $P_2(\omega) > 0$, the inequality (4.11) is strict unless $Q_1(\omega) = 0$, and similarly in the case $P_1(\omega) > 0$, $P_2(\omega) = 0$. If $P_1(\omega) > 0$ and $P_2(\omega) > 0$, then taking $g(t) = \log t$ in (4.9) and using the joint concavity of G gives that (4.11) holds and that the inequality is strict unless $P_1(\omega)/Q_1(\omega) = P_2(\omega)/Q_2(\omega)$. Summing (4.11) over ω we derive the statement. The discussion of the cases where the equality holds in (4.8) is simple and is left to the reader. $\qquad\square$

The relative entropy is super-additive in the following sense:

Proposition 4.6 *For any P and $Q = Q_l \otimes Q_r$ in $\mathcal{P}(\Omega_l \times \Omega_r)$,*

$$S(P_l|Q_l) + S(P_r|Q_r) \leq S(P|Q).   \qquad(4.12)$$

Moreover, if the r.h.s. in (4.12) is finite, the equality holds iff $P = P_l \otimes P_r$.

Proof We may assume that $P \ll Q$, in which case one easily verifies that $P_l \ll Q_l$ and $P_r \ll Q_r$. One computes

$$S(P|Q) - S(P_l|Q_l) - S(P_r|Q_r) = S(P_l) + S(P_r) - S(P),$$

and the result follows from Proposition 3.2. $\qquad\square$

In general, for $P, Q \in \mathcal{P}(\Omega_l \times \Omega_r)$ it is *not* true that $S(P|Q) \geq S(P_l|Q_l) + S(P_r|Q_r)$ even if $P = P_l \otimes P_r$.

Exercise 4.3 Find an example of faithful $P = P_l \otimes P_r$, $Q \in \mathcal{P}(\Omega_l \times \Omega_r)$ where $|\Omega_l| = |\Omega_r| = 2$ such that

$$S(P|Q) < S(P_l|Q_l) + S(P_r|Q_r).$$

Let $\Omega = (\omega_1, \cdots, \omega_L)$, $\widehat{\Omega} = \{\hat{\omega}_1, \cdots, \hat{\omega}_{\hat{L}}\}$ be two finite sets. A matrix of real numbers $[\Phi(\omega, \hat{\omega})]_{(\omega, \hat{\omega}) \in \Omega \times \widehat{\Omega}}$ is called *stochastic* if $\Phi(\omega, \hat{\omega}) \geq 0$ for all pairs $(\omega, \hat{\omega})$ and

$$\sum_{\hat{\omega} \in \widehat{\Omega}} \Phi(\omega, \hat{\omega}) = 1$$

for all $\omega \in \Omega$. A stochastic matrix induces a map $\Phi : \mathcal{P}(\Omega) \to \mathcal{P}(\widehat{\Omega})$ by

$$\Phi(P)(\hat{\omega}) = \sum_{\omega \in \Omega} P(\omega) \Phi(\omega, \hat{\omega}).$$

We shall refer to Φ as the *stochastic map* induced by the stochastic matrix $[\Phi(\omega, \hat{\omega})]$. One can interpret the elements of Ω and $\widehat{\Omega}$ as states of two stochastic systems and $P(\omega)$ as probability that the state ω is realized. $\Phi(\omega, \hat{\omega})$ is interpreted as the *transition probability*, i.e. the probability that in a unit of time the system will make a transition from the state ω to the state $\hat{\omega}$. With this interpretation, the probability that the state $\hat{\omega}$ is realized after the transition has taken place is $\Phi(P)(\hat{\omega})$.

Note that if $[\Phi(\omega, \hat{\omega})]_{(\omega, \hat{\omega}) \in \Omega \times \widehat{\Omega}}$ and $[\widehat{\Phi}(\hat{\omega}, \hat{\hat{\omega}})]_{(\hat{\omega}, \hat{\hat{\omega}}) \in \widehat{\Omega} \times \widehat{\widehat{\Omega}}}$ are stochastic matrices, then their product is also stochastic matrix and that the induced stochastic map is $\widehat{\Phi} \circ \Phi$. Another elementary property of stochastic maps is:

Proposition 4.7 $d_V(\Phi(P), \Phi(Q)) \le d_V(P, Q)$.

Exercise 4.4 Prove Proposition 4.7. When the equality holds?

The following result is deeper.

Proposition 4.8

$$S(\Phi(P)|\Phi(Q)) \le S(P|Q). \tag{4.13}$$

Remark 4.9 In information theory, the inequality (4.13) is sometimes called the *data processing inequality*. We shall refer to it as the *stochastic monotonicity*. If the relative entropy is interpreted as a measure of *distinguishability* of two probability measures, then the inequality asserts that probability measures are less distinguishable after an application of a stochastic map.

Proof We start with the so-called *log-sum* inequality: If $a_j, b_j, j = 1, \cdots, M$, are non-negative numbers, then

$$\sum_{j=1}^{M} a_j \log \frac{a_j}{b_j} \ge \sum_{j=1}^{M} a_j \log \frac{\sum_{k=1}^{M} a_k}{\sum_{k=1}^{M} b_k}, \tag{4.14}$$

with the usual convention that $0 \log 0/x = 0$. If $b_j = 0$ and $a_j > 0$ for some j, then l.h.s is ∞ and there is nothing to prove. If $a_j = 0$ for all j again there is nothing to prove. Hence, without loss of generality we may assume that $\sum_j a_j > 0$, $\sum b_j > 0$, and $b_j = 0 \Rightarrow a_j = 0$. Set $p = (p_1, \cdots, p_M)$, $p_k = a_k / \sum_j a_j$,

$q = (q_1, \cdots, q_M)$, $q_k = b_k / \sum_j b_j$. Then the inequality (4.14) is equivalent to

$$S(p|q) \geq 0.$$

This observation and Proposition 4.1 prove (4.14).

We now turn to the proof. Clearly, we need only to consider the case $P \ll Q$. Then

$$S(\Phi(P)|\Phi(Q)) = \sum_{\hat\omega \in \hat\Omega} \Phi(P)(\hat\omega) \log \frac{\Phi(P)(\hat\omega)}{\Phi(Q)(\hat\omega)}$$

$$= \sum_{\hat\omega \in \hat\Omega} \sum_{\omega \in \Omega} P(\omega)\Phi(\omega, \hat\omega) \log \frac{\sum_{\omega' \in \Omega} P(\omega')\Phi(\omega', \hat\omega)}{\sum_{\omega' \in \Omega} Q(\omega')\Phi(\omega', \hat\omega)}$$

$$\leq \sum_{\hat\omega \in \hat\Omega} \sum_{\omega \in \Omega} P(\omega)\Phi(\omega, \hat\omega) \log \frac{P(\omega)}{Q(\omega)}$$

$$= S(P|Q),$$

where the third step follows from the log-sum inequality. $\square$

Exercise 4.5 A stochastic matrix $[\Phi(\omega, \hat\omega)]$ is called doubly stochastic if

$$\sum_{\omega \in \Omega} \Phi(\omega, \hat\omega) = \frac{|\Omega|}{|\hat\Omega|}$$

for all $\hat\omega \in \hat\Omega$. Prove that $S(P) \leq S(\Phi(P))$ for all $P \in \mathcal{P}(\Omega)$ iff $[\Phi(\omega, \hat\omega)]$ is doubly stochastic.

Hint: Use that $\Phi(P_{ch}) = \hat{P}_{ch}$ iff $[\Phi(\omega, \hat\omega)]$ is doubly stochastic.

Exercise 4.6 Suppose that $\Omega = \hat\Omega$. Let $\gamma = \min_{(\omega_1, \omega_2)} \Phi(\omega_1, \omega_2)$ and suppose that $\gamma > 0$.

1. Show that $S(\Phi(P)|\Phi(Q)) = S(P|Q)$ iff $P = Q$.
2. Show that

$$d_V(\Phi(P), \Phi(Q)) \leq (1 - \gamma)d_V(P, Q).$$

3. Using Part 2 show that there exists unique probability measure $\overline{Q}$ such that $\Phi(\overline{Q}) = \overline{Q}$. Show that $\overline{Q}$ is faithful and that for any $P \in \mathcal{P}(\Omega)$,

$$d_V(\Phi^n(P), \overline{Q}) \leq (1 - \gamma)^n d_V(P, \overline{Q}),$$

where $\Phi^2 = \Phi \circ \Phi$, etc.

Hint: Follow the proof of the Banach fixed point theorem.

Exercise 4.7 The stochastic monotonicity yields the following elegant proof of Theorem 4.2.

1. Let $P, Q \in \mathcal{P}(\Omega)$ be given, where $|\Omega| \geq 2$. Let $T = \{\omega : P(\omega) \geq Q(\omega)\}$ and

$$p = (p_1, p_2) = (P(T), P(T^c)), \qquad q = (q_1, q_2) = (Q(T), Q(T^c)),$$

   be probability measures on $\widehat{\Omega} = \{1, 2\}$. Find a stochastic map $\Phi : \mathcal{P}(\Omega) \to \mathcal{P}(\widehat{\Omega})$ such that $\Phi(P) = p$, $\Phi(Q) = q$.
2. Since $S(P|Q) \geq S(p|q)$ and $d_V(P, Q) = d_V(p, q)$, observe that to prove Theorem 4.2 it suffices to show that for all $p, q \in \mathcal{P}(\widehat{\Omega})$,

$$S(p|q) \geq \frac{1}{2} d_V(p, q)^2. \tag{4.15}$$

3. Show that (4.15) is equivalent to the inequality

$$x \log \frac{x}{y} + (1 - x) \log \frac{1 - x}{1 - y} \geq 2(x - y)^2, \tag{4.16}$$

   where $0 \leq y \leq x \leq 1$. Complete the proof by establishing (4.16).

Hint: Fix $x > 0$ and consider the function

$$F(y) = x \log \frac{x}{y} + (1 - x) \log \frac{1 - x}{1 - y} - 2(x - y)^2$$

on $]0, x]$. Since $F(x) = 0$, it suffices to show that $F'(y) \leq 0$ for $y \in]0, x[$. Direct computation gives $F'(y) \leq 0 \Leftrightarrow y(1 - y) \leq \frac{1}{4}$ and the statement follows.

The log-sum inequality used in the proof Proposition 4.8 leads to the following refinement of Proposition 4.4.

Proposition 4.10 *Let* $P_1, \cdots, P_n, Q_1, \cdots, Q_n \in \mathcal{P}(\Omega)$ *and* $p = (p_1, \cdots, p_n)$, $q = (q_1, \cdots, q_n) \in \mathcal{P}_n$. *Then*

$$S(p_1 P_1 + \cdots + p_n P_n | q_1 Q_1 + \cdots + q_n Q_n) \leq p_1 S(P_1|Q_1) + \cdots + p_n S(P_n|Q_n) + S(p|q). \tag{4.17}$$

If the r.h.s. in (4.17) *is finite, then the equality holds iff for all* j, k *such that* $q_j > 0, q_k > 0$,

$$\frac{p_j P_j(\omega)}{q_j Q_j(\omega)} = \frac{p_k P_k(\omega)}{q_k Q_k(\omega)}$$

holds for all $\omega \in \operatorname{supp} Q_k \cap \operatorname{supp} Q_j$.

Exercise 4.8 Deduce Proposition 4.10 from the log-sum inequality.

4.2   Variational Principles

The relative entropy is characterized by the following variational principle.

Proposition 4.11

$$S(P|Q) = \sup_{X:\Omega \to \mathbb{R}} \left(\int_\Omega X \mathrm{d}P - \log \int_{\mathrm{supp}P} e^X \mathrm{d}Q \right). \tag{4.18}$$

If $S(P|Q) < \infty$, then the supremum is achieved, and each maximizer is equal to $S_{P|Q} + $ const on $\mathrm{supp}P$ and is arbitrary otherwise.

Proof Suppose that $Q(\omega_0) = 0$ and $P(\omega_0) > 0$ for some $\omega_0 \in \Omega$. Set $X_n(\omega) = n$ if $\omega = \omega_0$ and zero otherwise. Then

$$\int_\Omega X_n \mathrm{d}P = n P(\omega_0), \qquad \int_{\mathrm{supp}P} e^{X_n} \mathrm{d}Q = Q(\mathrm{supp}P).$$

Hence, if P is not absolutely continuous w.r.t. Q the relation (4.18) holds since both sides are equal to ∞.

Suppose now that $P \ll Q$. For given $X : \Omega \to \mathbb{R}$ set

$$Q_X(\omega) = \frac{e^{X(\omega)} Q(\omega)}{\sum_{\omega' \in \mathrm{supp}P} e^{X(\omega')} Q(\omega')}$$

if $\omega \in \mathrm{supp}P$ and zero otherwise. $Q_X \in \mathcal{P}(\Omega)$ and

$$S(P|Q_X) = S(P|Q) - \left(\int_\Omega X \mathrm{d}P - \log \int_{\mathrm{supp}P} e^X \mathrm{d}Q \right).$$

Hence,

$$S(P|Q) \geq \int_\Omega X \mathrm{d}P - \log \int_{\mathrm{supp}P} e^X \mathrm{d}Q$$

with equality iff $P = Q_X$. Obviously, $P = Q_X$ iff $X = S_{P|Q} + $ const on $\mathrm{supp}P$ and is arbitrary otherwise. $\square$

Exercise 4.9 Show that

$$S(P|Q) = \sup_{X:\Omega \to \mathbb{R}} \left(\int_\Omega X \mathrm{d}P - \log \int_\Omega e^X \mathrm{d}Q \right). \tag{4.19}$$

When is the supremum achieved? Use (4.19) to prove that the map $(P, Q) \mapsto S(P|Q)$ is jointly convex.

Proposition 4.12 *The following dual variational principle holds: for $X : \Omega \to \mathbb{R}$ and $Q \in \mathcal{P}(\Omega)$,*

$$\log \int_\Omega e^X dQ = \max_{P \in \mathcal{P}(\Omega)} \left(\int_\Omega X dP - S(P|Q) \right).$$

The maximizer is unique and is given by

$$P_{X,Q}(\omega) = \frac{e^{X(\omega)} Q(\omega)}{\sum_{\omega' \in \Omega} e^{X(\omega')} Q(\omega')}.$$

Proof For any $P \ll Q$,

$$\log \int_\Omega e^X dQ - \int_\Omega X dP + S(P|Q) = S(P|P_{X,Q}),$$

and the result follows from Proposition 4.1.   □

Setting $Q = P_{\mathrm{ch}}$ in Propositions 4.11 and 4.12, we derive the variational principle for entropy and the respective dual variational principle.

Proposition 4.13

(1)

$$S(P) = \inf_{X:\Omega \to \mathbb{R}} \left(\log \left(\sum_{\omega \in \Omega} e^{X(\omega)} \right) - \int_\Omega X dP \right).$$

The infimum is achieved if P is faithful and $X = -S_P + \text{const.}$
(2) For any $X : \Omega \to \mathbb{R}$,

$$\log \left(\sum_{\omega \in \Omega} e^{X(\omega)} \right) = \max_{P \in \mathcal{P}(\Omega)} \left(\int_\Omega X dP + S(P) \right).$$

The maximizer is unique and is given by

$$P(\omega) = \frac{e^{X(\omega)}}{\sum_{\omega' \in \Omega} e^{X(\omega')}}.$$

4.3   Stein's Lemma

Let $P, Q \in \mathcal{P}(\Omega)$ and let P_N, Q_N be the induced product probability measures on Ω^N. For $\gamma \in]0, 1[$ the Stein exponents are defined by

$$s_N(\gamma) = \min \left\{ Q_N(T) \mid T \subset \Omega^N, \; P_N(T) \geq \gamma \right\}. \tag{4.20}$$

The following result is often called *Stein's Lemma*.

Theorem 4.14

$$\lim_{N \to \infty} \frac{1}{N} \log s_N(\gamma) = -S(P|Q).$$

Remark 4.15 If $Q = P_{\text{ch}}$, then Stein's Lemma reduces to Proposition 3.3. In fact, the proofs of the two results are very similar.

Proof We deal first with the case $S(P|Q) < \infty$. Set $S_{P|Q}(\omega) = 0$ for $\omega \notin \text{supp} P$ and

$$\mathcal{S}_N(\omega = (\omega_1, \cdots, \omega_N)) = \sum_{j=1}^{N} S_{P|Q}(\omega_j).$$

For given $\epsilon > 0$ let

$$R_{N,\epsilon} = \left\{ \omega \in \Omega^N \mid \frac{\mathcal{S}_N(\omega)}{N} \geq S(P|Q) - \epsilon \right\}.$$

By the LLN,

$$\lim_{N \to \infty} P_N(R_{N,\epsilon}) = 1,$$

and so for N large enough, $P_N(R_{N,\epsilon}) \geq \gamma$. We also have

$$Q_N(R_{N,\epsilon}) = Q_N \left\{ e^{\mathcal{S}_N(\omega)} \geq e^{NS(P|Q) - N\epsilon} \right\} \leq e^{N\epsilon - NS(P|Q)} \mathbb{E}_{Q_N}(e^{\mathcal{S}_N}).$$

Since

$$\mathbb{E}_{Q_N}(e^{\mathcal{S}_N}) = \left(\int_\Omega \Delta_{P|Q} dQ \right)^N = 1,$$

we derive

$$\limsup_{N\to\infty} \frac{1}{N} \log s_N(\gamma) \le -S(P|Q) + \epsilon.$$

Since $\epsilon > 0$ is arbitrary,

$$\limsup_{N\to\infty} \frac{1}{N} \log s_N(\gamma) \le -S(P|Q).$$

To prove the lower bound, let $U_{N,\gamma}$ be the set for which the minimum in (4.20) is achieved. Let $\epsilon > 0$ be given and let

$$D_{N,\epsilon} = \left\{ \omega \in \Omega^N \ \middle| \ \frac{S_N(\omega)}{N} \le S(P|Q) + \epsilon \right\}.$$

Again, by the LLN,

$$\lim_{N\to\infty} P_N(D_{N,\epsilon}) = 1,$$

and so for N large enough, $P_N(D_{N,\epsilon}) \ge \gamma$. We then have

$$P_N(U_{N,\gamma} \cap D_{N,\epsilon}) = \int_{U_{N,\gamma} \cap D_{N,\epsilon}} \Delta_{P_N|Q_N} dQ_N = \int_{U_{N,\gamma} \cap D_{N,\epsilon}} e^{S_N} dQ_N$$

$$\le e^{NS(P|Q)+N\epsilon} Q_N(U_{N,\gamma} \cap D_{N,\epsilon})$$

$$\le e^{NS(P|Q)+N\epsilon} Q_N(U_{N,\gamma}).$$

Since

$$\liminf_{N\to\infty} P_N(U_{N,\gamma} \cap D_{N,\epsilon}) \ge \gamma,$$

we have

$$\liminf_{N\to\infty} \frac{1}{N} s_N(\gamma) \ge -S(P|Q) - \epsilon.$$

Since $\epsilon > 0$ is arbitrary,

$$\liminf_{N\to\infty} \frac{1}{N} s_N(\gamma) \ge -S(P|Q).$$

This proves Stein's Lemma in the case $S(P|Q) < \infty$.

We now deal with the case $S(P|Q) = \infty$. For $0 < \delta < 1$ set $Q_\delta = (1-\delta)Q + \delta P$. Obviously, $S(P|Q_\delta) < \infty$. Let $s_{N,\delta}(\gamma)$ be the Stein exponent of the pair (P, Q_δ). Then

$$s_{N,\delta}(\gamma) \geq (1 - \delta)^N s_N(\gamma),$$

and

$$-S(P|Q_\delta) = \lim_{N\to\infty} \frac{1}{N} \log s_{N,\delta}(\gamma) \geq \log(1 - \delta) + \liminf_{N\to\infty} \frac{1}{N} \log s_N(\gamma).$$

The lower semicontinuity of relative entropy gives $\lim_{\delta\to 0} S(P|Q_\delta) = \infty$, and so

$$\lim_{N\to\infty} \frac{1}{N} \log s_N(\gamma) = \infty = -S(P|Q).$$

$\square$

Exercise 4.10 Prove the following variant of Stein's Lemma. Let

$$\underline{s} = \inf_{(T_N)} \left\{ \liminf_{N\to\infty} \frac{1}{N} Q_N(T_N) \mid \lim_{N\to\infty} P_N(T_N^c) = 0 \right\},$$

$$\overline{s} = \inf_{(T_N)} \left\{ \limsup_{N\to\infty} \frac{1}{N} Q_N(T_N) \mid \lim_{N\to\infty} P_N(T_N^c) = 0 \right\},$$

where the infimum is taken over all sequences $(T_N)_{N\geq 1}$ of sets such that $T_N \subset \Omega^N$ for all $N \geq 1$. Then

$$\underline{s} = \overline{s} = -S(P|Q).$$

4.4 Fluctuation Relation

Let Ω be a finite set and $P \in \mathcal{P}_f(\Omega)$. Let $\Theta : \Omega \to \Omega$ be a bijection such that

$$\Theta^2(\omega) = \Theta \circ \Theta(\omega) = \omega \tag{4.21}$$

for all ω. We set $P_\Theta(\omega) = P(\Theta(\omega))$. Obviously, $P_\Theta \in \mathcal{P}_f(\Omega)$. The relative entropy function

$$S_{P|P_\Theta}(\omega) = \log \frac{P(\omega)}{P_\Theta(\omega)}$$

satisfies

$$S_{P|P_\Theta}(\Theta(\omega)) = -S_{P|P_\Theta}(\omega), \tag{4.22}$$

and so the set of values of $S_{P|P_\Theta}$ is symmetric with respect to the origin. On the other hand,

$$S(P|P_\Theta) = \mathbb{E}_P(S(P|P_\Theta)) \geq 0$$

with equality iff $P = P_\Theta$. Thus, the probability measure P "favours" positive values of $S_{P|P_\Theta}$. Proposition 4.16 below is a refinement of this observation.

Let Q be the probability distribution of the random variable $S(P|P_\Theta)$ w.r.t. P. We recall that Q is defined by

$$Q(s) = P\left\{\omega \mid S_{P|P_\Theta}(\omega) = s\right\}.$$

Obviously, $Q(s) \neq 0$ iff $Q(-s) \neq 0$.

The following result is known as the *fluctuation relation*.

Proposition 4.16 *For all s,*

$$Q(-s) = e^{-s} Q(s).$$

Proof For any α,

$$\mathbb{E}_P\left(e^{-\alpha S_{P|P_\Theta}}\right) = \sum_{\omega \in \Omega} [P_\Theta(\omega)]^\alpha [P(\omega)]^{1-\alpha}$$

$$= \sum_{\omega \in \Omega} [P_\Theta(\Theta(\omega))]^\alpha [P(\Theta(\omega))]^{1-\alpha}$$

$$= \sum_{\omega \in \Omega} [P(\omega)]^\alpha [P_\Theta(\omega)]^{1-\alpha}$$

$$= \mathbb{E}_P\left(e^{-(1-\alpha)S_{P|P_\Theta}}\right).$$

Hence, if $S = \{s \mid Q(s) \neq 0\}$,

$$\sum_{s \in S} e^{-\alpha s} Q(s) = \sum_{s \in S} e^{-(1-\alpha)s} Q(s) = \sum_{s \in S} e^{(1-\alpha)s} Q(-s),$$

and so

$$\sum_{s \in S} e^{-\alpha s} (Q(s) - e^s Q(-s)) = 0. \tag{4.23}$$

Since (4.23) holds for all real α, we must have that $Q(s) - e^s Q(-s) = 0$ for all $s \in \mathcal{S}$, and the statement follows. $\square$

Remark 4.17 The assumption that P is faithful can be omitted if one assumes in addition that Θ preserves $\mathrm{supp} P$. If this is the case, one can replace Ω with $\mathrm{supp} P$, and the above proof applies.

Exercise 4.11 Prove that the fluctuation relation implies (4.22).

Exercise 4.12 This exercise is devoted to a generalization of the fluctuation relation which has also found fundamental application in physics. Consider a family $\{P_X\}_{X \in \mathbb{R}^n}$ of probability measures on Ω indexed by vectors $X = (X_1, \cdots, X_n) \in \mathbb{R}^n$. Set

$$\mathcal{E}_X(\omega) = \log \frac{P_X(\omega)}{P_X(\Theta_X(\omega))},$$

where Θ_X satisfies (4.21). Suppose that $\mathcal{E}_0 = 0$ and consider a decomposition

$$\mathcal{E}_X = \sum_{k=1}^n X_k \mathfrak{F}_{X,k}, \tag{4.24}$$

where the random variables $\mathfrak{F}_{X,k}$ satisfy

$$\mathfrak{F}_{X,k} \circ \Theta_X = -\mathfrak{F}_{X,k}. \tag{4.25}$$

We denote by $\mathcal{Q}_X$ the probability distribution of the vector random variable $(\mathfrak{F}_{X,1}, \cdots, \mathfrak{F}_{X,n})$ with respect to P_X: for $s = (s_1, \cdots, s_n) \in \mathbb{R}^n$,

$$\mathcal{Q}_X(s) = P_X \left\{ \omega \in \Omega \mid \mathcal{F}_{X,1} = s_1, \cdots, \mathcal{F}_{X,n} = s_n \right\}.$$

We also denote $\mathcal{S} = \{s \in \mathbb{R}^n \mid \mathcal{Q}_X(s) \neq 0\}$ and, for $Y = (Y_1, \cdots, Y_n) \in \mathbb{R}^n$, set

$$G(X, Y) = \sum_{s \in \mathcal{S}} e^{-\sum_k s_k Y_k} \mathcal{Q}_X(s).$$

1. Prove that a decomposition (4.24) satisfying (4.25) always exists and that, except in trivial cases, is never unique.
2. Prove that $\mathcal{Q}_X(s) \neq 0$ iff $\mathcal{Q}_X(-s) \neq 0$.
3. Prove that

$$G(X, Y) = G(X, X - Y).$$

4. Prove that

$$\mathcal{Q}_X(-s) = e^{-\sum_k s_k X_k} \mathcal{Q}_X(s).$$

4.5 Jensen-Shannon Entropy and Metric

The Jensen-Shannon entropy of two probability measures $P, Q \in \mathcal{P}(\Omega)$ is

$$S_{JS}(P|Q) = S(M(P, Q)) - \frac{1}{2}S(P) - \frac{1}{2}S(Q)$$

$$= \frac{1}{2}\left(S\left(P|M(P, Q)\right) + S\left(Q|M(P, Q)\right)\right),$$

where

$$M(P, Q) = \frac{P + Q}{2}.$$

The Jensen-Shannon entropy can be viewed as a measure of concavity of the entropy. Obviously, $S_{JS}(P|Q) \geq 0$ with equality iff $P = Q$. In addition:

Proposition 4.18

(1)

$$S_{JS}(P|Q) \leq \log 2,$$

with equality iff $P \perp Q$.
(2)

$$\frac{1}{8}d_V(P, Q)^2 \leq S_{JS}(P|Q) \leq d_V(P, Q) \log\sqrt{2}.$$

The first inequality is saturated iff $P = Q$ and the second iff $P = Q$ or $P \perp Q$.

Proof Part (1) follows from

$$S_{JS}(P|Q) = \frac{1}{2}\sum_{\omega \in \Omega}\left(P(\omega)\log\left(\frac{2P(\omega)}{P(\omega) + Q(\omega)}\right) + Q(\omega)\log\left(\frac{2Q(\omega)}{P(\omega) + Q(\omega)}\right)\right)$$

$$\leq \frac{1}{2}\sum_{\omega \in \Omega}(P(\omega) + Q(\omega))\log 2$$

$$= \log 2.$$

To prove (2), we start with the lower bound:

$$S_{JS}(P|Q) = \frac{1}{2}S(P|M(P, Q)) + \frac{1}{2}S(Q|M(P|Q))$$

$$\geq \frac{1}{4}d_V(P, M(P, Q))^2 + \frac{1}{4}d_V(Q, M(P, Q))^2$$

$$= \frac{1}{8}\left(\sum_{\omega \in \Omega} |P(\omega) - Q(\omega)|\right)^2 = \frac{1}{8}d_V(P|Q)^2,$$

where the inequality follows from Theorem 4.2.

To prove the upper bound, set $S_+ = \{\omega \mid P(\omega) \geq Q(\omega)\}$, $S_- = \{\omega \mid P(\omega) < Q(\omega)\}$. Then

$$S_{JS}(P|Q) = \frac{1}{2}\sum_{\omega \in S_+}\left(P(\omega)\log\left(\frac{2P(\omega)}{P(\omega) + Q(\omega)}\right) - Q(\omega)\log\left(\frac{P(\omega) + Q(\omega)}{2Q(\omega)}\right)\right)$$

$$+ \frac{1}{2}\sum_{\omega \in S_-}\left(Q(\omega)\log\left(\frac{2Q(\omega)}{P(\omega) + Q(\omega)}\right)\right.$$

$$\left. -P(\omega)\log\left(\frac{P(\omega) + Q(\omega)}{2P(\omega)}\right)\right)$$

$$\leq \frac{1}{2}\sum_{\omega \in S_+}(P(\omega) - Q(\omega))\log\left(\frac{2P(\omega)}{P(\omega) + Q(\omega)}\right)$$

$$+ \frac{1}{2}\sum_{\omega \in S_-}(Q(\omega) - P(\omega))\log\left(\frac{2Q(\omega)}{P(\omega) + Q(\omega)}\right)$$

$$\leq \frac{1}{2}\sum_{\omega \in S_-}(P(\omega) - Q(\omega))\log 2 + \frac{1}{2}\sum_{\omega \in S_-}(Q(\omega) - P(\omega))\log 2$$

$$= d_V(P, Q)\log\sqrt{2}.$$

In the first inequality we have used that for $P(\omega) \neq 0$ and $Q(\omega) \neq 0$,

$$\frac{P(\omega) + Q(\omega)}{2P(\omega)} \geq \frac{2Q(\omega)}{P(\omega) + Q(\omega)},$$

and the same inequality with P and Q interchanged.

The cases where equality holds in Parts (1) and (2) are easily identified from the above argument and we leave the formal proof as an exercise for the reader. □

Set

$$d_{JS}(P, Q) = \sqrt{S_{JS}(P, Q)}.$$

Theorem 4.19 d_{JS} is a metric on $\mathcal{P}(\Omega)$.

Remark 4.20 If $|\Omega| \geq 2$, then S_{JS} is not a metric on $\mathcal{P}(\Omega)$. To see that, pick $\omega_1, \omega_2 \in \Omega$ and define $P, Q, R \in \mathcal{P}(\Omega)$ by $P(\omega_1) = 1$, $Q(\omega_2) = 1$, $R(\omega_1) = R(\omega_2) = \frac{1}{2}$. Then

$$S_{JS}(P|Q) = \log 2 > \frac{3}{2} \log \frac{4}{3} = S_{JS}(P|R) + S_{JS}(R|Q).$$

Remark 4.21 In the sequel we shall refer to d_{SJ} as the *Jensen-Shannon* metric.

Proof Note that only the triangle inequality needs to be proved. Set $\mathbb{R}_+ =]0, \infty[$. For $p, q \in \mathbb{R}_+$ let

$$L(p, q) = p \log \left(\frac{2p}{p+q} \right) + q \log \left(\frac{2q}{p+q} \right).$$

Since the function $F(x) = x \log x$ is strictly convex, writing

$$L(p, q) = (p+q) \left[\frac{1}{2} F \left(\frac{2p}{p+q} \right) + \frac{1}{2} F \left(\frac{2q}{p+q} \right) \right]$$

and applying the Jensen inequality to the expression in the brackets, we derive that $L(p, q) \geq 0$ with equality iff $p = q$. Our goal is to prove that for all $p, q, r \in \mathbb{R}_+$,

$$L(p, q) \leq \sqrt{L(p, r)} + \sqrt{L(r, q)}. \tag{4.26}$$

This yields the triangle inequality for d_{JS} as follows. If $P, Q, R \in \mathcal{P}_f(\Omega)$, (4.26) and Minkowski's inequality give

$$d_{JS}(P, Q) = \left(\sum_{\omega \in \Omega} \sqrt{L(P(\omega), Q(\omega))}^2 \right)^{\frac{1}{2}}$$

$$\leq \left(\sum_{\omega \in \Omega} \left(\sqrt{L(P(\omega), R(\omega))} + \sqrt{L(R(\omega), Q(\omega))} \right)^2 \right)^{\frac{1}{2}}$$

$$\leq \left(\sum_{\omega \in \Omega} \sqrt{L(P(\omega), R(\omega))}^2 \right)^{\frac{1}{2}} + \left(\sum_{\omega \in \Omega} \sqrt{L(R(\omega), Q(\omega))}^2 \right)^{\frac{1}{2}}$$

$$= d_{JS}(P, R) + d_{JS}(R, Q).$$

This yields the triangle inequality on $\mathcal{P}_f(\Omega)$. Since the map $(P, Q) \mapsto d_{JS}(P, Q)$ is continuous, the triangle inequality extends to $\mathcal{P}(\Omega)$.

The proof of (4.26) is an elaborate calculus exercise. The relation is obvious if $p = q$. Since $L(p, q) = L(q, p)$, it suffices to consider the case $p < q$. We fix such p and q and set

$$f(r) = \sqrt{L(p, r)} + \sqrt{L(r, q)}.$$

Then

$$f'(r) = \frac{1}{2\sqrt{L(p, r)}} \log\left(\frac{2r}{p + r}\right) + \frac{1}{2\sqrt{L(r, q)}} \log\left(\frac{2r}{r + q}\right).$$

Define $g : \mathbb{R}_+ \setminus \{1\} \to \mathbb{R}$ by

$$g(x) = \frac{1}{\sqrt{L(x, 1)}} \log\left(\frac{2}{x + 1}\right),$$

One easily verifies that

$$f'(r) = \frac{1}{2\sqrt{r}}\left(g\left(\frac{p}{r}\right) + g\left(\frac{q}{r}\right)\right). \tag{4.27}$$

We shall need the following basic properties of g, clearly displayed in the graph below:

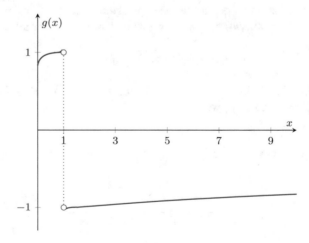

(a) $g > 0$ on $]0, 1[$, $g < 0$ on $]1, \infty[$.
(b) $\lim_{x \uparrow 1} g(x) = 1$, $\lim_{x \downarrow 1} g(x) = -1$. This follows from $\lim_{x \to 1}[g(x)]^2 = 1$, which can be established by applying l'Hopital's rule twice.

(c) $g'(x) > 0$ for $x \in \mathbb{R}_+ \setminus \{1\}$. To prove this one computes

$$g'(x) = -\frac{h(x)}{(x+1)L(x,1)^{3/2}},$$

where

$$h(x) = 2x \log\left(\frac{2x}{x+1}\right) + 2 \log\left(\frac{2}{x+1}\right) + (x+1) \log\left(\frac{2x}{x+1}\right) \log\left(\frac{2}{x+1}\right).$$

One further computes

$$h'(x) = \log\left(\frac{2x}{x+1}\right) \log\left(\frac{2}{x+1}\right) + \log\left(\frac{2x}{x+1}\right) + \frac{1}{x} \log\left(\frac{2}{x+1}\right),$$

$$h''(x) = -\frac{1}{x+1} \log\left(\frac{2x}{x+1}\right) - \frac{1}{x^2(x+1)} \log\left(\frac{2}{x+1}\right).$$

Note that $h(1) = h'(1) = h''(1) = 0$. The inequality $\log t \geq (t-1)/t$, which holds for all $t > 0$, gives

$$h''(x) \leq -\frac{1}{x+1}\left(1 - \frac{x+1}{2x}\right) - \frac{1}{x^2(x+1)}\left(1 - \frac{x+1}{2}\right) = -\frac{(x-1)^2}{2x^2(x+1)}.$$

Hence $h''(x) < 0$ for $x \in \mathbb{R}_+ \setminus \{1\}$, and the statement follows.

(d) Note that (a), (b), and (c) give that $0 < g(x) < 1$ on $]0, 1[$ and $-1 < g(x) < 0$ on $]1, \infty[$.

If follows from (a) that $f'(r) < 0$ for $r \in]0, p[$, $f'(r) > 0$ for $r > q$, and so $f(r)$ is decreasing on $]0, p[$ and increasing on $]q, \infty[$. Hence, for $r < p$ and $r > q$, $f(r) > f(p)$, which gives (4.26) for those r's. To deal with the case $p < r < q$, set $m(r) = g(p/r) + g(q/r)$. It follows from (b) that $m'(r) < 0$ for $p < r < q$, while (b) and (d) give $m(p+) = 1 + g(q/p) > 0$, $m(q-) = -1 + g(p/q) < 0$. Hence $f'(r)$ has precisely one zero r_m in the interval $]p, q[$. Since $f'(p+) > 0$, $f'(q-) > 0$, $f(r)$ is increasing in $[p, r_m]$ and decreasing on $[r_m, q]$. On the first interval, $f(r) \geq f(p)$, and on the second interval $f(r) \geq f(q)$, which gives that (4.26) also holds for $p < r < q$.   □

The graph of $r \mapsto f(r)$ is plotted below for $p = \frac{1}{10}$ and $q = \frac{2}{3}$. In this case $r_m \approx 0.28$.

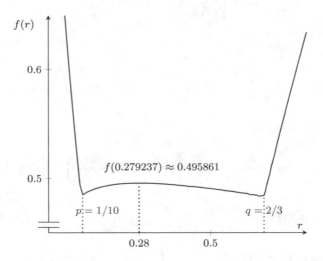

4.6 Rényi's Relative Entropy

Let Ω be a finite set and $P, Q \in \mathcal{P}(\Omega)$. For $\alpha \in]0, 1[$ we set

$$S_\alpha(P|Q) = \frac{1}{\alpha - 1} \log \left(\sum_{\omega \in \Omega} P(\omega)^\alpha Q(\omega)^{1-\alpha} \right).$$

$S_\alpha(P|Q)$ is called Rényi's relative entropy of P with respect to Q. Note that

$$S_\alpha(P|P_{\text{ch}}) = S_\alpha(P) + \log |\Omega|.$$

Proposition 4.22

(1) $S_\alpha(P|Q) \geq 0$.
(2) $S_\alpha(P|Q) = \infty$ iff $P \perp Q$ and $S_\alpha(P|Q) = 0$ iff $P = Q$.
(3)

$$S_\alpha(P|Q) = \frac{\alpha}{1 - \alpha} S_{1-\alpha}(Q|P).$$

(4)

$$\lim_{\alpha \uparrow 1} S_\alpha(P|Q) = S(P|Q).$$

(5) Suppose that $P \not\perp Q$. Then the function $]0, 1[\ni \alpha \mapsto S_\alpha(P|Q)$ is strictly increasing

(6) The map $(P, Q) \mapsto S_\alpha(P|Q) \in [0, \infty]$ is continuous and jointly convex.

(7) Let $\Phi : \mathcal{P}(\Omega) \to \mathcal{P}(\hat{\Omega})$ be a stochastic map. Then for all $P, Q \in \mathcal{P}(\Omega)$,

$$S_\alpha(\Phi(P)|\Phi(Q)) \le S_\alpha(P|Q).$$

(8) If $S(P|Q) < \infty$, then $\alpha \mapsto S_\alpha(P|Q)$ extends to a real-analytic function on $\mathbb{R}$.

Proof Obviously, $S_\alpha(P|Q) = \infty$ iff $P \perp Q$. In what follows, if $P \not\perp Q$, we set

$$T = \operatorname{supp} P \cap \operatorname{supp} Q.$$

An application of Jensen's inequality gives

$$\sum_{\omega \in \Omega} P(\omega)^\alpha Q(\omega)^{1-\alpha} = Q(T) \sum_{\omega \in T} \left(\frac{P(\omega)}{Q(\omega)} \right)^\alpha \frac{Q(\omega)}{Q(T)}$$

$$\le Q(T) \left(\sum_{\omega \in T} \frac{P(\omega)}{Q(\omega)} \frac{Q(\omega)}{Q(T)} \right)^\alpha$$

$$= Q(T)^{1-\alpha} P(T)^\alpha.$$

Hence, $\sum_{\omega \in \Omega} P(\omega)^\alpha Q(\omega)^{1-\alpha} \le 1$ with the equality iff $P = Q$, and Parts (1), (2) follow.

Part (3) is obvious. To prove (4), note that

$$\lim_{\alpha \uparrow 1} \sum_\omega P(\omega)^\alpha Q(\omega)^{1-\alpha} = P(T),$$

and that $P(T) = 1$ iff $P \ll Q$. Hence, if P is not absolutely continuous with respect to Q, then $\lim_{\alpha \uparrow 1} S_\alpha(P|Q) = \infty = S(P|Q)$. If $P \ll Q$, an application of L'Hopital rule gives $\lim_{\alpha \uparrow 1} S_\alpha(P|Q) = S(P|Q)$.

To prove (5), set

$$F(\alpha) = \log \left(\sum_{\omega \in \Omega} P(\omega)^\alpha Q(\omega)^{1-\alpha} \right),$$

and note that $\mathbb{R} \ni \alpha \mapsto F(\alpha)$ is a real-analytic strictly convex function satisfying $F(0) \le 0$, $F(1) \le 0$. We have

$$\frac{dS_\alpha(P|Q)}{d\alpha} = \frac{F'(\alpha)(\alpha - 1) - (F(\alpha) - F(1))}{(\alpha - 1)^2} - \frac{F(1)}{(\alpha - 1)^2}.$$

By the mean-value theorem, $F(\alpha) - F(1) = (\alpha - 1)F'(\zeta_\alpha)$ for some $\zeta_\alpha \in]\alpha, 1[$. Since F' is strictly increasing, $F'(\alpha) < F'(\zeta_\alpha)$ and

$$\frac{dS_\alpha(P|Q)}{d\alpha} > 0$$

for $\alpha \in]0, 1[$.

The continuity part of (6) is obvious. The proof of the joint convexity is the same as the proof of Proposition 4.4 (one now takes $g(t) = t^\alpha$) and is left as an exercise for the reader.

We now turn to Part (7). First, we have

$$[\Phi(P)(\hat{\omega})]^\alpha [\Phi(Q)(\hat{\omega})]^{1-\alpha} \geq \sum_\omega P(\omega)^\alpha Q(\omega)^{1-\alpha} \Phi(\omega, \hat{\omega}).$$

This inequality is obvious if the r.h.s. is equal to zero. Otherwise, let

$$R = \{\omega \mid P(\omega)Q(\omega)\Phi(\omega, \hat{\omega}) > 0\}.$$

Then

$$[\Phi(P)(\hat{\omega})]^\alpha [\Phi(Q)(\hat{\omega})]^{1-\alpha} \geq \left(\sum_{\omega \in R} P(\omega)\Phi(\omega, \hat{\omega})\right)^\alpha \left(\sum_{\omega \in R} Q(\omega)\Phi(\omega, \hat{\omega})\right)^{1-\alpha}$$

$$= \left(\frac{\sum_{\omega \in R} P(\omega)\Phi(\omega, \hat{\omega})}{\sum_{\omega \in R} Q(\omega)\Phi(\omega, \hat{\omega})}\right)^\alpha \sum_{\omega \in R} Q(\omega)\Phi(\omega, \hat{\omega})$$

$$\geq \sum_\omega P(\omega)^\alpha Q(\omega)^{1-\alpha} \Phi(\omega, \hat{\omega}),$$

where in the last step we have used the joint concavity of the function $(x, y) \mapsto x(y/x)^\alpha$ (recall proof of Proposition 4.4). Hence,

$$\sum_{\hat{\omega}} [\Phi(P)(\hat{\omega})]^\alpha [\Phi(Q)(\hat{\omega})]^{1-\alpha} \geq \sum_{\hat{\omega}} \sum_\omega P(\omega)^\alpha Q(\omega)^{1-\alpha} \Phi(\omega, \hat{\omega})$$

$$= \sum_\omega P(\omega)^\alpha Q(\omega)^{1-\alpha},$$

and Part (7) follows.

It remains to prove Part (8). For $\alpha \in \mathbb{R} \setminus \{1\}$ set

$$\mathfrak{S}_\alpha(P|Q) = \frac{1}{\alpha - 1} \log\left(\sum_{\omega \in T} P(\omega)^\alpha Q(\omega)^{1-\alpha}\right).$$

Obviously, $\alpha \mapsto \mathfrak{S}_\alpha(P|Q)$ is real-analytic on $\mathbb{R} \setminus \{1\}$. Since

$$\lim_{\alpha \uparrow 1} \mathfrak{S}_\alpha(P|Q) = \lim_{\alpha \downarrow 1} \mathfrak{S}_\alpha(P|Q) = S(P|Q),$$

$\alpha \mapsto \mathfrak{S}_\alpha(P|Q)$ extends to a real-analytic function on $\mathbb{R}$ with $\mathfrak{S}_1(P|Q) = S(P|Q)$. Finally, Part (8) follows from the observation that $S_\alpha(P|Q) = \mathfrak{S}_\alpha(P|Q)$ for $\alpha \in]0, 1[$.

$\qquad\qquad\qquad\qquad\qquad\qquad\qquad\qquad\qquad\qquad\qquad\qquad\qquad\qquad\qquad$ $\square$

Following on the discussion at the end of Sect. 3.6, we set

$$\widehat{S}_\alpha(P|Q) = \log \left(\sum_{\omega \in T} P(\omega)^\alpha Q(\omega)^{1-\alpha} \right), \qquad \alpha \in \mathbb{R}.$$

If $P \ll Q$, then

$$\widehat{S}_\alpha(P|Q) = \log \mathbb{E}_Q(e^{\alpha S_{P|Q}}), \tag{4.28}$$

and so $\widehat{S}_\alpha(P|Q)$ is the cumulant generating function for the relative entropy function $S_{P|Q}$ defined on the probability space (T, P). The discussion at the end of Sect. 3.6 can be now repeated verbatim (we will return to this point in Sect. 5.1). Whenever there is no danger of the confusion, we shall also call $\widehat{S}_\alpha(P|Q)$ Rényi's relative entropy of the pair (P, Q). Note that

$$\widehat{S}_\alpha(P_{\text{ch}}|P) = \widehat{S}_\alpha(P) - \alpha \log |\Omega|. \tag{4.29}$$

Some care is needed in transposing the properties listed in Proposition 4.22 to $\widehat{S}_\alpha(P|Q)$. This point is discussed in the Exercise 4.14.

Exercise 4.13

1. Describe the subset of $\mathcal{P}(\Omega) \times \mathcal{P}(\Omega)$ on which the function $(P, Q) \mapsto S_\alpha(P|Q)$ is strictly convex.
2. Describe the subset of $\mathcal{P}(\Omega) \times \mathcal{P}(\Omega)$ on which $S_\alpha(\Phi(P)|\Phi(Q)) < S_\alpha(P|Q)$.
3. Redo the Exercise 4.2 in Sect. 4.1 and reprove Proposition 4.8 following the proofs of Parts (7) and (8) of Proposition 4.22. Describe the subset of $\mathcal{P}(\Omega)$ on which

$$S(\Phi(P)|\Phi(Q)) < S(P|Q).$$

Exercise 4.14 Prove the following properties of $\widehat{S}_\alpha(P|Q)$.

1. $\widehat{S}_\alpha(P|Q) = -\infty$ iff $P \perp Q$.
   In the remaining statements we shall suppose that $P \not\perp Q$.

2. The function $\mathbb{R} \ni \alpha \mapsto \widehat{S}_\alpha(P|Q)$ is real-analytic and convex. This function is trivial (i.e., identically equal to zero) iff $P = Q$. If P/Q not constant on $T = \mathrm{supp} P \cap \mathrm{supp} Q$, then the function $\alpha \mapsto \widehat{S}_\alpha(P|Q)$ is strictly convex.
3. If $Q \ll P$, then

$$\frac{d\widehat{S}_\alpha(P|Q)}{d\alpha}\Big|_{\alpha=0} = -S(Q|P).$$

If $P \ll Q$, then

$$\frac{d\widehat{S}_\alpha(P|Q)}{d\alpha}\Big|_{\alpha=1} = S(P|Q).$$

4. If P and Q are mutually absolutely continuous, then $\widehat{S}_0(P|Q) = \widehat{S}_1(P|Q) = 0$, $\widehat{S}_\alpha(P|Q) \leq 0$ for $\alpha \in [0, 1]$, and $\widehat{S}_\alpha(P|Q) \geq 0$ for $\alpha \notin [0, 1]$. Moreover,

$$\widehat{S}_\alpha(P|Q) \geq \max\{-\alpha S(Q|P), (\alpha - 1)S(P|Q)\}.$$

5. For $\alpha \in]0, 1[$ the function $(P, Q) \mapsto \widehat{S}_\alpha(P|Q)$ is continuous and jointly concave. Moreover, for any stochastic matrix Φ,

$$\widehat{S}_\alpha(\Phi(P)|\Phi(Q)) \geq \widehat{S}_\alpha(P|Q).$$

Exercise 4.15 Prove that the fluctuation relation of Sect. 4.4 is equivalent to the following statement: for all $\alpha \in \mathbb{R}$,

$$\widehat{S}_\alpha(P|P_\Theta) = \widehat{S}_{1-\alpha}(P|P_\Theta).$$

4.7 Hypothesis Testing

Let Ω be a finite set and P, Q two distinct probability measures on Ω. We shall assume that P and Q are faithful.

Suppose that we know a priori that a probabilistic experiment is with probability p described by P and with probability $1 - p$ by Q. By performing an experiment we wish to decide with minimal error probability what is the correct probability measure. For example, suppose that we are given two coins, one fair ($P(\text{Head}) = P(\text{Tail}) = 1/2$) and one unfair ($Q(\text{Head}) = s$, $Q(\text{Tail}) = 1 - s$, $s > 1/2$). We pick coin randomly (hence $p = 1/2$). The experiment is a coin toss. After tossing a coin we wish to decide with minimal error probability whether we picked the fair or the unfair coin. The correct choice is obvious: if the outcome is Head, pick Q, if the outcome is Tail, pick P.

The following procedure is known as *hypothesis testing*. A *test T* is a subset of Ω. On the basis of the outcome of the experiment with respect to T one chooses between P or Q. More precisely, if the outcome of the experiment is in T, one chooses Q (Hypothesis I: Q is correct) and if the outcome is not in T, one chooses P (Hypothesis II: P is correct). $P(T)$ is the conditional error probability of accepting I if II is true and $Q(T^c)$ is the conditional error probability of accepting II if I is true. The average error probability is

$$D_p(P, Q, T) = pP(T) + (1 - p)Q(T^c),$$

and we are interested in minimizing $D_p(P, Q, T)$ w.r.t. T. Let

$$D_p(P, Q) = \inf_T D_p(P, Q, T).$$

The Bayesian distinguishability problem is to identify tests T such that $D_p(P, Q, T) = D_p(P, Q)$. Let

$$T_{\text{opt}} = \{\omega \mid pP(\omega) \le (1 - p)Q(\omega)\}.$$

Proposition 4.23

(1) T_{opt} is a minimizer of the function $T \mapsto D_p(P, Q, T)$. If T is another minimizer, then $T \subset T_{\text{opt}}$ and $pP(\omega) = (1 - p)Q(\omega)$ for $\omega \in T_{\text{opt}} \setminus T$.
(2)

$$D_p(P, Q) = \int_\Omega \min\{1 - p, p\Delta_{P|Q}(\omega)\}\mathrm{d}Q.$$

(3) For $\alpha \in]0, 1[$,

$$D_p(P, Q) \le p^\alpha(1 - p)^{1-\alpha}e^{\widehat{S}_\alpha(P|Q)}.$$

(4)

$$D_p(P, Q) \ge \int_\Omega \frac{p\Delta_{P|Q}}{1 + \frac{p}{1-p}\Delta_{P|Q}}\mathrm{d}Q.$$

Remark 4.24 Part (1) of this proposition is called Neyman-Pearson lemma. Part (3) is called Chernoff bound.

Proof

$$D_p(P, Q, T) = 1 - p - \sum_{\omega \in T} ((1 - p)Q(\omega) - pP(\omega))$$

$$\geq 1 - p - \sum_{\omega \in T_{\text{opt}}} ((1 - p)Q(\omega) - pP(\omega)),$$

and Part (1) follows. Part (2) is a straightforward computation. Part (3) follows from (2) and the bound $\min\{x, y\} \leq x^\alpha y^{1-\alpha}$ that holds for $x, y \geq 0$ and $\alpha \in]0, 1[$. Part (4) follows from (2) and the obvious estimate

$$\min\{1 - p, p\Delta_{P|Q}(\omega)\} \geq \frac{p\Delta_{P|Q}}{1 + \frac{p}{1-p}\Delta_{P|Q}}.$$

$\square$

Obviously, the errors are smaller if the hypothesis testing is based on repeated experiments. Let P_N and Q_N be the respective product probability measures on Ω^N.

Theorem 4.25

$$\lim_{N \to \infty} \frac{1}{N} \log D_p(P_N, Q_N) = \min_{\alpha \in [0,1]} \widehat{S}_\alpha(P|Q).$$

Proof By Part (2) of the last proposition, for any $\alpha \in]0, 1[$,

$$D_p(P_N, Q_N) \leq p^\alpha (1 - p)^{1-\alpha} e^{\widehat{S}_\alpha(P_N|Q_N)} = p^\alpha (1 - p)^{1-\alpha} e^{N\widehat{S}_\alpha(P|Q)},$$

and so

$$\frac{1}{N} \log D_p(P_N, Q_N) \leq \min_{\alpha \in [0,1]} \widehat{S}_\alpha(P|Q).$$

This yields the upper bound:

$$\limsup_{N \to \infty} \frac{1}{N} \log D_p(P_N|Q_N) \leq \min_{\alpha \in [0,1]} \widehat{S}_\alpha(P|Q).$$

To prove the lower bound we shall make use of the lower bound in Cramér's theorem (Corollary 2.18). Note first that the function

$$x \mapsto \frac{px}{1 + \frac{p}{1-p}x}$$

is increasing on $\mathbb{R}_+$. Let $\theta > 0$ be given. By Part (4) of the last proposition,

$$D_p(P_N, Q_N) \geq \frac{pe^{N\theta}}{1 + \frac{p}{1-p}e^{N\theta}} Q_N \left\{ \omega \in \Omega^N \mid \Delta_{P_N|Q_N}(\omega) \geq e^{N\theta} \right\}.$$

Hence,

$$\liminf_{N\to\infty} \frac{1}{N} \log D_p(P_N|Q_N)$$

$$\geq \liminf_{N\to\infty} \frac{1}{N} \log Q_N \left\{ \omega \in \Omega^N \mid \log \Delta_{P_N|Q_N}(\omega) \geq N\theta \right\}. \qquad (4.30)$$

Let $X = \log \Delta_{P|Q}$ and $S_N(\omega) = \sum_{k=1}^N X(\omega_k)$. Note that $S_N = \log \Delta_{P_N|Q_N}$. The cumulant generating function of X w.r.t. Q is

$$\log \mathbb{E}_Q(e^{\alpha X}) = \widehat{S}_\alpha(P|Q).$$

Since $\mathbb{E}_Q(X) = -S(Q|P) < 0$ and $\theta > 0$, it follows from Corollary 2.18 that

$$\lim_{N\to\infty} \frac{1}{N} \log Q_N \left\{ \omega \in \Omega^N \mid \log \Delta_{P_N|Q_N}(\omega) \geq N\theta \right\} \geq -I(\theta) \qquad (4.31)$$

Since

$$\frac{d\widehat{S}_\alpha}{d\alpha}\Big|_{\alpha=0} = -S(Q|P) < 0, \qquad \frac{d\widehat{S}_\alpha}{d\alpha}\Big|_{\alpha=1} = S(P|Q) > 0,$$

the rate function $I(\theta)$ is continuous around zero, and it follows from (4.30) and (4.31) that

$$\liminf_{N\to\infty} \frac{1}{N} \log D_p(P_N|Q_N) \geq -I(0) = -\sup_{\alpha\in\mathbb{R}}(-\widehat{S}_\alpha(P|Q)).$$

Since $\widehat{S}_\alpha(P|Q) \leq 0$ for $\alpha \in [0, 1]$ and $\widehat{S}_\alpha(P|Q) \geq 0$ for $\alpha \notin [0, 1]$,

$$-\sup_{\alpha\in\mathbb{R}}(-\widehat{S}_\alpha(P|Q)) = \min_{\alpha\in[0,1]} \widehat{S}_\alpha(P|Q),$$

and the lower bound follows:

$$\liminf_{N\to\infty} \frac{1}{N} \log D_p(P_N|Q_N) \geq \min_{\alpha\in[0,1]} \widehat{S}_\alpha(P|Q).$$

$\square$

4.8 Asymmetric Hypothesis Testing

We continue with the framework and notation of the previous section. The asymmetric hypothesis testing concerns individual error probabilities $P_N(T_N)$ (type I-error) and $Q_N(T_N^c)$ (type II-error). For $\gamma \in]0, 1[$ the Stein error exponents are defined by

$$s_N(\gamma) = \min\left\{P(T_N) \mid T_N \subset \Omega^N, \ Q(T_N^c) \leq \gamma\right\}.$$

Theorem 4.14 gives

$$\lim_{N\to\infty} \frac{1}{N} \log s_N(\gamma) = -S(Q|P).$$

The Hoeffding error exponents are similar to Stein's exponents, but with a tighter constraint on the family $(T_N)_{N\geq 1}$ of tests which are required to ensure exponential decay of type-II errors with a minimal rate $s > 0$. They are defined as

$$\overline{h}(s) = \inf_{(T_N)}\left\{\limsup_{N\to\infty} \frac{1}{N} \log P_N(T_N) \ \middle| \ \limsup_{N\to\infty} \frac{1}{N} \log Q_N(T_N^c) \leq -s\right\},$$

$$\underline{h}(s) = \inf_{(T_N)}\left\{\liminf_{N\to\infty} \frac{1}{N} \log P_N(T_N) \ \middle| \ \limsup_{N\to\infty} \frac{1}{T} \log Q_N(T_N^c) \leq -s\right\},$$

$$h(s) = \inf_{(T_N)}\left\{\lim_{N\to\infty} \frac{1}{N} \log P_N(T_N) \ \middle| \ \limsup_{N\to\infty} \frac{1}{N} \log Q_N(T_N^c) \leq -s\right\},$$

where in the last case the infimum is taken over all sequences of tests $(T_N)_{N\geq 1}$ for which the limit

$$\lim_{N\to\infty} \frac{1}{N} \log P_N(T_N)$$

exists. The analysis of these exponents is centred around the function

$$\psi(s) = \inf_{\alpha\in[0,1[} \frac{s\alpha + \widehat{S}_\alpha(Q|P)}{1-\alpha}, \qquad s \geq 0.$$

We first describe some basic properties of ψ.

Proposition 4.26

(1) ψ is continuous on $[0, \infty[$, $\psi(0) = -S(Q|P)$ and $\psi(s) = 0$ for $s \geq S(P|Q)$.
(2) ψ is strictly increasing and strictly concave on $[0, S(P|Q)]$, and real analytic on $]0, S(P|Q)[$.

(3)

$$\lim_{s \downarrow 0} \psi'(s) = \infty, \qquad \lim_{s \uparrow S(P|Q)} \psi'(s) = \left[\widehat{S}_\alpha''(Q|P)\big|_{\alpha=0} \right]^{-1}.$$

(4) For $\theta \in \mathbb{R}$ set

$$\varphi(\theta) = \sup_{\alpha \in [0,1]} \left(\theta\alpha - \widehat{S}_\alpha(Q|P) \right), \qquad \hat{\varphi}(\theta) = \varphi(\theta) - \theta.$$

Then for all $s \geq 0$,

$$\psi(s) = -\varphi(\hat{\varphi}^{-1}(s)). \tag{4.32}$$

Proof Throughout the proof we shall often use Part 3 of the Exercise 4.14. We shall prove Parts (1)–(3) simultaneously. Set

$$F(\alpha) = \frac{s\alpha + \widehat{S}_\alpha(Q|P)}{1 - \alpha}.$$

Then

$$F'(\alpha) = \frac{G(\alpha)}{(1 - \alpha)^2},$$

where $G(\alpha) = s + \widehat{S}_\alpha(Q|P) + (1 - \alpha)\widehat{S}_\alpha'(Q|P)$. Furthermore, $G'(\alpha) = (1 - \alpha)\widehat{S}_\alpha''(Q|P)$ and so $G'(\alpha) > 0$ for $\alpha \in [0, 1[$. Note that $G(0) = s - S(P|Q)$ and $G(1) = s$. It follows that if $s = 0$, then $G(\alpha) < 0$ for $\alpha \in [0, 1[$ and $F(\alpha)$ is decreasing on $[0, 1[$. Hence,

$$\psi(0) = \lim_{\alpha \to 1} \frac{\widehat{S}_\alpha(Q|P)}{1 - \alpha} = -S(Q|P).$$

On the other hand, if $0 < s < S(P|Q)$, then $G(0) < 0$, $G(1) > 0$, and so there exists unique $\alpha_*(s) \in]0, 1[$ such that

$$G(\alpha_*(s)) = 0. \tag{4.33}$$

In this case,

$$\psi(s) = \frac{s\alpha_*(s) + \widehat{S}_{\alpha_*(s)}(Q|P)}{1 - \alpha_*(s)} = -s - \widehat{S}_{\alpha_*(s)}'(Q|P). \tag{4.34}$$

If $s \geq S(P|Q)$, then $G(\alpha) \geq 0$ for $\alpha \in [0, 1[$, and $\psi(s) = F(0) = 0$. The analytic implicit function theorem yields that $s \mapsto \alpha_*(s)$ is analytic on $]0, S(P|Q)[$, and so

ψ is real-analytic on $]0, S(P|Q)[$. The identity

$$0 = G(\alpha_*(s)) = s + \widehat{S}_{\alpha_*(s)}(Q|P) + (1 - \alpha_*(s))\widehat{S}'_{\alpha_*(s)}(Q|P), \qquad (4.35)$$

which holds for $s \in]0, S(P|Q)[$, gives that

$$\alpha'_*(s) = -\frac{1}{(1 - \alpha_*(s))G'(\alpha_*(s))}, \qquad (4.36)$$

and so $\alpha'_*(s) < 0$ for $s \in]0, S(P|Q)[$. One computes

$$\psi'(s) = \frac{\alpha_*(s) - s\alpha'_*(s)}{(1 - \alpha_*(s))^2}, \qquad (4.37)$$

and so ψ is strictly increasing on $]0, S(P|Q)[$ and hence on $[0, S(P|Q)]$. Since $\alpha_*(s)$ is strictly decreasing on $]0, S(P|Q)[$, the limits

$$\lim_{s \downarrow 0} \alpha_*(s) = x, \qquad \lim_{s \uparrow S(P|Q)} \alpha_*(s) = y,$$

exist. Obviously, $x, y \in [0, 1]$, $x > y$, and the definition of G and α_* give that

$$\widehat{S}_x(Q|P) + (1 - x)\widehat{S}'_x(Q|P) = 0, \quad S(P|Q) + \widehat{S}_y(Q|P) + (1 - y)\widehat{S}'_y(Q|P) = 0. \qquad (4.38)$$

We proceed to show that $x = 1$ and $y = 0$. Suppose that $x < 1$. The mean value theorem gives that for some $z \in]x, 1[$

$$-\widehat{S}_x(Q|P) = \widehat{S}_1(Q|P) - \widehat{S}_x(Q|P) = (1 - x)\widehat{S}'_z(Q|P) > (1 - x)\widehat{S}'_x(Q|P), \qquad (4.39)$$

where we used that $\alpha \mapsto \widehat{S}'_\alpha(Q|P)$ is strictly increasing. Obviously, (4.39) contradicts the first equality in (4.38), and so $x = 1$. Similarly, if $y > 0$,

$$S(P|Q) + \widehat{S}_y(Q|P) + (1 - y)\widehat{S}'_y(Q|P) > S(P|Q) + \widehat{S}_y(Q|P) + (1 - y)\widehat{S}'_0(Q|P)$$

$$= \widehat{S}_y(Q|P) - y\widehat{S}'_0(Q|P) > 0,$$

contradicting the second equality in (4.38). Since $x = 1$ and $y = 0$, (4.36) and (4.37) yield Part (3). Finally, to prove that ψ is strictly concave on $[0, S(P|Q)]$ (in view of real analyticity of ψ on $]0, S(P|Q)[$), it suffices to show that ψ' is not constant on $]0, S(P|Q)[$. That follows from Part (3), and the proofs of Parts (1)–(3) are complete.

We now turn to Part (4). The following basic properties of the "restricted Legendre transform" φ are easily proven following the arguments in Sect. 2.5 and we leave the details as an exercise for the reader: φ is continuous, non-negative, and convex on $\mathbb{R}$, $\varphi(\theta) = 0$ for $\theta \leq -S(P|Q)$, φ is real analytic, strictly increasing, and strictly convex on $]-S(P|Q), S(Q|P)[$, and $\varphi(\theta) = \theta$ for $\theta \geq S(Q|P)$. The properties of $\hat{\varphi}$ are now deduced from those of φ and we mention the following: $\hat{\varphi}$ is convex, continuous, and decreasing, $\hat{\varphi}(\theta) = \theta$ for $\theta \leq -S(P|Q)$, and $\varphi(\theta) = 0$ for $\theta \geq S(Q|P)$. Moreover, the map $\hat{\varphi} :]-\infty, S(Q|P)] \rightarrow [0, \infty[$ is a bijection, and we denote by $\hat{\varphi}^{-1}$ its inverse. For $s \geq S(P|Q)$, $\hat{\varphi}^{-1}(s) = -s$ and $\varphi(-s) = 0$, and so (4.32) holds for $s \geq S(P|Q)$. Since $\hat{\varphi}^{-1}(0) = S(Q|P)$ and $\varphi(S(Q|P)) = S(Q|P)$, (4.32) also holds for $s = 0$.

It remains to consider the case $s \in]0, S(P|Q)[$. The map $\hat{\varphi} :]-S(P|Q), S(Q|P)[\rightarrow]0, S(P|Q)[$ is a strictly decreasing bijection. Since

$$-\varphi(\hat{\varphi}^{-1}(s)) = -s - \hat{\varphi}^{-1}(s),$$

it follows from (4.34) that it suffices to show that

$$\hat{\varphi}^{-1}(s) = \widehat{S}'_{\alpha_*(s)}(Q|P),$$

or equivalently, that

$$\varphi(\widehat{S}'_{\alpha_*(s)}(Q|P)) = -s - \widehat{S}'_{\alpha_*(s)}(Q|P). \tag{4.40}$$

Since on $]-S(P|Q), S(Q|P)[$ the function φ coincides with the Legendre transform of $\widehat{S}_\alpha(P|Q)$, it follows from Part (1) of Proposition 2.8 that

$$\varphi(\widehat{S}'_{\alpha_*(s)}(Q|P)) = \alpha_*(s)\widehat{S}'_{\alpha_*(s)}(Q|P) - \widehat{S}_{\alpha_*(s)}(Q|P),$$

and (4.40) follows from (4.35).   □

Exercise 4.16 Prove the properties of φ and $\hat{\varphi}$ that were stated and used in the proof of Part (4) of Proposition 4.26.

The next result sheds additional light on the function ψ. For $\alpha \in [0, 1]$ we define $R_\alpha \in \mathcal{P}(\Omega)$ by

$$R_\alpha(\omega) = \frac{Q(\omega)^\alpha P(\omega)^{1-\alpha}}{\sum_{\omega'} Q(\omega')^\alpha P(\omega')^{1-\alpha}}.$$

Proposition 4.27

(1) For all $s \geq 0$,

$$\psi(s) = -\inf\{S(R|P) \mid R \in \mathcal{P}(\Omega), S(R|Q) \leq s\}. \tag{4.41}$$

(2) For any $s \in]0, S(P|Q)[$,

$$S(R_{\alpha_*(s)}|Q) = s, \qquad S(R_{\alpha_*(s)}|P) = -\psi(s),$$

where $\alpha_*(s)$ is given by (4.33).

Proof Denote by $\phi(s)$ the r.h.s. in (4.41). Obviously, $\phi(0) = -S(Q|P)$ and $\phi(s) = 0$ for $s \geq S(P|Q)$. So we need to prove that $\psi(s) = \phi(s)$ for $s \in]0, S(P|Q)[$.

For any $R \in \mathcal{P}(\Omega)$ and $\alpha \in [0, 1]$,

$$S(R|R_\alpha) = \alpha S(R|Q) + (1 - \alpha)S(R|P) + \widehat{S}_\alpha(Q|P).$$

If R is such that $S(R|Q) \leq s$ and $\alpha \in [0, 1[$, then

$$\frac{S(R|R_\alpha)}{1 - \alpha} \leq \frac{\alpha s + \widehat{S}_\alpha(Q|P)}{1 - \alpha} + S(R|P).$$

Since $S(R|R_\alpha) \geq 0$,

$$\inf_{\alpha \in [0,1[} \frac{\alpha s + \widehat{S}_\alpha(Q|P)}{1 - \alpha} + S(R|P) \geq 0.$$

This gives that $\phi(s) \leq \psi(s)$. If Part (2) holds, then also $\phi(s) \geq \psi(s)$ for all $s \in]0, S(P|Q)[$, and we have the equality $\phi = \psi$. To prove Part (2), a simple computation gives

$$S(R_\alpha|Q) = -(1 - \alpha)\widehat{S}'_\alpha(Q|P) - \widehat{S}_\alpha(Q|P), \qquad S(R_\alpha|Q) = S(R_\alpha|P) + \widehat{S}'_\alpha(Q|P).$$

After setting $\alpha = \alpha_*(s)$ in these equalities, Part (2) follows from (4.35) and (4.34).

$\square$

The main result of this section is

Theorem 4.28 For all $s > 0$,

$$\overline{h}(s) = \underline{h}(s) = h(s) = \psi(s). \tag{4.42}$$

Proof Note that the functions $\overline{h}, \underline{h}, h$ are non-negative and increasing on $]0, \infty[$ and that

$$\underline{h}(s) \leq \overline{h}(s) \leq h(s) \tag{4.43}$$

for all $s > 0$.

We shall prove that for all $s \in]0, S(P|Q)[$,

$$h(s) \leq \psi(s), \qquad \underline{h}(s) \geq \psi(s). \tag{4.44}$$

In view of (4.43), that proves (4.42) for $s \in]0, S(P|Q)[$. Assuming that (4.44) holds, the relations $h(s) \leq h(S(P|Q)) \leq 0$ for $s \in]0, S(P|Q)[$ and

$$\lim_{s \uparrow S(P|Q)} h(s) = \lim_{s \uparrow S(P|Q)} \psi(s) = 0$$

give that $h(S(P|Q)) = 0$. Since h is increasing, $h(s) = 0$ for $s \geq S(P|Q)$ and so $h(s) = \psi(s)$ for $s \geq S(P|Q)$. In the same way one shows that $\overline{h}(s) = \underline{h}(s) = \psi(s)$ for $s \geq S(P|Q)$.

We now prove the first inequality in (4.44). Recall that the map $\hat{\varphi} :] - S(P|Q), S(Q|P)[\to]0, S(P|Q)[$ is a bijection. Fix $s \in]0, S(P|Q)[$ and let $\theta \in] - S(P|Q), S(Q|P[$ be such that $\hat{\varphi}(\theta) = s$. Let

$$T_N(\theta) = \left\{ \omega \in \Omega^N \mid Q_N(\omega) \geq e^{N\theta} P_N(\omega) \right\}. \qquad (4.45)$$

Then

$$P_N(T_N(\theta)) = P_N \left\{ \omega = (\omega_1, \cdots, \omega_N) \in \Omega^N \mid \frac{1}{N} \sum_{j=1}^{N} S_{Q|P}(\omega_j) \geq \theta \right\}.$$

Since the cumulant generating function for $S_{Q|P}$ with respect to P is $\widehat{S}_\alpha(Q|P)$, and the rate function I for $S_{Q|P}$ with respect to P coincides with φ on $]S(P|Q), S(Q|P)[$, it follows from Part (1) of Corollary 2.18 that

$$\lim_{N \to \infty} \frac{1}{N} \log P_N(T_N(\theta)) = -\varphi(\theta). \qquad (4.46)$$

Similarly,

$$Q_N([T_N(\theta)]^c) = Q_N \left\{ \omega = (\omega_1, \cdots, \omega_N) \in \Omega^N \mid \frac{1}{N} \sum_{j=1}^{N} S_{Q|P}(\omega_j) < \theta \right\}.$$

The cumulant generating function for $S_{Q|P}$ with respect to Q is $\widehat{S}_{\alpha+1}(Q|P)$, and the rate function for $S_{Q|P}$ with respect to Q on $]S(P|Q), S(Q|P)[$ is $\hat{\varphi}$. Part (2) of Corollary 2.18 yields

$$\lim_{N \to \infty} \frac{1}{N} \log Q_N([T_N(\theta)]^c) = -\hat{\varphi}(\theta). \qquad (4.47)$$

The relations (4.46) and (4.47) yield that $h(\hat{\varphi}(\theta)) \leq -\varphi(-\theta)$. Since $\hat{\varphi}(\theta) = s$, the first inequality (4.44) follows from Part (4) of Proposition 4.26.

We now turn to the second inequality in (4.44). For $\theta \in] - S(P|Q), S(Q|P)[$ and $T_N \subset \Omega^N$ we set

$$D_N(T_N, \theta) = Q_N([T_N]^c) + e^{\theta N} P_N(T_N).$$

Arguing in the same way as in the proof of Parts (1)–(3) of Proposition 4.23, one shows that for any T_N,

$$D_N(T_N, \theta) \geq D_N(T_N(\theta), \theta).$$

The relations (4.46) and (4.47) yield

$$\lim_{N \to \infty} \frac{1}{N} \log D_N(T_N(\theta), \theta)) = -\hat{\varphi}(\theta).$$

Fix now $s \in]0, S(P|Q)[$ and let $\theta \in] - S(P|Q), S(Q|P)[$ be such that $\hat{\varphi}(\theta) = s$. Let $(T_N)_{N \geq 1}$ be a sequence of tests such that

$$\limsup_{N \to \infty} \frac{1}{N} \log Q_N(T_N^c) \leq -s.$$

Then, for any θ' satisfying $\theta < \theta' < S(Q|P)$ we have

$$
\begin{aligned}
-\hat{\varphi}(\theta') &= \lim_{N \to \infty} \frac{1}{N} \log \left(Q_N([T_N(\theta')]^c) + e^{\theta' N} P_N(T_N(\theta')) \right) \\[2mm]
&\leq \liminf_{N \to \infty} \frac{1}{N} \log \left(Q_N(T_N^c) + e^{\theta' N} P_N(T_N) \right) \\[2mm]
&\leq \max \left(\liminf_{N \to \infty} \frac{1}{N} \log Q_N(T_N^c), \theta' + \liminf_{N \to \infty} \frac{1}{N} \log P_N(T_N) \right) \\[2mm]
&\leq \max \left(-\hat{\varphi}(\theta), \theta' + \liminf_{N \to \infty} \frac{1}{N} \log P_N(T_N) \right).
\end{aligned}
\tag{4.48}
$$

Since $\hat{\varphi}$ is strictly decreasing on $] - S(P|Q), S(Q|P)[$ we have that $-\hat{\varphi}(\theta') > -\varphi(\theta)$, and (4.48) gives

$$\liminf_{N \to \infty} \frac{1}{N} \log P_N(T_N) \geq -\theta' - \hat{\varphi}(\theta') = -\varphi(\theta').$$

Taking $\theta' \downarrow \theta$, we derive

$$\liminf_{N \to \infty} \frac{1}{N} \log P_N(T_N) \geq -\varphi(\theta) = -\varphi(\hat{\varphi}^{-1}(s)) = \psi(s),$$

and so $\underline{h}(s) \geq \psi(s)$. $\square$

Remark 4.29 Theorem 4.28 and its proof give the following. For any sequence of tests $(T_N)_{N \geq 1}$ such that

$$\limsup_{N \to \infty} \frac{1}{N} \log Q_N(T_N^c) \leq -s \qquad (4.49)$$

one has

$$\liminf_{N \to \infty} \frac{1}{N} \log P_N(T_N) \geq \psi(s).$$

On the other hand, if $s \in]0, S(P|Q)[$, $\hat{\varphi}(\theta) = s$, and $T_N(\theta)$ is defined by (4.45), then

$$\limsup_{N \to \infty} \frac{1}{N} \log Q_N([T_N(\theta)]^c) = -s \qquad \text{and} \qquad \lim_{N \to \infty} \frac{1}{N} \log P_N(T_N(\theta)) = \psi(s).$$

Exercise 4.17 Set

$$\overline{h}(0) = \inf_{(T_N)} \left\{ \limsup_{N \to \infty} \frac{1}{N} \log P_N(T_N) \,\middle|\, \limsup_{N \to \infty} \frac{1}{N} \log Q_N(T_N^c) < 0 \right\},$$

$$\underline{h}(0) = \inf_{(T_N)} \left\{ \liminf_{N \to \infty} \frac{1}{N} \log P_N(T_N) \,\middle|\, \limsup_{N \to \infty} \frac{1}{N} \log Q_N(T_N^c) < 0 \right\},$$

$$h(0) = \inf_{(T_N)} \left\{ \lim_{N \to \infty} \frac{1}{N} \log P_N(T_N) \,\middle|\, \limsup_{N \to \infty} \frac{1}{N} \log Q_N(T_N^c) < 0 \right\},$$

where in the last case the infimum is taken over all sequences of tests $(T_N)_{N \geq 1}$ for which the limit

$$\lim_{N \to \infty} \frac{1}{N} \log P_N(T_N)$$

exists. Prove that

$$\overline{h}(0) = \underline{h}(0) = h(0) = -S(Q|P).$$

Compare with Exercise 4.10.

4.9 Notes and References

The relative entropy $S(P|P_{ch})$ already appeared in Shannon's work [44]. The definition (4.1) is commonly attributed to Kullback and Leibler [33], and the relative entropy is sometimes called the Kullback-Leibler divergence. From a historical perspective, it is interesting to note that the symmetrized relative entropy $S(P|Q) + S(Q|P)$ was introduced by Jeffreys in [29] (see Equation (1)) in 1946.

The basic properties of the relative entropy described in Sect. 4.1 are so well-known that it is difficult to trace the original sources. The statement of Proposition 4.1 is sometimes called Gibbs's inequality and sometimes Shannon's inequality. For the references regarding Theorem 4.2 and Exercise 4.7 see Exercise 17 in Chapter 3 of [11] (note the typo regarding the value of the constant c).

The variational principles discussed in Sect. 4.2 are of fundamental importance in statistical mechanics and we postpone their discussion to Part II of the lecture notes.

The attribution of Theorem 4.14 to statistician Charles Stein appears to be historically inaccurate; for a hilarious account of the events that has led to this, see the footnote on the page 85 of [30]. Theorem 4.14 was proven by Hermann Chernoff in [7]. To avoid further confusion, we have used the usual terminology. To the best of my knowledge, the Large Deviations arguments behind the proof of Stein's Lemma, which were implicit in the original work [7], were brought to the surface for the first time in [3, 47], allowing for a substantial generalization of the original results.[4] Our proof follows [47].

The Fluctuation Relation described in Sect. 4.4 is behind the spectacular developments in non-equilibrium statistical mechanics mentioned in the Introduction. We will return to this topic in Part II of the lecture notes.

The choice of the name for Jensen-Shannon entropy (or divergence) and metric is unclear; see [37]. To the best of my knowledge, Theorem 4.19 was first proven in [16, 40]. Our proof follows closely [16]. For additional information, see [20].

The definition of the Rényi relative entropy is usually attributed to [41], although the "un-normalized" $\widehat{S}_\alpha(P|Q)$ already appeared in the work of Chernoff [7] in 1952.

The hypothesis testing is an essential procedure in statistics. Its relevance to modern developments in non-equilibrium statistical mechanics will be discussed in Part II of the lecture notes. Theorem 4.25 is due to Chernoff [7]. As in the case of Stein's Lemma, the LDP based proof allows to considerably generalize the original result. The Hoeffding error exponents were first introduced and studied in [25] and the previous remarks regarding the proof applies to them as well. For additional information about hypothesis testing, see [35].

---

[4]By this I mean that essentially the same argument yields the proof of Stein's Lemma in a very general probabilistic setting.

5   Why is the Relative Entropy Natural?

5.1   Introduction

This chapter is a continuation of Sect. 3.4 and concerns naturalness of the relative entropy.

1. Operational Interpretation   Following on Shannon's quote in Sect. 3.7, Stein's Lemma gives an operational interpretation of the relative entropy $S(P|Q)$. Chernoff and Hoeffding error exponents, Theorems 4.25 and 4.28, give an operational interpretation of Rényi's relative entropy $\widehat{S}_\alpha(P|Q)$ and, via formula (4.29), of Rényi's entropy $\widehat{S}_\alpha(P)$ as well. Note that this operational interpretation of Rényi's entropies is rooted in the LDPs for respective entropy functions which are behind the proofs of Theorems 4.25 and 4.28.

2. Axiomatic Characterizations   Recall that $\mathcal{A}(\Omega) = \{(P, Q) \in \mathcal{P}(\Omega) \mid P \ll Q\}$. Set $\mathcal{A} = \cup_\Omega \mathcal{A}(\Omega)$. The axiomatic characterizations of relative entropy concern choice of a function $\mathfrak{S} : \mathcal{A} \to \mathbb{R}$ that should qualify as a measure of *entropic distinguishability* of a pair $(P, Q) \in \mathcal{A}$. The goal is to show that intuitive natural demands uniquely specify $\mathfrak{S}$ up to a choice of units, namely that for some $c > 0$ and all $(P, Q) \in \mathcal{A}$, $\mathfrak{S}(P, Q) = cS(P|Q)$.

We list basic properties that any candidate $\mathfrak{S}$ for relative entropy should satisfy. The obvious ones are

$$\mathfrak{S}(P, P) = 0, \qquad \mathfrak{S}(P, Q) \geq 0, \qquad \exists\, (P, Q) \text{ such that } \mathfrak{S}(P, Q) > 0. \tag{5.1}$$

Another obvious requirement is that if $|\Omega_1| = |\Omega_2|$ and $\theta : \Omega_1 \to \Omega_2$ is a bijection, then for any $(P, Q) \in \mathcal{A}$,

$$\mathfrak{S}(P, Q) = \mathfrak{S}(P \circ \theta, Q \circ \theta).$$

In other words, the distinguishability of a pair (P, Q) should not depend on the labeling of the elementary events. This requirement gives that $\mathfrak{S}$ is completely specified by its restriction $\mathfrak{S} : \cup_{L \geq 1} \mathcal{A}_L \to [0, \infty[$, where

$$\mathcal{A}_L = \{((p_1, \cdots, p_L), (q_1, \cdots, q_L)) \in \mathcal{P}_L \times \mathcal{P}_L \mid q_k = 0 \Rightarrow p_k = 0\},$$

and that this restriction satisfies

$$\mathfrak{S}((p_1, \cdots, p_L), (q_1, \cdots, q_L))) = \mathfrak{S}((p_{\pi(1)}, \cdots, p_{\pi(L)}), (q_{\pi(1)}, \cdots q_{\pi(L)})) \tag{5.2}$$

for any $L \geq 1$ and any permutation π of $\{1, \cdots, L\}$. In the proofs of Theorems 5.1 and 5.3 we shall assume that (5.1) and (5.2) are satisfied.

Split Additivity Characterization This axiomatic characterization is the relative entropy analog of Theorem 3.4, and has its roots in the identity (recall Proposition 4.10)

$$S(p_1 P_1 + \cdots + p_n P_n | q_1 Q_1 + \cdots + q_n Q_n) = p_1 S(P_1 | Q_1) + \cdots + p_n S(P_n | Q_n)$$
$$+ S((p_1, \cdots, p_n) | (q_1, \cdots q_n))$$

which holds if $(\operatorname{supp} P_j \cup \operatorname{supp} Q_j) \cap (\operatorname{supp} P_k \cup \operatorname{supp} Q_k) = \emptyset$ for all $j \neq k$.

Theorem 5.1 Let $\mathfrak{S} : \mathcal{A} \to [0, \infty[$ be a function such that:

(a) $\mathfrak{S}$ is continuous on $\mathcal{A}_2$.
(b) For any finite collection of disjoint sets Ω_j, $j = 1, \cdots, n$, any $(P_j, Q_j) \in \mathcal{A}(\Omega_j)$, and any $p = (p_1, \cdots, p_n)$, $q = (q_1, \cdots, q_n) \in \mathcal{P}_n$,

$$\mathfrak{S}\left(\bigoplus_{k=1}^{n} p_k P_k, \bigoplus_{k=1}^{n} q_k Q_k \right) = \sum_{k=1}^{n} p_k \mathfrak{S}(P_k, Q_k) + \mathfrak{S}(p|q). \tag{5.3}$$

Then there exists $c > 0$ such that for all $(P, Q) \in \mathcal{A}$,

$$\mathfrak{S}(P, Q) = c S(P|Q). \tag{5.4}$$

Remark 5.2 If the positivity and non-triviality assumptions are dropped, then the proof gives that (5.4) holds for some $c \in \mathbb{R}$.

Exercise 5.1 Following on Remark 3.6, can you verbalize the split-additivity property (5.3)?

We shall prove Theorem 5.1 in Sect. 5.2. The vanishing assumption $\mathfrak{S}(P, P) = 0$ for all P plays a very important role in the argument. Note that

$$\mathfrak{S}(P, Q) = - \sum_{\omega} P(\omega) \log Q(\omega)$$

satisfies (a) and (b) of Theorem 5.1 and assumptions (5.1) apart from $\mathfrak{S}(P, P) = 0$.

Stochastic Monotonicity + Super Additivity Characterization This characterization is related to Theorem 3.7, although its proof is both conceptually different and technically simpler. The characterization asserts that two intuitive requirements, the stochastic monotonicity (Proposition 4.8) and super-additivity (Proposition 4.12) uniquely specify relative entropy.

Theorem 5.3 Let $\mathfrak{S} : \mathcal{A} \to [0, \infty[$ be a function such that:

(a) $\mathfrak{S}$ is continuous on $\mathcal{A}_L$ for all $L \geq 1$.
(b) For any $P, Q \in \mathcal{A}(\Omega)$ and any stochastic map $\Phi : \mathcal{P}(\Omega) \to \mathcal{P}(\hat{\Omega})$ (note that $(\Phi(P), \Phi(Q)) \in \mathcal{A}(\hat{\Omega})$),

$$\mathfrak{S}(\Phi(P), \Phi(Q)) \leq \mathfrak{S}(P, Q). \tag{5.5}$$

(c) For any P and $Q = Q_l \otimes Q_r$ in $\mathcal{A}(\Omega_l \times \Omega_r)$,

$$\mathfrak{S}(P_l, Q_l) + \mathfrak{S}(P_r, Q_r) \leq \mathfrak{S}(P, Q), \tag{5.6}$$

with the equality iff $P = P_l \otimes P_r$.

Then there exists $c > 0$ such that for all $(P, Q) \in \mathcal{A}$,

$$\mathfrak{S}(P, Q) = cS(P|Q). \tag{5.7}$$

We shall prove Theorem 5.3 in Sect. 5.3. Note that neither assumptions (a) $\wedge$ (b) nor (a) $\wedge$ (c) are sufficient to deduce (5.7): (a) and (b) hold for the Rényi relative entropy $(P, Q) \mapsto S_\alpha(P, Q)$ if $\alpha \in]0, 1[$ ((c) fails here), while (a) and (c) hold for the entropy $(P, Q) \mapsto S(P)$ ((b) fails here, recall Exercise 4.5).

4. Sanov's Theorem This result is a deep refinement of Crámer's theorem and the basic indicator of the central role the relative entropy plays in the theory of Large Deviations. We continue with our framework: Ω is a finite set and P a given probability measure on Ω. We shall assume that P is faithful.

To avoid confusion, we shall occasionally denote the generic element of Ω with a letter a (and list the elements of Ω as $\Omega = \{a_1, \cdots, a_L\}$). For $\omega \in \Omega$ we denote by $\delta_\omega \in \mathcal{P}(\Omega)$ the pure probability measure concentrated at ω: $\delta_\omega(a) = 1$ if $a = \omega$ and zero otherwise. For $\omega = (\omega_1, \cdots, \omega_N)$ we set

$$\delta_\omega = \frac{1}{N} \sum_{k=1}^{N} \delta_{\omega_k}.$$

Obviously, $\delta_\omega \in \mathcal{P}(\Omega)$ and

$$\delta_\omega(a) = \frac{\text{the number of times } a \text{ appears in the sequence } \omega = (\omega_1, \cdots, \omega_N)}{N}.$$

Sanov's theorem concerns the statistics of the map $\Omega^N \ni \omega \mapsto \delta_\omega \in \mathcal{P}(\Omega)$ w.r.t. the product probability measure P_N. The starting point is the corresponding law of large numbers.

Proposition 5.4 *For any $\epsilon > 0$,*

$$\lim_{N \to \infty} P_N \left\{ \omega \in \Omega^N \,|\, d_V(\delta_\omega, P) \geq \epsilon \right\} = 0.$$

Sanov's theorem concerns fluctuations in the above LLN, or more precisely, for a given $\Gamma \subset \mathcal{P}(\Omega)$, it estimates the probabilities

$$P_N \left\{ \omega \in \Omega^N \,|\, \delta_\omega \in \Gamma \right\}$$

in the limit of large N.

Theorem 5.5 For any closed set $\Gamma \subset \mathcal{P}(\Omega)$,

$$\limsup_{N \to \infty} \frac{1}{N} \log P_N \left\{ \omega \in \Omega^N \,|\, \delta_\omega \in \Gamma \right\} \leq - \inf_{Q \in \Gamma} S(Q|P),$$

and for any open set $\Gamma \subset \mathcal{P}(\Omega)$,

$$\liminf_{N \to \infty} \frac{1}{N} \log P_N \left\{ \omega \in \Omega^N \,|\, \delta_\omega \in \Gamma \right\} \geq - \inf_{Q \in \Gamma} S(Q|P).$$

We shall prove Proposition 5.4 and Theorem 5.5 in Sect. 5.4 where the reader can also find additional information about Sanov's theorem.

5.2   Proof of Theorem 5.1

The function

$$F(t) = \mathfrak{S}((1, 0), (t, 1 - t)), \qquad t \in]0, 1],$$

will play an important role in the proof. Obviously, F is continuous on $]0, 1]$ and $F(1) = 0$.

We split the proof into five steps.

Step 1 Let $(P, Q) \in \mathcal{A}(\Omega)$, where $\Omega = \{\omega_1, \cdots, \omega_n\}$, and suppose that $P(\omega_j) = 0$ for $j > k$. Set $\Omega_1 = \{\omega_1, \cdots, \omega_k\}$, $P_1(\omega_j) = P(\omega_j)$, and

$$Q_1(\omega_j) = \frac{Q(\omega_j)}{Q(\omega_1) + \cdots + Q(\omega_k)}.$$

It is obvious that $(P_1, Q_1) \in \mathcal{A}(\Omega_1)$. We then have

$$\mathfrak{S}(P, Q) = F(q_1 + \cdots + q_k) + \mathfrak{S}(P_1, Q_1). \tag{5.8}$$

Note that if $k = n$, then (5.8) follows from $F(1) = 1$. Otherwise, write $\Omega = \Omega_1 \oplus \Omega_2$, with $\Omega_2 = \{\omega_{k+1}, \cdots, \omega_n\}$. Take any $P_2 \in \mathcal{P}(\Omega_2)$, write

$$(P, Q) = (1 \cdot P_1 \oplus 0 \cdot P_2, t Q_1 \oplus (1 - t) Q_2),$$

where $t = q_1 + \cdots + q_k$, Q_2 is arbitrary if $t = 1$, and $Q_2(\omega_j) = Q(\omega_j)/(1 - t)$ if $t < 1$, and observe that the statement follows from (5.5).

Step 2 $F(ts) = F(t) + F(s)$ for all $s, t \in]0, 1]$.
  Consider $\mathfrak{S}((1, 0, 0), (ts, t(1 - s), 1 - t))$. Applying Step 1 with $k = 1$ we get

$$\mathfrak{S}((1, 0, 0), (ts, t(1 - s), 1 - t)) = F(ts) + \mathfrak{S}((1), (1)) = F(ts).$$

Applying Step 1 with $k = 2$ gives

$$\mathfrak{S}((1, 0, 0), (ts, t(1 - s), 1 - t)) = F(t) + \mathfrak{S}((1, 0), (s, 1 - s)) = F(t) + F(s),$$

and the statement follows.

Step 3 For some $c \in \mathbb{R}$, $F(t) = -c \log t$ for all $t \in]0, 1]$.
  Set $H(s) = F(e^{-s})$. Then H is continuous on $[0, \infty[$ and satisfies $H(s_1 + s_2) = H(s_1) + H(s_2)$. It is now a standard exercise to show that $H(s) = cs$ where $c = H(1)$. Setting $t = e^{-s}$ gives $F(t) = -c \log t$.
  This is the only point where the regularity assumption (a) has been used (implying the continuity of F), and so obviously (a) can be relaxed.[5] Note that (5.1) implies $c \geq 0$.

Step 4 We now prove that for any $n \geq 2$ and any pair $(p, q) \in \mathcal{A}_n$ of faithful probability measures,

$$\mathfrak{S}(p, q) = c S(p|q), \tag{5.9}$$

where c is the constant from Step 3.
  Let $p = (p_1, \cdots, p_n)$, $q = (q_1, \cdots, q_n)$, and choose $t \in]0, 1]$ such that $q_k - t p_k \geq 0$ for all k. Set

$$K = \mathfrak{S}((p_1, \cdots, p_n, 0, \cdots, 0), (t p_1, \cdots, t p_n, q_1 - t p_1, \cdots, q_n - t p_n)).$$

---

[5]It suffices that F is Borel measurable.

It follows from Steps 1 and 3 that

$$K = F(t) + \mathfrak{S}(p, p) = -c \log t. \tag{5.10}$$

On the other hand, (5.2) and (5.3) yield

$$K = \mathfrak{S}((p_1, 0, \cdots, p_n, 0), (tp_1, q_1 - tp_1, \cdots, tp_n, q_n - tp_n))$$
$$= \mathfrak{S}((p_1(1, 0), \cdots, p_n(1, 0)),$$
$$\left(q_1 \left(\frac{tp_1}{q_1}, 1 - \frac{tp_1}{q_1}\right)\right), \cdots, q_n \left(\frac{tp_n}{q_n}, 1 - \frac{tp_n}{q_n}\right)\right)$$
$$= \sum_{k=1}^{n} p_k F\left(\frac{tp_k}{q_k}\right) + \mathfrak{S}(p, q),$$

and it follows from Step 3 that

$$K = -c \log t - cS(p|q) + \mathfrak{S}(p, q). \tag{5.11}$$

Comparing (5.10) and (5.11) we derive (5.9).

Step 5 We now show that (5.9) also holds for non-faithful p's and complete the proof of Theorem 5.1. By (5.2) we may assume that $p_j > 0$ for $j \le k$ and $p_j = 0$ for $j > k$, where $k < n$. Then, setting $s = q_1 + \cdots q_k$, Steps 1 and 3 yield

$$\mathfrak{S}(p, q) = -c \log s + \mathfrak{S}((p_1, \cdots, p_k), (q_1/s, \cdots, q_k/s)),$$

and it follows from Step 4 that

$$\mathfrak{S}(p, q) = -c \log s + cS((p_1, \cdots, p_k)|(q_1/s, \cdots, q_k/s)).$$

On the other hand, a direct computation gives

$$S(p|q) = -\log s + S((p_1, \cdots, p_k)|(q_1/s, \cdots, q_k/s)),$$

and so $\mathfrak{S}(p, q) = cS(p|q)$.

The non-triviality assumption that $\mathfrak{S}$ is not vanishing on $\mathcal{A}$ gives that $c > 0$.

5.3   Proof of Theorem 5.3

We shall need the following preliminary result which is of independent interest and which we will prove at the end of this section. Recall that if P is a probability measure on Ω, then $P_N = P \otimes \cdots \otimes P$ is the product probability measure on $\Omega^N = \Omega \times \cdots \times \Omega$.

Proposition 5.6 *Suppose that $(P, Q) \in \mathcal{A}(\Omega)$ and $(\widehat{P}, \widehat{Q}) \in \mathcal{A}(\widehat{\Omega})$ are such that $S(P|Q) > S(\widehat{P}|\widehat{Q})$. Then there exists a sequence of stochastic maps $(\Phi_N)_{N \geq 1}$, $\Phi_N : \mathcal{P}(\Omega^N) \to \mathcal{P}(\widehat{\Omega}^N)$ such that $\Phi_N(Q_N) = \widehat{Q}_N$ for all $N \geq 1$ and*

$$\lim_{N \to \infty} d_V(\Phi_N(P_N), \widehat{P}_N) = 0.$$

We now turn to the proof of Theorem 5.3. Recall our standing assumptions (5.1). Let $(P^{(0)}, Q^{(0)}) \in \mathcal{A}$ be such that $\mathfrak{S}(P^{(0)}, Q^{(0)}) > 0$, and let $c > 0$ be such that

$$\mathfrak{S}(P^{(0)}, Q^{(0)}) = cS(P^{(0)}|Q^{(0)}).$$

Let $(P, Q) \in \mathcal{A}$, $P \neq Q$, be given and let L, M, L', M' be positive integers such that

$$\frac{L'}{M'} S(P^{(0)}|Q^{(0)}) < S(P|Q) < \frac{L}{M} S(P^{(0)}|Q^{(0)}). \qquad (5.12)$$

We work first with the r.h.s. of this inequality which can be rewritten as

$$S(P_M|Q_M) < S(P_L^{(0)}|Q_L^{(0)}).$$

It follows from Proposition 5.6 that there exists a sequence of stochastic maps $(\Phi_N)_{N \geq 1}$ such that $\Phi_N(Q_{LN}^{(0)}) = Q_{MN}$ and

$$\lim_{N \to \infty} d_V(\Phi_N(P_L^{(0)}), P_{MN}) = 0. \qquad (5.13)$$

We now turn to $\mathfrak{S}(P, Q)$ and note that

$$M\mathfrak{S}(P, Q) = \mathfrak{S}(P_M, Q_M) = \frac{1}{N}\mathfrak{S}(P_{MN}, Q_{MN})$$

$$= \frac{1}{N}\left[\mathfrak{S}(P_{MN}, Q_{MN}) - \mathfrak{S}(\Phi_N(P_L^{(0)}), Q_{MN})\right]$$

$$+ \frac{1}{N}\mathfrak{S}(\Phi_N(P_L^{(0)}), \Phi_N(Q_{LN}^{(0)}))$$

$$\leq \frac{1}{N}\left[\mathfrak{S}(P_{MN}, Q_{MN}) - \mathfrak{S}(\Phi_N(P_L^{(0)}), Q_{MN})\right]$$

$$+ \frac{1}{N}\mathfrak{S}(P_{LN}^{(0)}, Q_{LN}^{(0)})$$

$$= \frac{1}{N}\left[\mathfrak{S}(P_{MN}, Q_{MN}) - \mathfrak{S}(\Phi_N(P_L^{(0)}), Q_{MN})\right]$$

$$+ L\mathfrak{S}(P_L^{(0)}, Q^{(0)}). \qquad (5.14)$$

Write $Q_{MN} = Q_M \otimes \cdots \otimes Q_M$ and denote by $R_{k,N}$ the marginal of $\Phi_N(P_L^{(0)})$ with the respect to the k-th component of this decomposition. Assumption (c) gives

$$\frac{1}{N}\left[\mathfrak{S}(P_{MN}, Q_{MN}) - \mathfrak{S}(\Phi_N(P_L^{(0)}), Q_{MN})\right] \leq \frac{1}{N}\sum_{k=1}^{N}[\mathfrak{S}(P_M, Q_M)$$
$$-\mathfrak{S}(R_{k,N}, Q_M)\big]. \quad (5.15)$$

One easily shows that (5.13) implies that for any k,

$$\lim_{N\to\infty} d_V(R_{k,N}, P_M) = 0. \quad (5.16)$$

It then follows from (5.15) that

$$\limsup_{N\to\infty} \frac{1}{N}\left[\mathfrak{S}(P_{MN}, Q_{MN}) - \mathfrak{S}(\Phi_N(P_L^{(0)}), Q_{MN})\right] \leq 0. \quad (5.17)$$

Returning to (5.14), (5.17) yields

$$\mathfrak{S}(P, Q) \leq \frac{L}{M}\mathfrak{S}(P^{(0)}, Q^{(0)}) = \frac{L}{M}cS(P^{(0)}|Q^{(0)}). \quad (5.18)$$

Since the only constraint regarding the choice of L and M is that (5.12) holds, we derive from (5.18) that

$$\mathfrak{S}(P, Q) \leq cS(P|Q).$$

Starting with the l.h.s. of the inequality (5.12) and repeating the above argument one derives that $\mathfrak{S}(P, Q) \geq cS(P|Q)$. Hence, $\mathfrak{S}(P, Q) = cS(P|Q)$ for all $(P, Q) \in \mathcal{A}$ with $P \neq Q$. Since this relation holds trivially for $P = Q$, the proof is complete. $\square$

Exercise 5.2 Prove that (5.13) implies (5.16).

Proof of Proposition 5.6 The statement is trivial if $\widehat{P} = \widehat{Q}$, so we assume that $\widehat{P} \neq \widehat{Q}$ (hence $S(\widehat{P}|\widehat{Q}) > 0$). Let $t, \hat{t}$ be such that

$$S(\widehat{P}|\widehat{Q}) < \hat{t} < t < S(P|Q).$$

It follows from Stein's Lemma that one can find a sequence of sets $(T_N)_{N\geq 1}$, $T_N \subset \Omega_N$, such that

$$\lim_{N\to\infty} P_N(T_N) = 1, \qquad Q_N(T_N) \leq C_1 e^{-Nt},$$

for some constant $C_1 > 0$. Let $\Psi_N : \mathcal{P}(\Omega) \to \mathcal{P}(\{0, 1\})$ be a stochastic map induced by the matrix

$$\Psi_N(\omega, 0) = \chi_{T_N}(\omega), \qquad \Psi_N(\omega, 1) = \chi_{T_N^c}(\omega),$$

where χ_{T_N} and $\chi_{T_N^c}$ are the characteristic functions of T_N and its complement T_N^c. It follows that

$$\Psi_N(P_N) = (p_N, \overline{p}_N), \qquad \Psi(Q_N) = (q_N, \overline{q}_N),$$

where

$$p_N = P_N(T_N), \qquad q_N = Q(T_N).$$

Obviously $\overline{p}_N = 1 - p_N, \overline{q}_N = 1 - q_N$.

It follows again from Stein's Lemma that one can find a sequence of sets $(\widehat{T}_N)_{N \geq 1}, \widehat{T}_N \subset \widehat{\Omega}_N$, such that

$$\lim_{N \to \infty} \widehat{P}_N(\widehat{T}_N) = 1, \qquad Q_N(\widehat{T}_N^c) > C_2 e^{-N\hat{\imath}},$$

for some constant $C_2 > 0$. We now construct a stochastic map $\widehat{\Psi}_N : \mathcal{P}(\{0, 1\}) \to \mathcal{P}(\widehat{\Omega})$ as follows. Let $\delta_0 = (1, 0)$, $\delta_1 = (0, 1)$. We set first

$$\widehat{\Psi}_N(\delta_0)(\omega) = \frac{\widehat{P}_N(\omega)}{\sum_{\omega' \in \widehat{T}_N} \widehat{P}_N(\omega')} \qquad \text{if } \omega \in \widehat{T}_N,$$

$\widehat{\Psi}_N(\delta_0)(\omega) = 0$ otherwise, and observe that

$$d_V(\widehat{\Psi}_N(\delta_0), \widehat{P}_N) \leq \widehat{P}_N(\widehat{T}_N^c) + \frac{1 - \widehat{P}_N(\widehat{T}_N)}{\widehat{P}_N(\widehat{T}_N)}.$$

Hence,

$$\lim_{N \to \infty} d_V(\widehat{\Psi}_N(\delta_0), \widehat{P}_N) = 0.$$

Let

$$D_N(\omega) = \widehat{Q}_N(\omega) - q_N \Phi_N(\delta_0)(\omega).$$

If $\omega \notin \widehat{T}_N$, then obviously $D_N(\omega) = \widehat{Q}_N(\omega) \geq 0$, and if $\omega \in \widehat{T}_N$,

$$D_N(\omega) \geq C_2^{-\hat{\imath}N} - c_1 e^{-tN}.$$

Since $0 < \hat{t} < t$, there is N_0 such that for $N \geq N_0$ and all $\omega \in \widehat{\Omega}$, $D_N(\omega) \geq 0$. From now on we assume that $N \geq N_0$, set

$$\widehat{\Psi}_N(\delta_1) = \frac{1}{q_N}(Q_N - q_N \Phi_N(\delta_0)),$$

and define $\widehat{\Psi}_N : \mathcal{P}(\{0, 1\}) \to \mathcal{P}(\widehat{\Omega})$ by

$$\widehat{\Psi}_N(p, q) = p \Psi(\delta_0) + q \Psi(\delta_1).$$

The map $\widehat{\Psi}_N$ is obviously stochastic and

$$\widehat{\Psi}_N(q_N, \overline{q}_N) = \widehat{Q}_N.$$

Moreover,

$$d_V(\widehat{\Psi}_N(p_N, \overline{p}_N), \widehat{P}_N) \leq d_V(\widehat{\Psi}_N(p_N, \overline{p}_N), \widehat{\Psi}_N(\delta_0)) + d_V(\widehat{\Psi}_N(\delta_0), \widehat{P}_N)$$

$$\leq 2(1 - p_N) + d_V(\widehat{\Psi}_N(\delta_0), \widehat{P}_N),$$

and so

$$\lim_{N \to \infty} d_V(\widehat{\Psi}_N(p_N, \overline{p}_N), \widehat{P}_N) = 0.$$

For $N < N_0$ we take for Φ_N an arbitrary stochastic map satisfying $\Phi_N(Q_N) = \widehat{Q}_N$ and for $N \geq N_0$ we set $\Phi_N = \widehat{\Psi}_N \circ \Psi_N$. Then $\Phi_N(Q_N) = \widehat{Q}_N$ for all $N \geq 1$ and

$$\lim_{N \to \infty} d_V(\Phi_N(P_N), \widehat{P}_N) = 0,$$

proving the proposition.   □

Exercise 5.3 Write down the stochastic matrix that induces $\widehat{\Psi}_N$.

5.4   Sanov's Theorem

We start with

Proof of Proposition 5.4. Recall that $L = |\Omega|$. We have

$$d_V(\delta_\omega, P) = \sum_{a \in \Omega} \left| \frac{\sum_{k=1}^N \delta_{\omega_k}(a)}{N} - P(a) \right|,$$

and

$$\left\{ \omega \in \Omega^N \mid d_V(\delta_\omega, P) \geq \epsilon \right\} \subset \bigcup_{a \in \Omega} \left\{ \omega \in \Omega^N \mid \left| \frac{\sum_{k=1}^N \delta_{\omega_k}(a)}{N} - P(a) \right| \geq \frac{\epsilon}{L} \right\}.$$

Hence,

$$P_N \left\{ \omega \in \Omega^N \mid d_V(\delta_\omega, P) \geq \epsilon \right\} \leq \sum_{a \in \Omega} P_N \left\{ \omega \in \Omega^N \mid \left| \frac{\sum_{k=1}^N \delta_{\omega_k}(a)}{N} - P(a) \right| \geq \frac{\epsilon}{L} \right\}.$$

$$(5.19)$$

For given $a \in \Omega$, consider a random variable $X : \Omega \to \mathbb{R}$ defined by $X(\omega) = \delta_\omega(a)$. Obviously, $\mathbb{E}(X) = P(a)$ and the LLN yields that

$$\lim_{N \to \infty} P_N \left\{ \omega \in \Omega^N \mid \left| \frac{\sum_{k=1}^N \delta_{\omega_k}(a)}{N} - P(a) \right| \geq \frac{\epsilon}{L} \right\} = 0.$$

The proposition follows by combining this observation with inequality (5.19). □

We now turn to the proof of Sanov's theorem. Recall the assumption that P is faithful. We start with the upper bound.

Proposition 5.7 *Suppose that $\Gamma \subset \mathcal{P}(\Omega)$ is a closed set. Then*

$$\limsup_{N \to \infty} \frac{1}{N} \log P_N \left\{ \omega \in \Omega^N \mid \delta_\omega \in \Gamma \right\} \leq - \inf_{Q \in \Gamma} S(Q|P).$$

Remark 5.8 Recall that the map $\mathcal{P}(\Omega) \ni Q \mapsto S(Q|P) \in [0, \infty[$ is continuous (P is faithful). Since Γ is compact, there exists $Q_m \in \mathcal{P}(\Omega)$ such that

$$\inf_{Q \in \Gamma} S(Q|P) = S(Q_m|P).$$

Proof Let $\epsilon > 0$ be given. Let $Q \in \Gamma$. By Exercise 4.9,

$$S(Q|P) = \sup_{X:\Omega \to \mathbb{R}} \left(\int_\Omega X \, dQ - \log \int_\Omega e^X \, dP \right).$$

Hence, we can find X such that

$$S(Q|P) - \epsilon < \int_\Omega X \, dQ - \log \int_\Omega e^X \, dP.$$

Let

$$U_\epsilon(Q) = \left\{ Q' \in \mathcal{P}(\Omega) \mid \left| \int_\Omega X \mathrm{d}Q - \int_\Omega X \mathrm{d}Q' \right| < \epsilon \right\}.$$

Since the map $\mathcal{P}(\Omega) \ni Q' \mapsto \int_\Omega X \mathrm{d}Q'$ is continuous, $U_\epsilon(Q)$ is an open subset of $\mathcal{P}(\Omega)$. We now estimate

$$P_N\left\{ \delta_\omega \in U_\epsilon(Q) \right\} = P_N\left\{ \left| \int_\Omega X \mathrm{d}Q - \int_\Omega X \mathrm{d}\delta_\omega \right| < \epsilon \right\}$$

$$\leq P_N\left\{ \int_\Omega X \delta_\omega > \int_\Omega X \mathrm{d}Q - \epsilon \right\}$$

$$= P_N\left\{ \sum_{k=1}^N X(\omega_k) > N \int_\Omega X \mathrm{d}Q - N\epsilon \right\}$$

$$= P_N\left\{ \mathrm{e}^{\sum_{k=1}^N X(\omega_k)} > \mathrm{e}^{N \int_\Omega X \mathrm{d}Q - N\epsilon} \right\}$$

$$\leq \mathrm{e}^{-N \int_\Omega X \mathrm{d}Q + N\epsilon} \mathbb{E}(\mathrm{e}^X)^N$$

$$= \mathrm{e}^{-N \int_\Omega X \mathrm{d}Q + N \log \int_\Omega \mathrm{e}^X \mathrm{d}P + N\epsilon}$$

$$\leq \mathrm{e}^{-NS(Q|P) + 2N\epsilon}$$

Since Γ is compact, we can find $Q_1, \cdots, Q_M \in \Gamma$ such that

$$\Gamma \subset \bigcup_{j=1}^M U_\epsilon(Q_j).$$

Then

$$P_N\left\{ \delta_\omega \in \Gamma \right\} \leq \sum_{j=1}^M P_N\left\{ \delta_\omega \in U_\epsilon(Q_j) \right\}$$

$$\leq \mathrm{e}^{2N\epsilon} \sum_{j=1}^M \mathrm{e}^{-NS(Q_j|P)}$$

$$\leq \mathrm{e}^{2N\epsilon} M \mathrm{e}^{-N \inf_{Q \in \Gamma} S(Q|P)}.$$

Hence

$$\limsup_{N\to\infty} \frac{1}{N} \log P_N \left\{ \omega \in \Omega^N \mid \delta_\omega \in \Gamma \right\} \leq - \inf_{Q\in\Gamma} S(Q|P) + 2\epsilon.$$

Since $\epsilon > 0$ is arbitrary, the statement follows. $\qquad\square$

We now turn to the lower bound.

Proposition 5.9 *For any open set $\Gamma \subset \mathcal{P}(\Omega)$,*

$$\liminf_{N\to\infty} \frac{1}{N} \log P_N \left\{ \omega \in \Omega^N \mid \delta_\omega \in \Gamma \right\} \geq - \inf_{Q\in\Gamma} S(Q|P).$$

Proof Let $Q \in \Gamma$ be faithful. Recall that $S_{Q|P} = \log \Delta_{Q|P}$ and

$$\int_\Omega S_{P|Q} d\delta_\omega = \frac{S_{Q|P}(\omega_1) + \cdots + S_{Q|P}(\omega_N)}{N}.$$

Let $\epsilon > 0$ and

$$R_{N,\epsilon} = \left\{ \delta_\omega \in \Gamma \mid \left| \int_\Omega S_{Q|P} d\delta_\omega - S(Q|P) \right| < \epsilon \right\}.$$

Then

$$P_N \left\{ \delta_\omega \in \Gamma \right\} \geq P_N(R_{N,\epsilon}) = \int_{R_{N,\epsilon}} \Delta_{P_N|Q_N} dQ_N = \int_{R_{N,\epsilon}} \Delta_{Q_N|P_N}^{-1} dQ_N$$

$$= \int_{R_{N,\epsilon}} e^{-\sum_{k=1}^N S_{Q|P}(\omega_k)} dQ_N$$

$$\geq e^{-NS(Q|P) - N\epsilon} Q_N(R_{N,\epsilon}).$$

Note that for ϵ small enough (Γ is open!)

$$R_{N,\epsilon} \supset \left\{ \omega \in \Omega^N \mid d_V(Q, \delta_\omega) < \epsilon \right\}$$

$$\cap \left\{ \omega \in \Omega^N \mid \left| \frac{S_{Q|P}(\omega_1) + \cdots + S_{Q|P}(\omega_N)}{N} - S(Q|P) \right| < \epsilon \right\}.$$

By the LLN,

$$\lim_{N\to\infty} Q_N(R_{N,\epsilon}) = 1.$$

Hence, for any faithful $Q \in \Gamma$,

$$\liminf_{N \to \infty} \frac{1}{N} \log P_N \left\{ \omega \in \Omega^N \mid \delta_\omega \in \Gamma \right\} \geq -S(Q|P). \tag{5.20}$$

Since Γ is open and the map $\mathcal{P}(\Omega) \ni Q \to S(Q|P)$ is continuous,

$$\inf_{Q \in \Gamma \cap \mathcal{P}_f(\Omega)} S(Q|P) = \inf_{Q \in \Gamma} S(Q|P). \tag{5.21}$$

The relations (5.20) and (5.21) imply

$$\liminf_{N \to \infty} \frac{1}{N} \log P_N \left\{ \omega \in \Omega^N \mid \delta_\omega \in \Gamma \right\} \geq - \inf_{Q \in \Gamma} S(Q|P).$$

$\square$

Exercise 5.4 Prove the identity (5.21).

A set $\Gamma \in \mathcal{P}(\Omega)$ is called *Sanov-nice* if

$$\inf_{Q \in \text{int } \Gamma} S(Q|P) = \inf_{Q \in \text{cl } \Gamma} S(Q|P),$$

where int/cl stand for the interior/closure. If Γ is Sanov-nice, then

$$\lim_{N \to \infty} \frac{1}{N} \log P_N \left\{ \omega \in \Omega^N \mid \delta_\omega \in \Gamma \right\} = - \inf_{Q \in \Gamma} S(Q|P).$$

Exercise 5.5 1. Prove that any open set $\Gamma \subset \mathcal{P}(\Omega)$ is Sanov-nice.
2. Suppose that $\Gamma \subset \mathcal{P}(\Omega)$ is convex and has non-empty interior. Prove that Γ is Sanov-nice.

We now show that Sanov's theorem implies Cramér's theorem. The argument we shall use is an example of the powerful *contraction principle* in theory of Large Deviations.

Suppose that in addition to Ω and P we are given a random variable $X : \Omega \to \mathbb{R}$. C and I denote the cumulant generating function and the rate function of X. Note that

$$\frac{S_N(\omega)}{N} = \frac{X(\omega_1) + \cdots + X(\omega_N)}{N} = \int_\Omega X d\delta_\omega.$$

Hence, for any $S \subset \mathbb{R}$,

$$\frac{S_N(\omega)}{N} \in S \iff \delta_\omega \in \Gamma_S,$$

where

$$\Gamma_S = \left\{ Q \in \mathcal{P}(\Omega) \mid \int_\Omega X dQ \in S \right\}.$$

Exercise 5.6 Prove that

$$\text{int } \Gamma_S = \Gamma_{\text{int}S}, \qquad \text{cl } \Gamma_S = \Gamma_{\text{cl}S}.$$

Sanov's theorem and the last exercise yield

Proposition 5.10 *For any* $S \subset \mathbb{R}$,

$$- \inf_{Q \in \Gamma_{\text{int}S}} S(Q|P) \leq \liminf_{N \to \infty} \frac{1}{N} \log P_N \left\{ \omega \in \Omega^N \mid \frac{S_N(\omega)}{N} \in S \right\}$$

$$\leq \limsup_{N \to \infty} \frac{1}{N} \log P_N \left\{ \omega \in \Omega^N \mid \frac{S_N(\omega)}{N} \in S \right\}$$

$$\leq - \inf_{Q \in \Gamma_{\text{cl}S}} S(Q|P),$$

To relate this result to Cramér's theorem we need:

Proposition 5.11 *For any* $S \subset \mathbb{R}$,

$$\inf_{\theta \in S} I(\theta) = \inf_{Q \in \Gamma_S} S(Q|P). \tag{5.22}$$

Proof Let $Q \in \mathcal{P}(\Omega)$. An application of Jensen's inequality gives that for all $\alpha \in \mathbb{R}$,

$$C(\alpha) = \log \left(\sum_{\omega \in \Omega} e^{\alpha X(\omega)} P(\omega) \right)$$

$$\geq \log \left(\sum_{\omega \in \text{supp}Q} e^{\alpha X(\omega)} \frac{P(\omega)}{Q(\omega)} Q(\omega) \right)$$

$$\geq \sum_{\omega \in \text{supp}Q} Q(\omega) \log \left[e^{\alpha X(\omega)} \frac{P(\omega)}{Q(\omega)} \right].$$

Hence,

$$C(\alpha) \geq \alpha \int_\Omega X dQ - S(Q|P). \tag{5.23}$$

If Q is such that $\theta_0 = \int_\Omega X \mathrm{d}Q \in S$, then (5.23) gives

$$S(Q|P) \geq \sup_{\alpha \in \mathbb{R}} (\alpha \theta_0 - C(\alpha)) = I(\theta_0) \geq \inf_{\theta \in S} I(\theta),$$

and so

$$\inf_{Q \in \Gamma_S} S(Q|P) \geq \inf_{\theta \in S} I(\theta). \tag{5.24}$$

On the other hand, if $\theta \in]m, M[$, where $m = \min_{\omega \in \Omega} X(\omega)$ and $M = \max_{\omega \in \Omega} X(\omega)$, and $\alpha = \alpha(\theta)$ is such that $C'(\alpha(\theta)) = \theta$, then, with Q_α defined by (2.3) (recall also the proof of Cramer's theorem), $\theta = \int_\Omega X \mathrm{d}Q_\alpha$ and $S(Q_\alpha|P) = \alpha\theta - C(\alpha) = I(\theta)$. Hence, if $S \subset]m, M[$, then for any $\theta_0 \in S$, $\inf_{Q \in \Gamma_S} S(Q|P) \leq I(\theta_0)$, and so

$$\inf_{Q \in \Gamma_S} S(Q|P) \leq \inf_{\theta \in S} I(\theta). \tag{5.25}$$

It follows from (5.24) and (5.25) that (5.22) holds for $S \subset]m, M[$. One checks directly that

$$I(m) = \inf_{Q: \int_\Omega X \mathrm{d}Q = m} S(Q|P), \qquad I(M) = \inf_{Q: \int_\Omega X \mathrm{d}Q = M} S(Q|P). \tag{5.26}$$

If $S \cap [m, M] = \emptyset$, then both sides in (5.22) are ∞ (by definition, $\inf \emptyset = \infty$). Hence,

$$\inf_{\theta \in S} I(\theta) = \inf_{\theta \in S \cap [m, M]} I(\theta) = \inf_{Q \in \Gamma_{S \cap [m, M]}} S(Q|P) = \inf_{Q \in \Gamma_S} S(Q|P),$$

and the statement follows.   □

Exercise 5.7 Prove the identities (5.26).

Propositions 5.10 and 5.11 yield the following generalization of Cramér's theorem:

Theorem 5.12 For any $S \subset \mathbb{R}$,

$$- \inf_{\theta \in \mathrm{int} S} I(\theta) \leq \liminf_{N \to \infty} \frac{1}{N} \log P_N \left\{ \omega \in \Omega^N \,\Big|\, \frac{S_N(\omega)}{N} \in S \right\}$$

$$\leq \limsup_{N \to \infty} \frac{1}{N} \log P_N \left\{ \omega \in \Omega^N \,\Big|\, \frac{S_N(\omega)}{N} \in S \right\} \leq - \inf_{\theta \in \mathrm{cl} S} I(\theta).$$

A set S is called *Cramer-nice* if

$$\inf_{\theta \in \text{int}S} I(\theta) = \inf_{\theta \in \text{cl}S} I(\theta).$$

Obviously, if S is Cramer-nice, then

$$\lim_{N \to \infty} \frac{1}{N} \log P_N \left\{ \omega \in \Omega^N \mid \frac{S_N(\omega)}{N} \in S \right\} = - \inf_{\theta \in S} I(\theta).$$

Exercise 5.8

1. Is it true that any open/closed interval is Cramér-nice?
2. Prove that any open set $S \subset]m, M[$ is Cramér-nice.
3. Describe all open sets that are Cramér-nice.

5.5 Notes and References

Theorem 5.1 goes back to the work of Hobson [24] in 1969. Following in Shannon's steps, Hobson has proved Theorem 5.1 under the additional assumptions that $\mathfrak{S}$ is continuous on $\mathcal{A}_L$ for all $L \geq 1$, and that the function

$$(n, n_0) \mapsto \mathfrak{S} \left(\left(\frac{1}{n}, \cdots, \frac{1}{n}, 0, \cdots, 0 \right), \left(\frac{1}{n_0}, \cdots, \frac{1}{n_0} \right) \right),$$

defined for $n \leq n_0$, is an increasing function of n_0 and a decreasing function of n. Our proof of Theorem 5.1 follows closely [36] where the reader can find additional information about the history of this result.

The formulation and the proof of Theorem 5.3 are based on the recent works [38, 53].

For additional information about axiomatizations of relative entropy, we refer the reader to Section 7.2 in [1].

Regarding Sanov's theorem, for the original references and additional information, we refer the reader to [9, 13]. In these monographs one can also find a purely combinatorial proof of Sanov's theorem and we urge the reader to study this alternative proof. As in the case of Cramér's theorem, the proof presented here has the advantage that it extends to a much more general setting that will be discussed in the Part II of the lecture notes.

6 Fisher Entropy

6.1 Definition and Basic Properties

Let Ω be a finite set and $[a, b]$ a bounded closed interval in $\mathbb{R}$. To avoid trivialities, we shall always assume that $|\Omega| = L > 1$. Let $\{P_\theta\}_{\theta \in [a,b]}$, $P_\theta \in \mathcal{P}_f(\Omega)$, be a family of faithful probability measures on Ω indexed by points $\theta \in [a, b]$. We shall assume that the functions $[a, b] \ni \theta \mapsto P_\theta(\omega)$ are C^2 (twice continuously differentiable) for all $\omega \in \Omega$. The expectation and variance with respect to P_θ are denoted by $\mathbb{E}_\theta$ and Var_θ. The entropy function is denoted by $S_\theta = -\log P_\theta$. The derivatives w.r.t. θ are denoted as $\dot{f}(\theta) = \partial_\theta f(\theta)$, $\ddot{f}(\theta) = \partial_\theta^2 f(\theta)$, etc. Note that

$$\dot{S}_\theta = -\frac{\dot{P}_\theta}{P_\theta}, \qquad \ddot{S}_\theta = -\frac{\ddot{P}_\theta}{P_\theta} + \frac{\dot{P}_\theta^2}{P_\theta^2}, \qquad \mathbb{E}_\theta(\dot{S}_\theta) = 0.$$

The Fisher entropy of P_θ is defined by

$$\mathcal{I}(\theta) = \mathbb{E}_\theta([\dot{S}_\theta]^2) = \sum_{\omega \in \Omega} \frac{[\dot{P}_\theta(\omega)]^2}{P_\theta(\omega)}.$$

Obviously,

$$\mathcal{I}(\theta) = \mathrm{Var}_\theta(\dot{S}_\theta) = \mathbb{E}_\theta(\ddot{S}_\theta).$$

Example 6.1 Let $X : \Omega \to \mathbb{R}$ be a random variable and

$$P_\theta(\omega) = \frac{e^{\theta X(\omega)}}{\sum_{\omega'} e^{\theta X(\omega')}}.$$

Then

$$\mathcal{I}(\theta) = \mathrm{Var}_\theta(X).$$

The Fisher entropy arises by considering local relative entropy distortion of P_θ. Fix $\theta \in I$ and set

$$L(\epsilon) = S(P_{\theta+\epsilon}|P_\theta), \qquad R(\epsilon) = S(P_\theta|P_{\theta+\epsilon}).$$

The functions $\epsilon \mapsto L(\epsilon)$ and $\epsilon \mapsto R(\epsilon)$ are well-defined in a neighbourhood of θ (relative to the interval $[a, b]$). An elementary computation yields:

Proposition 6.2

$$\lim_{\epsilon \to 0} \frac{1}{\epsilon^2} L(\epsilon) = \lim_{\epsilon \to 0} \frac{1}{\epsilon^2} R(\epsilon) = \frac{1}{2}\mathcal{I}(\theta).$$

In terms of the Jensen-Shannon entropy and metric we have

Proposition 6.3

$$\lim_{\epsilon \to 0} \frac{1}{\epsilon^2} S_{JS}(P_{\theta+\epsilon}, P_\theta) = \frac{1}{4}\mathcal{I}(\theta),$$

$$\lim_{\epsilon \to 0} \frac{1}{|\epsilon|} d_{JS}(P_{\theta+\epsilon}, P_\theta) = \frac{1}{2}\sqrt{\mathcal{I}(\theta)}.$$

Exercise 6.1 Prove Propositions 6.2 and 6.3.

Since the relative entropy is stochastically monotone, Proposition 6.2 implies that the Fisher entropy is also stochastically monotone. More precisely, let $[\Phi(\omega, \hat{\omega})]_{(\omega,\hat{\omega}) \in \Omega \times \hat{\Omega}}$ be a stochastic matrix and $\Phi : \mathcal{P}(\Omega) \to \mathcal{P}(\hat{\Omega})$ the induced stochastic map. Set

$$\widehat{P}_\theta = \Phi(P_\theta),$$

and note that $\widehat{P}_\theta$ is faithful. Let $\widehat{\mathcal{I}}(\theta)$ be the Fisher entropy of $\widehat{P}_\theta$. Then

$$\widehat{\mathcal{I}}(\theta) = \lim_{\epsilon \to 0} \frac{1}{\epsilon^2} S(\widehat{P}_{\theta+\epsilon} | \widehat{P}_\theta) \leq \lim_{\epsilon \to 0} \frac{1}{\epsilon^2} S(P_{\theta+\epsilon} | P_\theta) = \mathcal{I}(\theta).$$

The inequality $\widehat{\mathcal{I}}(\theta) \leq \mathcal{I}(\theta)$ can be directly proven as follows. Since the function $x \mapsto x^2$ is convex, the Jensen inequality yields

$$\left(\sum_\omega \Phi(\omega, \hat{\omega}) \dot{P}_\theta(\omega) \right)^2 = \left(\sum_\omega \Phi(\omega, \hat{\omega}) P_\theta(\omega) \frac{\dot{P}_\theta(\omega)}{P_\theta(\omega)} \right)^2$$

$$\leq \left(\sum_\omega \Phi(\omega, \hat{\omega}) \frac{[\dot{P}_\theta(\omega)]^2}{P_\theta(\omega)} \right) \left(\sum_\omega \Phi(\omega, \hat{\omega}) P_\theta(\omega) \right).$$

Hence,

$$
\widehat{\mathcal{I}}(\theta) = \sum_{\hat\omega} \left(\sum_{\omega} \Phi(\omega, \hat\omega) P_\theta(\omega) \right)^{-1} \left(\sum_{\omega} \Phi(\omega, \hat\omega) \dot P_\theta(\omega) \right)^2
$$

$$
\leq \sum_{\hat\omega} \sum_{\omega} \Phi(\omega, \hat\omega) P_\theta(\omega) \frac{[\dot P_\theta(\omega)]^2}{P_\theta(\omega)}
$$

$$
= \mathcal{I}(\theta).
$$

6.2   Entropic Geometry

We continue with the framework of the previous section. In this section we again identify $\mathcal{P}_{\mathrm{f}}(\Omega)$ with

$$
\mathcal{P}_{L,\mathrm{f}} = \left\{ (p_1, \cdots, p_L) \in \mathbb{R}^L \mid p_k > 0, \sum_k p_k = 1 \right\}.
$$

We view $\mathcal{P}_{L,\mathrm{f}}$ as a surface in $\mathbb{R}^L$ and write $p = (p_1, \cdots, p_L)$. The family $\{P_\theta\}_{\theta \in [a,b]}$ is viewed as a map (we will also call it a path)

$$
[a, b] \ni \theta \mapsto p_\theta = (p_{\theta 1}, \cdots, p_{\theta L}) \in \mathcal{P}_{L,\mathrm{f}},
$$

where $p_{\theta k} = P_\theta(\omega_k)$. For the purpose of this section it suffices to assume that all such path are C^1 (that is, continuously differentiable). The tangent vector $\dot p_\theta = (\dot p_{\theta 1}, \cdots, \dot p_{\theta L})$ satisfies $\sum_k \dot p_{\theta k} = 0$ and hence belongs to the hyperplane

$$
\mathcal{T}_L = \left\{ \zeta = (\zeta_1, \cdots, \zeta_L) \mid \sum_k \zeta_k = 0 \right\}.
$$

The tangent space of the surface $\mathcal{P}_{L,\mathrm{f}}$ is $T_L = \mathcal{P}_{L,\mathrm{f}} \times \mathcal{T}_L$.

A Riemannian structure (abbreviated RS) on $\mathcal{P}_{L,\mathrm{f}}$ is a family $g_L = \{g_{L,p}(\cdot, \cdot)\}_{p \in \mathcal{P}_L}$ of real inner products on $\mathcal{T}_L$ such that for all $\zeta, \eta \in \mathcal{T}_L$ the map

$$
\mathcal{P}_L \ni p \mapsto g_{L,p}(\zeta, \eta) \tag{6.1}
$$

is continuous. The geometric notions (angles, length of curves, curvature, etc.) on $\mathcal{P}_L$ are defined with respect to the RS (to define some of them one needs additional regularity of the maps (6.1)). For example, the energy of the path $\theta \mapsto p_\theta$ is

$$\mathcal{E}([p_\theta]) = \int_a^b g_{L,p_\theta}(\dot{p}_\theta, \dot{p}_\theta)d\theta,$$

and its length is

$$\mathcal{L}([p_\theta]) = \int_a^b \sqrt{g_{L,p_\theta}(\dot{p}_\theta, \dot{p}_\theta)}d\theta.$$

Jensen's inequality for integrals (which is proven by applying Jensen's inequality to Riemann sums) gives that

$$\mathcal{L}([p_\theta]) \geq [(b-a)\mathcal{E}([p_\theta])]^{1/2}. \tag{6.2}$$

The Fisher Riemannian structure (abbreviated FRS) is defined by

$$g_p^F(\zeta, \eta) = \sum_k \frac{1}{p_k} \zeta_k \eta_k.$$

In this case,

$$g_{p(\theta)}^F(\dot{p}_\theta, \dot{p}_\theta) = \mathcal{I}(\theta),$$

where $\mathcal{I}(\theta)$ is the Fisher entropy of P_θ. Hence.

$$\mathcal{E}([p_\theta]) = \int_a^b \mathcal{I}(\theta)d\theta, \qquad \mathcal{L}([p_\theta]) = \int_a^b \sqrt{\mathcal{I}(\theta)}d\theta.$$

We have the following general bounds:

Proposition 6.4

$$\int_a^b \mathcal{I}(\theta)d\theta \geq \frac{1}{b-a}d_V(p_a, p_b)^2, \qquad \int_a^b \sqrt{\mathcal{I}(\theta)}d\theta \geq d_V(p_a, p_b), \tag{6.3}$$

where d_V is the variational distance defined by (3.2).

Remark 6.5 The first inequality in (6.3) yields the "symmetrized" version of Theorem 4.2. Let $p, q \in \mathcal{P}_{L,f}$ and consider the path $p_\theta = \theta p + (1-\theta)q, \theta \in [0, 1]$. Then

$$\int_0^1 \mathcal{I}(\theta)d\theta = S(p|q) + S(q|p),$$

and the first inequality in (6.3) gives

$$S(p|q) + S(q|p) \geq d_V(p, q)^2.$$

Proof To prove the first inequality, note that Jensen's inequality gives

$$\mathcal{I}(\theta) = \sum_{k=1}^{L} \frac{\dot{p}_{\theta k}^2}{p_{\theta k}} = \sum_{k=1}^{L} \left[\frac{\dot{p}_{\theta k}}{p_{\theta k}} \right]^2 p_{\theta k} \geq \left(\sum_{k=1}^{L} |\dot{p}_{\theta k}| \right)^2. \tag{6.4}$$

Hence,

$$\int_a^b \mathcal{I}(\theta) d\theta \geq \int_a^b \left(\sum_{k=1}^{L} |\dot{p}_{\theta k}| \right)^2 d\theta \geq \frac{1}{b-a} \left(\sum_{k=1}^{L} \int_a^b |\dot{p}_{\theta k}| d\theta \right)^2,$$

where the second inequality follows from Jensen's integral inequality. The last inequality and

$$\int_a^b |\dot{p}_{\theta k}| d\theta \geq \left| \int_a^b \dot{p}_{\theta k} d\theta \right| = |p_{bk} - p_{ak}| \tag{6.5}$$

yield the statement.

Note that the first inequality in (6.3) and (6.2) implies the second. Alternatively, the second inequality follows immediately from (6.4) and (6.5).   □

The geometry induced by the FRS can be easily understood in terms of the surface

$$\mathfrak{S}_L = \{s = (s_1, \cdots, s_L) \in \mathbb{R}^L \,|\, s_k > 0, \sum_k s_k^2 = 1\}.$$

The respective tangent space is $\mathfrak{S}_L \times \mathbb{R}^{L-1}$ which we equip with the Euclidian RS

$$e_s(\zeta, \eta) = \sum_k \zeta_k \eta_k.$$

Note that $e_s(\zeta, \eta)$ does not depend on $s \in \mathfrak{S}_L$ and we will drop the subscript s. Let now $\theta \mapsto p_\theta = (p_{\theta_1}, \cdots, p_{\theta L})$ be a path connecting $p = (p_1, \cdots, p_L)$ and $q = (q_1, \cdots, q_L)$ in $\mathcal{P}_{L,\mathrm{f}}$. Then,

$$\theta \mapsto s_\theta = (\sqrt{p_{\theta_1}}, \cdots, \sqrt{p_{\theta L}})$$

is a path in $\mathfrak{S}_L$ connecting $s = (\sqrt{p_1}, \cdots, \sqrt{p_L})$ and $u = (\sqrt{q_1}, \cdots, \sqrt{q_L})$. The map $[p_\theta] \mapsto [s_\theta]$ is a bijective correspondences between all C^1-paths in $\mathcal{P}_{L,\mathrm{f}}$

connecting p and q and all C^1-paths in $\mathfrak{S}_L$ connecting s and u. Since

$$e(\dot{s}_\theta, \dot{s}_\theta) = \frac{1}{4} g^F_{p(\theta)}(\dot{p}_\theta, \dot{p}_\theta) = \frac{1}{4}\mathcal{I}(\theta),$$

the geometry on $\mathcal{P}_{L,f}$ induced by the FRS is identified with the Euclidian geometry of $\mathfrak{S}_L$ via the map $[p_\theta] \mapsto [s_\theta]$.

Exercise 6.2 The geodesic distance between $p, q \in \mathcal{P}_{L,f}$ w.r.t. the FRS is defined by

$$\gamma(p, q) = \inf \int_a^b \sqrt{g^F_{p(\theta)}(\dot{p}_\theta, \dot{p}_\theta)}\, d\theta, \qquad (6.6)$$

where inf is taken over all C^1-paths $[a, b] \ni \theta \mapsto p_\theta \in \mathcal{P}_{L,f}$ such that $p_a = p$ and $p_b = q$. Prove that

$$\gamma(p, q) = \arccos\left(\sum_{k=1}^L \sqrt{p_k q_k}\right).$$

Show that the r.h.s. in (6.6) has a unique minimizer and identify this minimizer.

The obvious hint for a solution of this exercise is to use the correspondence between the Euclidian geometry of the sphere and the FRS geometry of $\mathcal{P}_{L,f}$. We leave it to the interested reader familiar with basic notions of differential geometry to explore this connection further. For example, can you compute the sectional curvature of $\mathcal{P}_{L,f}$ w.r.t. the FRS?

6.3 Chentsov's Theorem

Let $(g_L)_{L\geq 2}$ be a sequence of RS, where g_L is a RS on $\mathcal{P}_{L,f}$. The sequence $(g_L)_{L\geq 2}$ is called stochastically monotone if for any $L, \widehat{L} \geq 2$ and any stochastic map $\Phi : \mathcal{P}_{L,f} \to \mathcal{P}_{\widehat{L},f}$,

$$g_{\widehat{L},\Phi(p)}(\Phi(\zeta), \Phi(\zeta)) \leq g_{L,p}(\zeta, \zeta)$$

for all $p \in \mathcal{P}_{L,f}$ and $\zeta \in \mathcal{T}_L$. Here we used that, in the obvious way, Φ defines a linear map $\Phi : \mathbb{R}^L \mapsto \mathbb{R}^{\widehat{L}}$ which maps $\mathcal{T}_L$ to $\mathcal{T}_{\widehat{L}}$.

Proposition 6.6 *The sequence* $(g^F_L)_{L\geq 1}$ *of the FRS is stochastically monotone.*

Proof The argument is a repetition of the direct proof of the inequality $\mathcal{I}(\theta) \leq \widehat{\mathcal{I}}(\theta)$ given in Sect. 6.1. The details are as follows.

Let $[\Phi(i,j)]_{1\leq i\leq L, 1\leq j\leq\widehat{L}}$ be a stochastic matrix defining $\Phi : \mathcal{P}_{L,\mathrm{f}} \to \mathcal{P}_{\widehat{L},\mathrm{f}}$, i.e., for any $v = (v_1, \cdots, v_L) \in \mathbb{R}^L$, $\Phi(v) \in \mathbb{R}^{\widehat{L}}$ is given by

$$(\Phi(v))_j = \sum_{i=1}^{L} \Phi(i,j)v_i.$$

For $p \in \mathcal{P}_L$ and $\zeta \in \mathcal{T}_L$ the convexity gives

$$\left(\sum_i \Phi(i,j)\zeta_i\right)^2 = \left(\sum_i \Phi(i,j)p_i\frac{\zeta_i}{p_i}\right)^2 \leq \left(\sum_i \Phi(i,j)\frac{\zeta_i^2}{p_i}\right)\left(\sum_i \Phi(i,j)p_i\right)$$

$$= \left(\sum_i \Phi(i,j)\frac{\zeta_i^2}{p_i}\right)(\Phi(p))_j.$$

Hence,

$$g_L^F(\Phi(\zeta), \Phi(\zeta)) = \sum_j \frac{1}{(\Phi(p))_j}\left(\sum_i \Phi(i,j)\zeta_i\right)^2$$

$$\leq \sum_j\sum_i \Phi(i,j)\frac{\zeta_i^2}{p_i} = \sum_i \frac{\zeta_i^2}{p_i} = g_{L,p}^F(\zeta, \zeta).$$

$$\square$$

The main result of this section is:

Theorem 6.7 Suppose that a sequence $(g_L)_{L\geq 2}$ is stochastically monotone. Then there exists a constant $c > 0$ such that $g_L = cg_L^F$ for all $L \geq 2$.

Proof We start the proof by extending each $g_{L,p}$ to a bilinear map $G_{L,p}$ on $\mathbb{R}^L \times \mathbb{R}^L$ as follows. Set $v_L = (1, \cdots, 1) \in \mathbb{R}^L$ and note that any $v \in \mathbb{R}^L$ can be uniquely written as $v = av_L + \zeta$, where $a \in \mathbb{R}$ and $\zeta \in \mathcal{T}_L$. If $v = av_L + \zeta$ and $w = a'v_L + \zeta'$, we set

$$G_{L,p}(v, w) = g_{L,p}(\zeta, \zeta').$$

The map $G_{L,p}$ is obviously bilinear, symmetric ($G_{L,p}(v, w) = G_{L,p}(w, v)$), and non-negative ($G_{L,p}(v, v) \geq 0$). In particular, the polarization identity holds:

$$G_{L,p}(v, w) = \frac{1}{4}\left(G_{L,p}(v + w, v + w) - G_{L,p}(v - w, v - w)\right). \tag{6.7}$$

Note however that $G_{L,p}$ is not an inner product since $G_{L,p}(v_L, v_L) = 0$.

In what follows $p_{L,\text{ch}}$ denotes the chaotic probability distribution in $\mathcal{P}_L$, i.e., $p_{L,\text{ch}} = (1/L, \cdots, 1/L)$. A basic observation is that if the stochastic map $\Phi : \mathcal{P}_{L,\text{f}} \to \mathcal{P}_{\widehat{L},\text{f}}$ is stochastically invertible (that is, there exists a stochastic map $\Psi : \mathcal{P}_{\widehat{L},\text{f}} \to \mathcal{P}_{L,\text{f}}$ such that $\Phi \circ \Psi(p) = p$ for all $p \in \mathcal{P}_{L,\text{f}}$) and $\Phi(p_{L,\text{ch}}) = p_{\widehat{L},\text{ch}}$, then for all $v, w \in \mathbb{R}^L$,

$$G_{\widehat{L}, p_{\widehat{L},\text{ch}}}(\Phi(v), \Phi(w)) = G_{L, p_{L,\text{ch}}}(v, w). \tag{6.8}$$

To prove this, note that since Φ preserves the chaotic probability distribution, we have that $\Phi(v_L) = L\widehat{L}^{-1}v_{\widehat{L}}$. Then, writing $v = av_L + \zeta$, we have

$$G_{L, p_{L,\text{ch}}}(v, v) = g_{L, p_{L,\text{ch}}}(\zeta, \zeta) \geq g_{\widehat{L}, p_{\widehat{L},\text{ch}}}(\Phi(\zeta), \Phi(\zeta))$$

$$= G_{\widehat{L}, p_{\widehat{L},\text{ch}}}\left(aL\widehat{L}^{-1}v_{\widehat{L}} + \Phi(\zeta), aL\widehat{L}^{-1}v_{\widehat{L}} + \Phi(\zeta)\right)$$

$$\tag{6.9}$$

$$= G_{\widehat{L}, p_{\widehat{L},\text{ch}}}(a\Phi(v_L) + \Phi(\zeta), a\Phi(v_L) + \Phi(\zeta))$$

$$= G_{\widehat{L}, p_{\widehat{L},\text{ch}}}(\Phi(v), \Phi(v)).$$

If $\Psi : \mathcal{P}_{\widehat{L},\text{f}} \to \mathcal{P}_{L,\text{f}}$ is the stochastic inverse of Φ, then $\Psi(p_{\widehat{L},\text{ch}}) = p_{L,\text{ch}}$ and so by repeating the above argument we get

$$G_{\widehat{L}, p_{\widehat{L},\text{ch}}}(\Phi(v), \Phi(v)) \geq G_{L, p_{L,\text{ch}}}(\Psi(\Phi(v)), \Psi(\Phi(v))) = G_{L, p_{L,\text{ch}}}(v, v).$$

$$\tag{6.10}$$

The inequalities (6.9) and (6.10) yield (6.8) in the case $v = w$. The polarization identity (6.7) then yields the statement for all vectors v and w.

We proceed to identify $G_{\widehat{L}, p_{\widehat{L},\text{ch}}}$ and $g_{\widehat{L}, p_{\widehat{L},\text{ch}}}$. The identity (6.8) will play a central role in this part of the argument. Let $e_{L,k}$, $k = 1, \cdots, L$, be the standard basis of $\mathbb{R}^L$. Let π be a permutation of $\{1, \cdots, L\}$. Then for all $1 \leq j, k \leq L$,

$$G_{p_{L,\text{ch}}}(e_{L,j}, e_{L,k}) = G_{p_{L,\text{ch}}}(e_{L,\pi(j)}, e_{L,\pi(k)}). \tag{6.11}$$

To establish (6.11), we use (6.8) with $\Phi : \mathcal{P}_{L,\text{f}} \to \mathcal{P}_{L,\text{f}}$ defined by

$$\Phi((p_1, \cdots p_L)) = (p_{\pi(1)}, \cdots, p_{\pi(L)}).$$

Note that Φ is stochastically invertible with the inverse

$$\Psi((p_1, \cdots p_L)) = (p_{\pi^{-1}(1)}, \cdots, p_{\pi^{-1}(L)}),$$

and that $\Phi(p_{L,\text{ch}}) = p_{L,\text{ch}}$. An immediate consequence of the (6.11) is that for all k, j,

$$G_{p_{L,\text{ch}}}(e_{L,j}, e_{L,j}) = G_{p_{L,\text{ch}}}(e_{L,k}, e_{L,k}), \tag{6.12}$$

and that for all pairs (j, k), (j', k') with $j \neq j'$ and $k \neq k'$,

$$G_{p_{L,\text{ch}}}(e_{L,j}, e_{L,k}) = G_{p_{L,\text{ch}}}(e_{L,j'}, e_{L,k'}). \tag{6.13}$$

We introduce the constants

$$c_L = G_{p_{L,\text{ch}}}(e_{L,j}, e_{L,j}), \qquad b_L = G_{p_{L,\text{ch}}}(e_{L,j}, e_{L,k}),$$

where $j \neq k$. By (6.12) and (6.13), these constants do not depend on the choice of j, k. We now show that there exist constants $c, b \in \mathbb{R}$ such that for all $L \geq 2$, $c_L = cL + b$ and $b_L = b$. To prove this, let $L, L' \geq 2$ and consider the stochastic map $\Phi : \mathcal{P}_{L,\text{f}} \to \mathcal{P}_{LL',\text{f}}$ defined by

$$\Phi((p_1, \cdots, p_L)) = \left(\frac{p_1}{L'}, \cdots, \frac{p_1}{L'}, \cdots, \frac{p_L}{L'}, \cdots, \frac{p_L}{L'} \right),$$

where each term p_k/L' is repeated L' times. This map is stochastically invertible with the inverse

$$\Psi\left((p_1^{(1)}, \cdots, p_{L'}^{(1)}, \cdots, p_1^{(L)}, \cdots, p_{L'}^{(L)}) \right) = \left(\sum_{k=1}^{L'} p_k^{(1)}, \cdots, \sum_{k=1}^{L'} p_k^{(L)} \right).$$

Since $\Phi(p_{L,\text{ch}}) = p_{LL',\text{ch}}$, (6.8) holds. Combining (6.8) with the definition b_L, we derive that

$$b_L = b_{LL'} = b_{L'}.$$

Set $b = b_L$. Then, for $L, L' \geq 2$, (6.8) and the definition of c_L give

$$c_L = \frac{1}{L'}c_{LL'} + \frac{L'(L'-1)}{(L')^2}b_{LL'} = \frac{1}{L'}c_{LL'} + \frac{L'(L'-1)}{(L')^2},$$

and so

$$c_L - b = \frac{1}{L'}(c_{LL'} - b).$$

Hence,

$$\frac{1}{L}(c_L - b) = \frac{1}{LL'}(c_{LL'} - b) = \frac{1}{L'}(c_{L'} - b),$$

and we conclude that

$$c_L = cL + b$$

for some $c \in \mathbb{R}$. It follows that for $v, w \in \mathbb{R}^L$,

$$G_{P_{L,\mathrm{ch}}}(v, w) = cL \sum_{k=1}^{L} v_k w_k + b \left(\sum_{k=1}^{L} v_k \right) \left(\sum_{k=1}^{L} w_k \right),$$

and that for $\zeta, \eta \in \mathcal{T}_L$,

$$g_{P_{L,\mathrm{ch}}}(\zeta, \eta) = cL \sum_{k=1}^{L} \zeta_k \eta_k. \qquad (6.14)$$

The last relation implies in particular that $c > 0$. Note that (6.14) can be written as $g_{L,P_{L,\mathrm{ch}}} = c g_{L,P_{\mathrm{ch}}}^F$, proving the statement of the theorem for the special values $p = p_{L,\mathrm{ch}}$.

The rest of the argument is based on the relation (6.14). By essentially repeating the proof of the identity (6.8) one easily shows that if $\Phi : \mathcal{P}_{L,\mathrm{f}} \to \mathcal{P}_{\widehat{L},\mathrm{f}}$ is stochastically invertible, then for all $p \in \mathcal{P}_{L,\mathrm{f}}$ and $\zeta, \eta \in \mathcal{T}_L$,

$$g_{L,\Phi(p)}(\Phi(\zeta), \Phi(\eta)) = g_{L,p}(\zeta, \eta). \qquad (6.15)$$

Let now $\overline{p} = (\overline{p}_1, \cdots, \overline{p}_L) \in \mathcal{P}_{L,\mathrm{f}}$ be such that all $\overline{p}_k$'s are rational numbers. We can write

$$\overline{p} = \left(\frac{\ell_1}{L'}, \cdots, \frac{\ell_L}{L'} \right).$$

where all ℓ_k's are integers ≥ 1 and $\sum_k \ell_k = L'$. Let $\Phi : \mathcal{P}_{L,\mathrm{f}} \to \mathcal{P}_{L',\mathrm{f}}$ be a stochastic map defined by

$$\Phi((p_1, \cdots, p_L)) = \left(\frac{p_1}{\ell_1}, \cdots, \frac{p_1}{\ell_1}, \cdots, \frac{p_L}{\ell_L}, \cdots, \frac{p_L}{\ell_L} \right),$$

where each term p_k/ℓ_k is repeated ℓ_k times. The map Φ is stochastically invertible and its inverse is

$$\Psi((p_1^{(1)}, \cdots, p_{\ell_1}^{(1)}, \cdots, p_1^{(\ell_L)}, \cdots, p_{\ell_L}^{(\ell_L)})) = \left(\sum_{k=1}^{\ell_1} p_k^{\ell_1}, \cdots, \sum_{k=1}^{\ell_L} p_k^{(\ell_L)} \right).$$

Note that $\Phi(\overline{p}) = p_{L',\text{ch}}$, and so

$$g_{L,\overline{p}}(\zeta,\eta) = g_{L',p_{L',\text{ch}}}(\Phi(\zeta),\Phi(\eta)) = c\sum_{k=1}^{L}\frac{L'}{\ell_k}\zeta_k\eta_k = cg^F_{L,\overline{p}}(\zeta,\eta). \qquad (6.16)$$

Since the set of all $\overline{p}$'s in $\mathcal{P}_{L,\text{f}}$ whose all components are rational is dense in $\mathcal{P}_{L,\text{f}}$ and since the map $p \mapsto g_{L,p}(\zeta,\eta)$ is continuous, it follows from (6.16) that for all $L \geq 2$ and all $p \in \mathcal{P}_{L,\text{f}}$,

$$g_{L,p} = cg^F_{L,p}.$$

This completes the proof of Chentsov's theorem.   □

6.4   Notes and References

The Fisher entropy (also often called Fisher information) was introduced by Fisher in [18, 19] and plays a fundamental role in statistics (this is the topic of the next chapter). Although Fisher's work precedes Shannon's by 23 years, it apparently played no role in the genesis of the information theory. The first mentioning of the Fisher entropy in the context of information theory goes back to [33] where Proposition 6.2 was stated.

The geometric interpretation of the Fisher entropy is basically built in its definition. We shall return to this point in the Part II of the lecture notes where the reader can find references to the vast literature on this topic.

Chentsov's theorem goes back to [6]. Our proof is based on the elegant arguments of Campbel [5].

7   Parameter Estimation

7.1   Introduction

Let $\mathcal{A}$ be a set and $\{P_\theta\}_{\theta \in \mathcal{A}}$ a family of probability measures on a finite set Ω. We shall refer to the elements of $\mathcal{A}$ as *parameters*. Suppose that a probabilistic experiment is described by one unknown member of this family. By performing a trial we wish to choose the unknown parameter θ such that P_θ is the most likely description of the experiment. To predict θ one chooses a function $\hat{\theta} : \Omega \to \mathcal{A}$ which, in the present context, is called an *estimator*. If the outcome of a trial is $\omega \in \Omega$, then the value $\theta = \hat{\theta}(\omega)$ is the prediction of the unknown parameter and the probability. Obviously, a reasonable estimator should satisfy a reasonable requirements, and we will return to this point shortly.

The hypothesis testing, described in Sect. 4.7, is the simplest non-trivial example of the above setting with $\mathcal{A} = \{0, 1\}$, $P_0 = P$ and $P_1 = Q$ (we also assume that the priors are $p = q = 1/2$.) The estimators are identified with characteristic functions $\hat{\theta} = \chi_T$, $T \subset \Omega$. With an obvious change of vocabulary, the mathematical theory described in Sect. 4.7 can be viewed as a theory of parameter estimation in the case where $\mathcal{A}$ has two elements.

Here we shall assume that $\mathcal{A}$ is a bounded closed interval $[a, b]$ and we shall explore the conceptual and mathematical aspects the continuous set of parameters brings to the problem of estimation. The Fisher entropy will play an important role in this development. We continue with the notation and assumptions introduced in the beginning of Sect. 6.1, and start with some preliminaries.

A *loss function* is a map $L : \mathbb{R} \times [a, b] \to \mathbb{R}_+$ such that $L(x, \theta) \geq 0$ and $L(x, \theta) = 0$ iff $x = \theta$. To a given loss function and the estimator $\hat{\theta}$, one associates the *risk function* by

$$R(\hat{\theta}, \theta) = E_\theta(L(\hat{\theta}, \theta)) = \sum_{\omega \in \Omega} L(\hat{\theta}(\omega), \theta) P_\theta(\omega).$$

Once a choice of the loss function is made, the goal is to find an estimator that will minimize the risk function subject to appropriate consistency requirements.

We shall work only with the quadratic loss function $L(x, \theta) = (x - \theta)^2$. In this case, the risk function is

$$E_\theta((\hat{\theta} - \theta)^2) = \text{Var}_\theta(\hat{\theta}).$$

7.2 Basic Facts

The following general estimate is known as the Cramér-Rao bound.

Proposition 7.1 *For any estimator $\hat{\theta}$ and all $\theta \in [a, b]$,*

$$\frac{[\dot{E}_\theta(\hat{\theta})]^2}{\mathcal{I}(\theta)} \leq E_\theta((\hat{\theta} - \theta)^2).$$

Proof

$$\dot{E}_\theta(\hat{\theta}) = \sum_{\omega \in \Omega} \hat{\theta}(\omega) \dot{P}_\theta(\omega) = \sum_{\omega \in \Omega} (\hat{\theta}(\omega) - \theta) \dot{P}_\theta(\omega).$$

Writing $\dot{P}_\theta(\omega) = \dot{P}_\theta(\omega)\sqrt{P_\theta(\omega)}/\sqrt{P_\theta(\omega)}$ and applying the Cauchy-Schwartz inequality one gets

$$|\dot{E}_\theta(\hat{\theta})| \le \left(\sum_{\omega\in\Omega}(\hat{\theta}(\omega) - \theta)^2 P_\theta(\omega)\right)^{1/2} \left(\sum_{\omega\in\Omega}\frac{[\dot{P}_\theta(\omega)]^2}{P_\theta(\omega)}\right)^{1/2}$$

$$= \left(E_\theta((\hat{\theta} - \theta)^2)\right)^{1/2}\sqrt{\mathcal{I}(\theta)}.$$

$\square$

As in the case of hypothesis testing, multiple trials improve the errors in the parameter estimation. Passing to the product space Ω^N and the product probability measure $P_{\theta N}$, and denoting by $E_{\theta N}$ the expectation w.r.t. $P_{\theta N}$, the Cramér-Rao bound takes the following form.

Proposition 7.2 *For any estimator* $\hat{\theta}_N : \Omega^N \to [a, b]$ *and all* $\theta \in [a, b]$,

$$\frac{1}{N}\frac{[\dot{E}_{N\theta}(\hat{\theta}_N)]^2}{\mathcal{I}(\theta)} \le E_{\theta N}((\hat{\theta}_N - \theta)^2).$$

Proof

$$\dot{E}_{\theta N}(\hat{\theta}_N) = \sum_{\omega=(\omega_1,\cdots,\omega_N)\in\Omega^N} \sum_{k=1}^{N}(\hat{\theta}_N(\omega) - \theta)P_\theta(\omega_1)\cdots\dot{P}_\theta(\omega_k)\cdots P_\theta(\omega_N)$$

$$= \sum_{\omega=(\omega_1,\cdots,\omega_N)\in\Omega^N}\left(\sum_{k=1}^{N}\frac{\dot{P}_\theta(\omega_k)}{P_\theta(\omega_k)}\right)(\hat{\theta}_N(\omega) - \theta)P_{\theta N}(\omega).$$

Applying the Cauchy-Schwarz inequality

$$\int_{\Omega^N} fg\,\mathrm{d}P_{\theta N} \le \left(\int_{\Omega^N} f^2\,\mathrm{d}P_{\theta N}\right)^{1/2}\left(\int_{\Omega^N} g^2\,\mathrm{d}P_{\theta N}\right)^{1/2}$$

with

$$f(\omega) = \sum_{k=1}^{N}\frac{\dot{P}_\theta(\omega_k)}{P_\theta(\omega_k)}, \qquad g(\omega) = \hat{\theta}_N(\omega) - \theta,$$

one gets

$$
|\dot{E}_{\theta N}(\hat{\theta})| \leq \left(\sum_{\omega \in \Omega^N} (\hat{\theta}_N(\omega) - \theta)^2 P_{\theta N}(\omega) \right)^{1/2}
$$

$$
\left(\sum_{\omega = (\omega_1, \cdots, \omega_N)} \sum_{k=1}^{N} \frac{[\dot{P}_\theta(\omega_k)]^2}{[P_\theta(\omega_k)]^2} P_{\theta N}(\omega) \right)^{1/2}
$$

$$
= \left(E_{\theta N}((\hat{\theta}_N - \theta)^2) \right)^{1/2} \sqrt{N\mathcal{I}(\theta)}.
$$

$\square$

We now describe the *consistency* requirement. In a nutshell, the consistency states that if the experiment is described by P_θ, then the estimator should statistically return the value θ. An ideal consistency would be $E_{\theta N}(\hat{\theta}_N) = \theta$ for all $\theta \in [a, b]$. However, it is clear that in our setting such estimator cannot exist. Indeed, using that $\hat{\theta}$ takes values in $[a, b]$, the relations $E_{aN}(\hat{\theta}_N) = a$ and $E_{bN}(\hat{\theta}_N) = b$ give that $\hat{\theta}_N(\omega) = a$ and $\hat{\theta}_N(\omega) = b$ for all $\omega \in \Omega^N$. Requiring $E_{\theta N}(\hat{\theta}_N) = \theta$ only for $\theta \in]a, b[$ does not help, and the remaining possibility is to formulate the consistency in an asymptotic setting.

Definition 7.3 A sequence of estimators $\hat{\theta}_N : \Omega^N \to [a, b]$, $N = 1, 2, \cdots$, is called consistent if

$$
\lim_{N \to \infty} E_{\theta N}(\hat{\theta}_N) = \theta
$$

for all $\theta \in [a, b]$, and uniformly consistent if

$$
\lim_{N \to \infty} \sup_{\theta \in [a,b]} E_{\theta N}(|\hat{\theta} - \theta|) = 0.
$$

Finally, we introduce the notion of *efficiency*.

Definition 7.4 Let $\hat{\theta}_N : \Omega^N \to [a, b]$, $N = 1, 2, \cdots$ be a sequence of estimators. A continuous function $\mathcal{E} :]a, b[\to \mathbb{R}_+$ is called the efficiency of $(\hat{\theta}_N)_{N \geq 1}$ if

$$
\lim_{N \to \infty} N E_{\theta N} \left((\hat{\theta} - \theta)^2 \right) = \mathcal{E}(\theta) \tag{7.1}
$$

for all $\theta \in]a, b[$. The sequence $(\hat{\theta}_N)_{N \geq 1}$ is called uniformly efficient if in addition for any $[a', b'] \subset]a, b[$,

$$
\limsup_{N \to \infty} \sup_{\theta \in [a',b']} \left| N E_{\theta N} \left((\hat{\theta} - \theta)^2 \right) - \mathcal{E}(\theta) \right| = 0. \tag{7.2}
$$

To remain on a technically elementary level, we will work only with uniformly efficient estimators. The reason for staying away from the boundary points a and b in the definition of efficiency is somewhat subtle and we will elucidate it in Remark 7.12.

Proposition 7.5 *Let $(\hat{\theta}_N)_{N \geq 1}$ be a uniformly efficient consistent sequence of estimators. Then its efficiency $\mathcal{E}$ satisfies*

$$\mathcal{E}(\theta) \geq \frac{1}{\mathcal{I}(\theta)}$$

for all $\theta \in {]a, b[}$.

Proof Fix $\theta_1, \theta_2 \in {]a, b[}, \theta_1 < \theta_2$. The consistency gives

$$\theta_2 - \theta_1 = \lim_{N \to \infty} \left[E_{\theta_2 N}(\hat{\theta}_N) - E_{\theta_1 N}(\hat{\theta}_N) \right]. \tag{7.3}$$

The Cramér-Rao bound yields the estimate

$$E_{\theta_2 N}(\hat{\theta}_N) - E_{\theta_1 N}(\hat{\theta}_N) = \int_{\theta_1}^{\theta_2} \dot{E}_{\theta N}(\hat{\theta}_N) d\theta \leq \int_{\theta_1}^{\theta_2} |\dot{E}_{\theta N}(\hat{\theta}_N)| d\theta$$

$$\leq \int_{\theta_1}^{\theta_2} \left[N \mathcal{I}(\theta) E_{\theta N} \left((\hat{\theta}_N - \theta)^2 \right) \right]^{1/2} d\theta. \tag{7.4}$$

Finally, the uniform efficiency gives

$$\lim_{N \to \infty} \int_{\theta_1}^{\theta_2} \left[N \mathcal{I}(\theta) E_{\theta N} \left((\hat{\theta}_N - \theta)^2 \right) \right]^{1/2} d\theta$$

$$= \int_{\theta_1}^{\theta_2} \lim_{N \to \infty} \left[N \mathcal{I}(\theta) E_{\theta N} \left((\hat{\theta}_N - \theta)^2 \right) \right]^{1/2} d\theta$$

$$= \int_{\theta_1}^{\theta_2} \sqrt{\mathcal{I}(\theta) \mathcal{E}(\theta)} d\theta. \tag{7.5}$$

Combining (7.3)–(7.5), we derive that

$$\theta_2 - \theta_1 \leq \int_{\theta_1}^{\theta_2} \sqrt{\mathcal{I}(\theta) \mathcal{E}(\theta)} d\theta$$

for all $a \leq \theta_1 < \theta_2 \leq b$. Hence, $\sqrt{\mathcal{I}(\theta)\mathcal{E}(\theta)} \geq 1$ for all $\theta \in {]a, b[}$, and the statement follows.   □

In Sect. 7.4 we shall construct a uniformly consistent and uniformly efficient sequence of estimators whose efficiency is equal to $1/\mathcal{I}(\theta)$ for all $\theta \in]a, b[$. This sequence of estimators saturates the bound of Proposition 7.5 and in that sense is the best possible one. In Remark 7.12 we shall also exhibit a concrete example of such estimator sequence for which the limit (7.1) also exists for $\theta = a$ and satisfies $\mathcal{E}(a) < 1/\mathcal{I}(a)$. This shows that Proposition 7.5 is an optimal result.

7.3 Two Remarks

The first remark is that the existence of a consistent estimator sequence obviously implies that

$$\theta_1 \neq \theta_2 \;\Rightarrow\; P_{\theta_1} \neq P_{\theta_2}. \tag{7.6}$$

In Sect. 7.4 we shall assume that (7.6) holds and refer to it as the *identifiability* property of our starting family of probability measures $\{P_\theta\}_{\theta \in [a,b]}$.

The second remark concerns the LLN adapted to the parameter setting, which will play a central role in the proofs of the next section. This variant of the LLN is of independent interest, and for this reason we state it and prove it separately.

Proposition 7.6 *Let $X_\theta : \Omega \to \mathbb{R}$, $\theta \in [a, b]$, be random variables such that the map $[a, b] \ni \theta \mapsto X_\theta(\omega)$ is continuous for all $\omega \in \Omega$. Set*

$$\mathcal{S}_{\theta N}(\omega = (\omega_1, \cdots, \omega_N)) = \sum_{k=1}^{N} X_\theta(\omega_k).$$

Then for any $\epsilon > 0$,

$$\lim_{N \to \infty} \sup_{\theta \in [a,b]} P_{\theta N} \left\{ \omega \in \Omega^N \mid \sup_{\theta' \in [a,b]} \left| \frac{\mathcal{S}_{\theta' N}(\omega)}{N} - E_\theta(X_{\theta'}) \right| \geq \epsilon \right\} = 0. \tag{7.7}$$

Moreover, (7.7) can be refined as follows. For any $\epsilon > 0$ there are constants $C_\epsilon > 0$ and $\gamma_\epsilon > 0$ such that for all $N \geq 1$,

$$\sup_{\theta \in [a,b]} P_{\theta N} \left\{ \omega \in \Omega^N \mid \sup_{\theta' \in [a,b]} \left| \frac{\mathcal{S}_{\theta' N}(\omega)}{N} - E_\theta(X_{\theta'}) \right| \geq \epsilon \right\} \leq C_\epsilon e^{-\gamma_\epsilon N}. \tag{7.8}$$

Remark 7.7 The point of this result is uniformity in θ and θ'. Note that

$$\lim_{N \to \infty} P_{\theta N} \left\{ \omega \in \Omega^N \mid \left| \frac{\mathcal{S}_{\theta' N}(\omega)}{N} - E_\theta(X_{\theta'}) \right| \geq \epsilon \right\} = 0$$

is the statement of the LLN, while

$$P_{\theta N} \left\{ \omega \in \Omega^N \mid \left| \frac{S_{\theta' N}(\omega)}{N} - E_\theta(X_{\theta'}) \right| \geq \epsilon \right\} \leq C_\epsilon e^{-\gamma_\epsilon N},$$

with C_ϵ and γ_ϵ depending on θ, θ', is the statement of the strong LLN formulated in Exercise 2.4.

Proof By uniform continuity, there exists $\delta > 0$ such that for all $u, v \in [a, b]$ satisfying $|u - v| < \delta$ one has

$$\sup_{u' \in [a,b]} |E_{u'}(X_u) - E_{u'}(X_v)| < \frac{\epsilon}{4} \quad \text{and} \quad \sup_{\omega \in \Omega} |X_u(\omega) - X_v(\omega)| < \frac{\epsilon}{4}.$$

Let $a = \theta'_0 < \theta'_1 < \cdots < \theta'_n = b$ be such that $\theta'_k - \theta'_{k-1} < \delta$. Then, for all $\theta \in [a, b]$,

$$\left\{ \omega \in \Omega^N \mid \sup_{\theta' \in [a,b]} \left| \frac{S_{\theta' N}(\omega)}{N} - E_\theta(X_{\theta'}) \right| \geq \epsilon \right\}$$

$$\subset \bigcup_{k=1}^{n} \left\{ \omega \in \Omega^N \mid \left| \frac{S_{\theta'_k N}(\omega)}{N} - E_\theta(X_{\theta'_k}) \right| \geq \frac{\epsilon}{2} \right\}. \tag{7.9}$$

It follows that (recall the proof of the LLN, Proposition 2.2)

$$P_{\theta N} \left\{ \omega \in \Omega^N \mid \sup_{\theta' \in [a,b]} \left| \frac{S_{\theta' N}}{N} - E_\theta(X_{\theta'}) \right| \geq \epsilon \right\}$$

$$\leq \sum_{k=1}^{n} P_{\theta N} \left\{ \omega \in \Omega^N \mid \left| \frac{S_{\theta'_k N}(\omega)}{N} - E_\theta(X_{\theta'_k}) \right| \geq \frac{\epsilon}{2} \right\}$$

$$\leq \frac{4}{\epsilon^2} \sum_{k=1}^{n} E_{\theta N} \left(\left| \frac{S_{\theta'_k N}}{N} - E_\theta(X_{\theta'_k}) \right|^2 \right)$$

$$\leq \frac{4}{\epsilon^2} \frac{1}{N} \sum_{k=1}^{n} E_\theta \left(|X_{\theta'_k} - E_\theta(X_{\theta'_k})|^2 \right). \tag{7.10}$$

Setting

$$C = \max_{1 \leq k \leq n} \max_{\theta, \theta' \in [a,b]} E_\theta \left(|X_{\theta'} - E_\theta(X_{\theta'})|^2 \right),$$

we derive that

$$\sup_{\theta\in[a,b]} P_{\theta N} \left\{ \omega \in \Omega^N \mid \sup_{\theta'\in[a,b]} \left| \frac{S_{\theta'N}(\omega)}{N} - E_\theta(X_{\theta'}) \right| \geq \epsilon \right\} \leq \frac{4}{\epsilon^2} \frac{Cn}{N},$$

and (7.7) follows.

The proof of (7.8) also starts with (7.9) and follows the argument of Proposition 2.14 (recall the Exercise 2.4). The details are as follows. Let $\alpha > 0$. Then for any θ and k,

$$P_{\theta N} \left\{ \omega \in \Omega^N \mid \frac{S_{\theta'_k N}(\omega)}{N} - E_\theta(X_{\theta'_k}) \geq \frac{\epsilon}{2} \right\}$$

$$= P_{\theta N} \left\{ \omega \in \Omega^N \mid S_{\theta'_k N}(\omega) \geq N\frac{\epsilon}{2} + N E_\theta(X_{\theta'_k}) \right\}$$

$$= P_{\theta N} \left\{ \omega \in \Omega^N \mid e^{\alpha S_{\theta'_k N}(\omega)} \geq e^{\alpha N\epsilon/2} e^{\alpha N E_\theta(X_{\theta'_k})} \right\}$$

$$\leq e^{-\alpha N\epsilon/2} e^{-\alpha N E_\theta(X_{\theta'_k})} E_{\theta N} \left(e^{\alpha S_{\theta'_k N}} \right)$$

$$\leq e^{-\alpha N\epsilon/2} e^{-\alpha N E_\theta(X_{\theta'_k})} e^{N C_\theta^{(k)}(\alpha)},$$

$$(7.11)$$

where

$$C_\theta^{(k)}(\alpha) = \log E_\theta \left(e^{\alpha X_{\theta'_k}} \right).$$

We write

$$C_\theta^{(k)}(\alpha) - \alpha E_\theta(X_{\theta'_k}) = \int_0^\alpha \left[\left(C_\theta^{(k)} \right)'(u) - E_\theta(X_{\theta'_k}) \right] du,$$

and estimate

$$|C_\theta^{(k)}(\alpha) - \alpha E_\theta(X_{\theta'_k})| \leq \alpha \sup_{u\in[0,\alpha]} \left| \left(C_\theta^{(k)} \right)'(u) - E_\theta(X_{\theta'_k}) \right|.$$

Since $\left(C_\theta^{(k)} \right)'(0) = E_\theta(X_{\theta'_k})$, the uniform continuity gives

$$\lim_{\alpha\to0} \sup_{\theta\in[a,b]} \sup_{u\in[0,\alpha]} \left| \left(C_\theta^{(k)} \right)'(u) - E_\theta(X_{\theta'_k}) \right| = 0.$$

It follows that there exists $\alpha_\epsilon^+ > 0$ such that for all $k = 1, \cdots, n$,

$$\sup_{\theta \in [a,b]} \left| C_\theta^{(k)}(\alpha_\epsilon^+) - \alpha_\epsilon^+ E_\theta(X_{\theta_k'}) \right| \leq \frac{\epsilon}{4},$$

and (7.11) gives that for all k,

$$\sup_{\theta \in [a,b]} P_{\theta N} \left\{ \omega \in \Omega^N \mid \frac{S_{\theta_k' N}(\omega)}{N} - E_\theta(X_{\theta_k'}) \geq \frac{\epsilon}{2} \right\} \leq e^{-\alpha_\epsilon^+ N\epsilon/4}.$$

Going back to first inequality in (7.10), we conclude that

$$\sup_{\theta \in [a,b]} P_{\theta N} \left\{ \omega \in \Omega^N \mid \sup_{\theta' \in [a,b]} \left(\frac{S_{\theta' N}(\omega)}{N} - E_\theta(X_{\theta'}) \right) \geq \epsilon \right\} \leq n e^{-\alpha_\epsilon^+ N\epsilon/4}. \tag{7.12}$$

By repeating the above argument (or by simply applying the final estimate (7.12) to the random variables $-X_\theta$), one derives

$$\sup_{\theta \in [a,b]} P_{\theta N} \left\{ \omega \in \Omega^N \mid \inf_{\theta' \in [a,b]} \left(\frac{S_{\theta' N}(\omega)}{N} - E_\theta(X_{\theta'}) \right) \leq -\epsilon \right\} \leq n e^{-\alpha_\epsilon^- N\epsilon/4} \tag{7.13}$$

for a suitable $\alpha_\epsilon^- > 0$. Finally, since

$$\left\{ \omega \in \Omega^N \mid \sup_{\theta' \in [a,b]} \left| \frac{S_{\theta' N}(\omega)}{N} - E_\theta(X_{\theta'}) \right| \geq \epsilon \right\}$$

$$\subset \left\{ \omega \in \Omega^N \mid \sup_{\theta' \in [a,b]} \left(\frac{S_{\theta' N}(\omega)}{N} - E_\theta(X_{\theta'}) \right) \geq \epsilon \right\}$$

$$\cup \left\{ \omega \in \Omega^N \mid \inf_{\theta' \in [a,b]} \left(\frac{S_{\theta' N}(\omega)}{N} - E_\theta(X_{\theta'}) \right) \leq -\epsilon \right\},$$

(7.8) follows from (7.12) and (7.13).   □

Exercise 7.1 Prove the relation (7.9).

7.4   The Maximum Likelihood Estimator

For each N and $\omega = (\omega_1, \cdots, \omega_N) \in \Omega^N$, consider the function

$$[a, b] \ni \theta \mapsto P_{\theta N}(\omega_1, \cdots, \omega_N) \in]0, 1[. \tag{7.14}$$

By continuity, this function achieves its global maximum on the interval $[a, b]$. We denote by $\hat{\theta}_{ML,N}(\omega)$ a point where this maximum is achieved (in the case where there are several such points, we select one arbitrarily but always choosing $\hat{\theta}_{ML,N}(\omega) \in]a, b[$ whenever such possibility exists). This defines a random variable

$$\hat{\theta}_{ML,N} : \Omega^N \to [a, b]$$

that is called the *maximum likelihood estimator* (abbreviated MLE) of order N. We shall also refer to the sequence $(\hat{\theta}_{ML,N})_{N \geq 1}$ as the MLE.

Note that maximizing (7.14) is equivalent to minimizing the entropy function

$$[a, b] \ni \theta \mapsto S_{\theta N}(\omega) = \sum_{k=1}^{N} -\log P_\theta(\omega_k).$$

Much of our analysis of the MLE will make use of this elementary observation and will be centred around the entropy function $S_{\theta N}$. We set

$$S(\theta, \theta') = E_\theta(S_{\theta'}) = -\sum_{\omega \in \Omega} P_\theta(\omega) \log P_{\theta'}(\omega).$$

Obviously, $S(\theta, \theta) = S(P_\theta)$ and

$$S(\theta, \theta') - S(\theta, \theta) = S(P_\theta | P_{\theta'}). \tag{7.15}$$

The last relation and the identifiability (7.6), which we assume throughout, give that

$$S(\theta, \theta') > S(\theta, \theta) \qquad \text{for} \qquad \theta \neq \theta'. \tag{7.16}$$

Applying Proposition 7.6 to $X_\theta = -\log P_\theta$, we derive

Proposition 7.8 *For any $\epsilon > 0$,*

$$\lim_{N \to \infty} \sup_{\theta \in [a,b]} P_{\theta N} \left\{ \omega \in \Omega^N \mid \sup_{\theta' \in [a,b]} \left| \frac{S_{\theta' N}(\omega)}{N} - S(\theta, \theta') \right| \geq \epsilon \right\} = 0.$$

Moreover, for any $\epsilon > 0$ there is $C_\epsilon > 0$ and $\gamma_\epsilon > 0$ such that for all $N \geq 1$,

$$\sup_{\theta \in [a,b]} P_{\theta' N} \left\{ \omega \in \Omega^N \,\Big|\, \sup_{\theta' \in [a,b]} \left| \frac{S_{\theta' N}(\omega)}{N} - S(\theta, \theta') \right| \geq \epsilon \right\} \leq C_\epsilon e^{-\gamma_\epsilon N}.$$

The first result of this section is:

Theorem 7.9 For any $\epsilon > 0$,

$$\lim_{N \to \infty} \sup_{\theta \in [a,b]} P_{\theta N} \left\{ \omega \in \Omega^N \,|\, |\hat{\theta}_{ML,N}(\omega) - \theta| \geq \epsilon \right\} = 0.$$

Moreover, for any $\epsilon > 0$ there exists $C_\epsilon > 0$ and $\gamma_\epsilon > 0$ such that for all $N \geq 1$,

$$\sup_{\theta \in [a,b]} P_{\theta N} \left\{ \omega \in \Omega^N \,|\, |\hat{\theta}_{ML,N}(\omega) - \theta| \geq \epsilon \right\} \leq C_\epsilon e^{-\gamma_\epsilon N}.$$

Proof Let

$$I_\epsilon = \{(u, v) \in [a, b] \times [a, b] \,|\, |u - v| \geq \epsilon\}.$$

It follows from (7.16) and continuity that

$$\delta = \sup_{(u,v) \in I_\epsilon} [S(u, v) - S(u, u)] > 0. \tag{7.17}$$

Fix $\theta \in [a, b]$ and set $I_\epsilon(\theta) = \{\theta' \in [a, b] \,|\, |\theta - \theta'| \geq \epsilon\}$. Let

$$A = \left\{ \omega \in \Omega^N \,\Big|\, \sup_{\theta' \in I_\epsilon(\theta)} \left| \frac{S_{\theta' N}(\omega)}{N} - S(\theta, \theta') \right| < \frac{\delta}{2} \right\},$$

$$B = \left\{ \omega \in \Omega^N \,\Big|\, \sup_{\theta' \in [a,b] \setminus I_\epsilon(\theta)} \left| \frac{S_{\theta' N}(\omega)}{N} - S(\theta, \theta') \right| < \frac{\delta}{2} \right\}.$$

For $\omega \in A$ and $\theta' \in I_\epsilon(\theta)$,

$$\frac{S_{\theta' N}(\omega)}{N} < S(\theta, \theta') + \frac{\delta}{2} \leq S(\theta, \theta) - \frac{\delta}{2}. \tag{7.18}$$

On the other hand, for $\omega \in B$ and $\theta \in [a, b] \setminus I_\epsilon(\theta)$,

$$\frac{S_{\theta' N}(\omega)}{N} > S(\theta, \theta') - \frac{\delta}{2} \geq S(\theta, \theta) - \frac{\delta}{2}. \tag{7.19}$$

Since $\hat{\theta}_{ML,N}(\omega)$ minimizes the map $[a,b] \ni \theta' \mapsto S_{\theta'N}(\omega)$,

$$\omega \in A \cap B \qquad \Rightarrow \qquad |\hat{\theta}_{ML,N}(\omega) - \theta| < \epsilon.$$

It follows that

$$\left\{ \omega \in \Omega^N \mid |\hat{\theta}_{ML,N}(\omega) - \theta| \geq \epsilon \right\} \subset A^c \cup B^c = \left\{ \omega \in \Omega^N \mid \sup_{\theta' \in [a,b]} \left| \frac{S_{\theta'N}(\omega)}{N} - S(\theta, \theta') \right| \geq \frac{\delta}{2} \right\},$$

and so

$$\sup_{\theta \in [a,b]} P_{\theta N} \left\{ \omega \in \Omega^N \mid |\hat{\theta}_{ML,N}(\omega) - \theta| \geq \epsilon \right\}$$

$$\leq \sup_{\theta \in [a,b]} P_{\theta N} \left\{ \omega \in \Omega^N \mid \sup_{\theta' \in [a,b]} \left| \frac{S_{\theta'N}(\omega)}{N} - S(\theta, \theta') \right| \geq \frac{\delta}{2} \right\}.$$

Since δ depends only on the choice of ϵ (recall (7.17)), the last inequality and Proposition 7.8 yield the statement. $\square$

Theorem 7.9 gives that the MLE is consistent in a very strong sense, and in particular that is uniformly consistent.

Corollary 7.10

$$\lim_{N \to \infty} \sup_{\theta \in [a,b]} E_{\theta N}(|\hat{\theta}_{ML,N} - \theta|) = 0.$$

Proof Let $\epsilon > 0$. Then

$$E_{\theta N}(|\hat{\theta}_{ML,N} - \theta|) = \int_{\Omega^N} |\hat{\theta}_{ML,N} - \theta| \, dP_{\theta N}$$

$$= \int_{|\hat{\theta}_{ML,N} - \theta| < \epsilon} |\hat{\theta}_{ML,N} - \theta| \, dP_{\theta N}$$

$$+ \int_{|\hat{\theta}_{ML,N} - \theta| \geq \epsilon} |\hat{\theta}_{ML,N} - \theta| \, dP_{\theta N}$$

$$\leq \epsilon + (b - a) P_{\theta N} \left\{ \omega \in \Omega^N \mid |\hat{\theta}_{ML,N}(\omega) - \theta| \geq \epsilon \right\}.$$

Hence,

$$\sup_{\theta \in [a,b]} E_{\theta N}(|\hat{\theta}_{ML,N} - \theta|) \leq \epsilon + (b-a) \sup_{\theta \in [a,b]} P_{\theta N}\left\{ \omega \in \Omega^N \mid |\hat{\theta}_{ML,N}(\omega) - \theta| \geq \epsilon \right\},$$

and the result follows from Proposition 7.9.   □

We note that so far all results of this section hold under the sole assumptions that the maps $[a, b] \ni \theta \mapsto P_\theta(\omega)$ are continuous for all $\omega \in \Omega$ and that the identifiability condition (7.6) is satisfied.

We now turn to study of the efficiency of the MLE and prove the second main result of this section. We strengthen our standing assumptions and assume that the maps $[a, b] \ni \theta \mapsto P_\theta(\omega)$ are C^3 for all $\omega \in \Omega$.

Theorem 7.11 Suppose that $[a', b'] \subset]a, b[$. Then

$$\lim_{N \to \infty} \sup_{\theta \in [a',b']} \left| N E_{\theta N}(|\hat{\theta}_{ML,N} - \theta|^2) - \frac{1}{\mathcal{I}(\theta)} \right| = 0.$$

Proof Recall that

$$[a, b] \ni \theta \mapsto S_{\theta N}(\omega = (\omega_1, \cdots, \omega_N)) = - \sum_{k=1}^{N} \log P_\theta(\omega_k)$$

achieves its minimum at $\hat{\theta}_{ML,N}(\omega)$ and that $\hat{\theta}_{ML,N}(\omega) \in]a, b[$ unless a strict minimum is achieved at either a or b. Let

$$B_N(a) = \left\{ \omega \in \Omega^N \mid \hat{\theta}_{ML,N}(\omega) = a \right\}, \qquad B_N(b) = \left\{ \omega \in \Omega^N \mid \hat{\theta}_{ML,N}(\omega) = b \right\},$$

and

$$\zeta = \min\left(\inf_{\theta \in [a',b']} S(P_\theta | P_a), \inf_{\theta \in [a',b']} S(P_\theta | P_b) \right).$$

Since the maps $\theta \mapsto S(P_\theta | P_a)$, $\theta \mapsto S(P_\theta | P_b)$ are continuous, the identifiability (7.6) yields that $\zeta > 0$. Then, for $\theta \in [a', b']$,

$$P_{\theta N}(B_N(a)) \leq P_{\theta N}\left\{ \omega \in \Omega^N \mid \frac{1}{N} \sum_{k=1}^{N} \log \frac{P_\theta(\omega_k)}{P_a(\omega_k)} < 0 \right\}$$

$$\leq P_{\theta N}\left\{ \omega \in \Omega^N \mid \frac{1}{N} \sum_{k=1}^{N} \log \frac{P_\theta(\omega_k)}{P_a(\omega_k)} - S(P_\theta | P_a) \leq -\zeta \right\},$$

and similarly,

$$P_{\theta N}(B_N(b)) \le P_{\theta N}\left\{\omega \in \Omega^N \mid \frac{1}{N}\sum_{k=1}^{N}\log\frac{P_\theta(\omega_k)}{P_b(\omega_k)} - S(P_\theta|P_b) \le -\zeta\right\}.$$

Proposition 7.7 now yields that for some constants $K_\zeta > 0$ and $k_\zeta > 0$,

$$\sup_{\theta \in [a',b']} P_{\theta N}(B_N(a) \cup B_N(b)) \le K_\zeta e^{-k_\zeta N}$$

for all $N \ge 1$. A simple but important observation is that if $\omega \notin B_N(a) \cup B_N(b)$, then $\hat{\theta}_{ML,N}(\omega) \in \,]a, b[$ and so

$$\dot{S}_{\hat{\theta}_{ML,N}(\omega)N}(\omega) = 0. \tag{7.20}$$

The Taylor expansion gives that for any $\omega \in \Omega^N$ and $\theta \in [a, b]$ there is $\theta'(\omega)$ between $\hat{\theta}_{ML,N}(\omega)$ and θ such that

$$\dot{S}_{\hat{\theta}_{ML,N}(\omega)N}(\omega) - \dot{S}_{\theta N}(\omega) = (\hat{\theta}_{ML,N}(\omega) - \theta)\left[\ddot{S}_{\theta N} + \frac{1}{2}(\hat{\theta}_{ML,N}(\omega) - \theta)\dddot{S}_{\theta'(\omega)N}\right]. \tag{7.21}$$

Write

$$E_{\theta N}\left(\left(\dot{S}_{\hat{\theta}_{ML,N}(\omega)N}(\omega) - \dot{S}_{\theta N}(\omega)\right)^2\right) = L_N(\theta) + E_{\theta N}\left(\left[\dot{S}_{\theta N}\right]^2\right),$$

where

$$L_N(\theta) = E_{\theta N}\left(\left[\dot{S}_{\hat{\theta}_{ML,N}(\omega)N}\right]^2\right) + 2E_{\theta N}\left(\dot{S}_{\hat{\theta}_{ML,N}(\omega)N}\dot{S}_{\theta N}\right). \tag{7.22}$$

It follows from (7.20) that in (7.22) $E_{\theta N}$ reduces to integration over $B_N(a) \cup B_N(b)$, and we arrive at the estimate

$$\sup_{\theta \in [a',b']} |L_N(\theta)| \le K N^2 \sup_{\theta \in [a',b']} P_{\theta N}(B_N(a) \cup B_N(b)) \le K N^2 K_\zeta e^{-k_\zeta N} \tag{7.23}$$

for some uniform constant $K > 0$, where by uniform we mean that K does not depend on N. It is easy to see that one can take

$$K = 3 \sup_{\theta \in [a,b], \omega \in \Omega}\left(\frac{\dot{P}_\theta(\omega)}{P_\theta(\omega)}\right)^2.$$

In Exercise 7.2 the reader is asked to estimate other uniform constant that will appear in the proof.

Squaring both sides in (7.21), taking the expectation, and dividing both sides with N^2, we derive the identity

$$\frac{1}{N^2}L_N(\theta) + \frac{1}{N^2}E_{\theta N}\left(\left[\dot{S}_{\theta N}\right]^2\right)$$
$$= E_{\theta N}\left((\hat{\theta}_{ML,N} - \theta)^2 \times \left[\frac{\ddot{S}_{\theta N}}{N} + \frac{1}{2N}(\hat{\theta}_{ML,N} - \theta)\dddot{S}_{\theta' N}\right]^2\right).$$

$$(7.24)$$

An easy computation gives

$$\frac{1}{N^2}E_{\theta N}\left(\left[\dot{S}_{\theta N}\right]^2\right) = \frac{1}{N}\mathcal{I}(\theta).$$

Regarding the right hand side in (7.24), we write it as

$$E_{\theta N}\left((\hat{\theta}_{ML,N} - \theta)^2 \left[\frac{\ddot{S}_{\theta N}}{N}\right]^2\right) + R_N(\theta),$$

where the remainder $R_N(\theta)$ can be estimated as

$$|R_N(\theta)| \le C_1 E_{\theta N}\left(|\hat{\theta}_{ML,N} - \theta|^3\right) \tag{7.25}$$

for some uniform constant $C_1 > 0$.

With these simplifications, an algebraic manipulation of the identity (7.24) gives

$$N E_{\theta N}\left((\hat{\theta}_{ML,N} - \theta)^2\right) - \frac{1}{\mathcal{I}(\theta)} = -D_N(\theta) - \frac{N R_N(\theta)}{\mathcal{I}(\theta)^2} + \frac{1}{N}\frac{L_N(\theta)}{\mathcal{I}(\theta)^2}, \tag{7.26}$$

where

$$D_N(\theta) = N E_{\theta N}\left((\hat{\theta}_{ML} - \theta)^2\left(\left[\frac{\ddot{S}_{\theta N}}{N}\right]^2 \frac{1}{\mathcal{I}(\theta)^2} - 1\right)\right). \tag{7.27}$$

Writing

$$\left[\frac{\ddot{S}_{\theta N}}{N}\right]^2 \frac{1}{\mathcal{I}(\theta)^2} - 1 = \frac{1}{\mathcal{I}(\theta)^2}\left(\frac{\ddot{S}_{\theta N}}{N} + \mathcal{I}(\theta)\right)\left(\frac{\ddot{S}_{\theta N}}{N} - \mathcal{I}(\theta)\right)$$

and using that $\mathcal{I}(\theta)$ is continuous and strictly positive on $[a, b]$, we derive the estimate

$$|D_N(\theta)| \leq C_2 N E_{\theta N}\left((\hat{\theta}_{ML} - \theta)^2 \left|\frac{\ddot{S}_{\theta N}}{N} - \mathcal{I}(\theta)\right|\right) \tag{7.28}$$

for some uniform constant $C_2 > 0$.

Fix $\epsilon > 0$, and choose $C_\epsilon > 0$ and $\gamma_\epsilon > 0$ such that

$$\sup_{\theta \in [a,b]} P_{\theta N}\left\{\omega \in \Omega^N \mid |\hat{\theta}_{ML,N}(\omega) - \theta| \geq \epsilon\right\} \leq C_\epsilon e^{-\gamma_\epsilon N}, \tag{7.29}$$

$$\sup_{\theta \in [a,b]} P_{\theta N}\left\{\omega \in \Omega^N \mid \left|\frac{\ddot{S}_{\theta N}}{N} - \mathcal{I}(\theta)\right| \geq \epsilon\right\} \leq C_\epsilon e^{-\gamma_\epsilon N}. \tag{7.30}$$

Here, (7.29) follows from Theorem 7.9, while (7.30) follows from Proposition 7.7 applied to $X_\theta = -\frac{d^2}{d\theta^2} \log P_\theta$ (recall that $E_\theta(X_\theta) = \mathcal{I}(\theta)$).

Let $\delta = \inf_{u \in [a,b]} \mathcal{I}(u)$. Then, for all $\theta \in [a, b]$,

$$\begin{aligned}
\frac{N|\mathcal{R}_N(\theta)|}{\mathcal{I}(\theta)^2} &\leq \frac{C_1 N}{\delta^2} \int_{\Omega^N} |\hat{\theta}_{ML,N} - \theta|^3 dP_{\theta N} \\
&= \frac{C_1 N}{\delta^2} \int_{|\hat{\theta}_{ML,N} - \theta| < \epsilon} |\hat{\theta}_{ML,N} - \theta|^3 dP_{\theta N} \\
&\quad + \frac{C_1 N}{\delta^2} \int_{|\hat{\theta}_{ML,N} - \theta| \geq \epsilon} |\hat{\theta}_{ML,N} - \theta|^3 dP_{\theta N} \\
&\leq \epsilon \frac{C_1}{\delta^2} N E_{\theta N}\left((\hat{\theta}_{ML,N} - \theta)^2\right) + \frac{C_1 (b - a)^3 N}{\delta^2} C_\epsilon e^{-\gamma_\epsilon N}.
\end{aligned} \tag{7.31}$$

Similarly, splitting the integral (that is, $E_{\theta N}$) on the r.h.s. of (7.28) into the sum of integrals over the sets

$$\left|\frac{\ddot{S}_{\theta N}}{N} - \mathcal{I}(\theta)\right| < \epsilon, \qquad \left|\frac{\ddot{S}_{\theta N}}{N} - \mathcal{I}(\theta)\right| \geq \epsilon,$$

we derive that for all $\theta \in [a, b]$,

$$|D_N(\theta)| \leq \epsilon C_2 N E_{\theta N}\left((\hat{\theta}_{ML,N} - \theta)^2\right) + C_2 C_3 N C_\epsilon e^{-\gamma_\epsilon N}, \tag{7.32}$$

where $C_3 > 0$ is a uniform constant. Returning to (7.26) and taking $\epsilon = \epsilon_0$ such that

$$\epsilon_0 \frac{C_1}{\delta^2} < \frac{1}{4}, \qquad \epsilon_0 C_2 < \frac{1}{4},$$

the estimates (7.23), (7.31), and (7.32) give that for all $\theta \in [a', b']$,

$$NE_{\theta N}\left((\hat{\theta}_{ML,N} - \theta)^2\right) \leq \frac{2}{\mathcal{I}(\theta)} + C'_{\epsilon_0} N e^{-\gamma_{\epsilon_0} N} + \frac{2K}{\delta^2} K_\zeta e^{-k_\zeta N},$$

where $C'_{\epsilon_0} > 0$ is a uniform constant (that of course depends on ϵ_0). It follows that

$$C' = \sup_{N \geq 1} \sup_{\theta \in [a',b']} NE_{\theta N}\left((\hat{\theta}_{ML,N} - \theta)^2\right) < \infty. \tag{7.33}$$

Returning to (7.31), (7.32), we then have that for any $\epsilon > 0$,

$$\sup_{\theta \in [a',b']} \frac{N|\mathcal{R}_N(\theta)|}{\mathcal{I}(\theta)^2} \leq \epsilon \frac{C_1}{\delta^2} C' + \frac{C_1(b-a)^3 N}{\delta^2} C_\epsilon e^{-\gamma_\epsilon N}, \tag{7.34}$$

$$\sup_{\theta \in [a',b']} |D_N(\theta)| \leq \epsilon C_2 C' + C_2 C_3 N C_\epsilon . e^{-\gamma_\epsilon N}. \tag{7.35}$$

Finally, returning once again to (7.26), we derive that for any $\epsilon > 0$,

$$\sup_{\theta \in [a',b']} \left| NE_{\theta N}\left((\hat{\theta}_{ML,N} - \theta)^2\right) - \frac{1}{\mathcal{I}(\theta)} \right| \leq \sup_{\theta \in [a',b']} |D_N(\theta)| + \sup_{\theta \in [a',b']} \frac{N|R_N(\theta)|}{\mathcal{I}(\theta)^2}$$

$$+ \sup_{\theta \in [a',b']} \frac{|L_N(\theta)|}{N\mathcal{I}(\theta)^2} \leq \epsilon C'' + C''_\epsilon N e^{-\gamma_\epsilon N} + K K_\zeta e^{-k_\zeta N},$$

where $C'' > 0$ is a uniform constant and $C''_\epsilon > 0$ depends only on ϵ. Hence,

$$\limsup_{N \to \infty} \sup_{\theta \in [a',b']} \left| NE_{\theta N}\left((\hat{\theta}_{ML,N} - \theta)^2\right) - \frac{1}{\mathcal{I}(\theta)} \right| \leq \epsilon C''.$$

Since $\epsilon > 0$ is arbitrary, the result follows. $\square$

Exercise 7.2 Write an explicit estimate for all uniform constants that have appeared in the above proof.

Remark 7.12 The proof of Theorem 7.11 hints at the special role the boundary points a and b of the chosen parameter interval may play in study of the efficiency. The MLE is selected with respect to the $[a, b]$ and $\hat{\theta}_{ML,N}(\omega)$ may take value a or b without the derivative $\dot{S}_{\hat{\theta}_{ML,N}(\omega)N}(\omega)$ vanishing. That forces the estimation of the probability of the set $B_N(a) \cup B_N(b)$ and the argument requires that θ stays away from the boundary points. If the parameter interval is replaced by a circle, there would be no boundary points and the above proof then gives that the uniform efficiency of the MLE holds with respect to the entire parameter set. One may wonder whether a different type of argument may yield the same result in the case of $[a, b]$. The following example shows that this is not the case.

Let $\Omega = \{0, 1\}$ and let $P_\theta(0) = 1 - \theta$, $P_\theta(1) = \theta$, where $\theta \in]0, 1[$. One computes $\mathcal{I}(\theta) = (\theta - \theta^2)^{-1}$. If $[a, b] \subset]0, 1[$ is selected as the estimation interval, the MLE $\theta_{ML,N}$ takes the following form:

$$\hat{\theta}_{ML,N}(\omega_1, \cdots, \omega_N) = \frac{\omega_1 + \cdots + \omega_N}{N} \qquad \text{if} \qquad \frac{\omega_1 + \cdots + \omega_N}{N} \in [a, b],$$

$$\hat{\theta}_{ML,N}(\omega_1, \cdots, \omega_N) = a \qquad \text{if} \qquad \frac{\omega_1 + \cdots + \omega_N}{N} < a,$$

$$\hat{\theta}_{ML,N}(\omega_1, \cdots, \omega_N) = b \qquad \text{if} \qquad \frac{\omega_1 + \cdots + \omega_N}{N} > b.$$

We shall indicate the dependence of $\hat{\theta}_{ML,N}$ on $[a, b]$ by $\hat{\theta}_{ML,N}^{[a,b]}$. It follows from Theorem 7.11 that

$$\lim_{N \to \infty} N E_{(\theta=1/2)N}\left(\left(\hat{\theta}_{ML,N}^{[\frac{1}{3}, \frac{2}{3}]} - \frac{1}{2}\right)^2\right) = \left[\mathcal{I}\left(\frac{1}{2}\right)\right]^{-1} = \frac{1}{4}.$$

On the other hand, a moment's reflection shows that

$$\frac{1}{2} E_{(\theta=1/2)N}\left(\left(\hat{\theta}_{ML,N}^{[\frac{1}{3}, \frac{2}{3}]} - \frac{1}{2}\right)^2\right) = E_{(\theta=1/2)N}\left(\left(\hat{\theta}_{ML,N}^{[\frac{1}{2}, \frac{2}{3}]} - \frac{1}{2}\right)^2\right),$$

and so

$$\lim_{N \to \infty} N E_{(\theta=1/2)N}\left(\left(\hat{\theta}_{ML,N}^{[\frac{1}{2}, \frac{2}{3}]} - \frac{1}{2}\right)^2\right) = \frac{1}{8}.$$

Thus, in this case even the bound of Proposition 7.5 fails at the boundary point $1/2$ at which the MLE becomes "superefficient". In general, such artificial boundary effects are difficult to quantify and we feel it is best that they are excluded from the theory. These observations hopefully elucidate our definition of efficiency which excludes the boundary points of the interval of parameters.

7.5 Notes and References

For additional information and references about parameter estimation the reader may consult [34, 51]. For additional information about the Cramér-Rao bound and its history, we refer the reader to the respective Wikipedia and Scholarpedia articles.

The modern theory of the MLE started with the seminal work of Fisher [18]; for the fascinating history of the subject, see [48]. Our analysis of the MLE follows the standard route, but I have followed no particular reference. In particular, I am not aware whether Theorem 7.11 as formulated has appeared previously in the literature.

References

1. J. Aczél, Z. Daróczy, *On Measures of Information and Their Characterizations* (Academic, Cambridge, 1975)
2. J. Aczél, B. Forte, C.T. Ng, Why the Shannon and Hartley entropies are 'natural'. Adv. Appl. Prob. **6**, 131–146 (1974)
3. V. Anantharam, A large deviations approach to error exponents in source coding and hypothesis testing. IEEE Trans. Inf. Theory **36**, 938–943 (1990)
4. P. Billingsley, *Ergodic Theory and Information* (Wiley, Hoboken, 1965)
5. L.L. Campbel, An extended Cencov characterization of the information metric. Proc. AMS **98**, 135–141 (1996)
6. N.N. Cencov, *Statistical Decision Rules and Optimal Inference*. Translations of Mathematical Monographs, vol. 53 (AMS, Providence, 1981)
7. H. Chernoff, A measure of asymptotic efficiency for tests of a hypothesis based on the sum of observations. Ann. Math. Stat. **23**, 493 (1952)
8. K.L. Chung, *A Course in Probability Theory* (Academic, Cambridge, 2001)
9. T.A. Cover, J.A. Thomas, *Elements of Information Theory* (Willey, Hoboken, 1991)
10. I. Csiszár, Axiomatic characterizations of information measures. Entropy **10**, 261–273 (2008)
11. I. Csiszár, J. Körner, *Information Theory* (Academic, Cambridge, 1981)
12. Z. Daróczy, Über nittelwerte und entropien vollständiger wahrscheinlichkeitsverteilungen. Acta Math. Acad. Sci. Hungar. **15**, 203–210 (1964)
13. A. Dembo, O. Zeitouni, *Large Deviations Techniques and Applications* (Springer, Berlin, 1998)
14. F. den Hollander, *Large Deviations* (AMS, Providence, 2000)
15. R.S. Ellis, *Entropy, Large Deviations, and Statistical Mechanics* (Springer, Berlin, 1985). Reprinted in the series Classics of Mathematics (2006)
16. D.M. Endres, J.E. Schindelin, A new metric for probability distributions. IEEE Trans. Inf. Theory **49**, 1858–1860 (2003)
17. D.K. Faddeev, On the concept of entropy of a finite probabilistic scheme. Usp. Mat. Nauk **1**, 227–231 (1956)
18. R.A. Fisher, On the mathematical foundations of theoretical statistics. Philos. Trans. R. Soc. Lond. Ser. A **222**, 309–368 (1921)
19. R.A. Fisher, Theory of statistical estimation. Proc. Camb. Philos. Soc. **22**, 700–725 (1925)
20. B. Fugledge, F. Topsoe, Jensen-Shannon divergence and Hilbert space embedding, in *Proceedings of International Symposium on Information Theory, ISIT* (2004)
21. S. Goldstein, D.A. Huse, J.L. Lebowitz, P. Sartori, On the nonequilibrium entropy of large and small systems (2018). Preprint. https://arxiv.org/pdf/1712.08961.pdf
22. R.M. Gray, *Entropy and Information Theory* (Springer, New York, 2011)
23. R.V.L. Hartley, Transmission of information. Bell Syst. Tech. J. **7**, 535–563 (1928)
24. A. Hobson, A new theorem of information theory. J. Stat. Phys. **1**, 383–391 (1969)
25. H. Hoeffding, Asymptotically optimal tests for multinomial distributions. Ann. Math. Stat. **36**, 369 (1965)
26. V. Jakšić, C.-A. Pillet, L. Rey-Bellet, Entropic fluctuations in statistical mechanics: I. Classical dynamical systems. Nonlinearity **24**, 699 (2011)
27. E.T. Jaynes, Information theory and statistical mechanics. Phys. Rev. **106**, 620–630 (1957)
28. E.T. Jaynes, Information theory and statistical mechanics II. Phys. Rev. **108**, 171 (1957)

29. H. Jeffreys, An invariant form for the prior probability in estimation problems. Proc. Roy. Soc. A **186**, 453–461 (1946)
30. D. Johnson, *Statistical Signal Processing* (2018). https://cpb-us-e1.wpmucdn.com/blogs.rice. edu/dist/7/3490/files/2018/01/notes-1311a3s.pdf
31. I. Kátai, A remark on additive arithmetical functions. Ann. Univ. Sci. Budapest Edtvds Sect. Math. **12**, 81–83 (1967)
32. A.Ya. Khinchin, *Mathematical Foundations of Information Theory* (Dover Publications, New York, 1957)
33. S. Kullback, R.A. Leibler, On information and sufficiency. Ann. Math. Stat. **22**, 79–86 (1951)
34. E.L. Lehmann, G. Cassela, *Theory of Point Estimation* (Springer, New York, 1998)
35. E.L. Lehmann, J.P. Romano, *Testing Statistical Hypotheses* (Springer, New York, 2005)
36. T. Leinster, A short characterization of relative entropy. J. Math. Phys. **60**, 023302 (2019)
37. J. Lin, Divergence measures based on the Shannon entropy. IEEE Trans. Inf. Theory **27**, 145–151 (1991)
38. K. Matsumoto, Reverse test and characterization of quantum relative entropy (2010). Preprint. https://arxiv.org/pdf/1010.1030.pdf
39. N. Merhav, *Statistical Physics and Information Theory*. Foundations and Trends in Communications and Information Theory, vol. 6 (Now Publishers, Hanover, 2009)
40. F. Österreicher, I. Vajda, A new class of metric divergences on probability spaces and its statistical applications. Ann. Inst. Stat. Math. **55**, 639–653 (2003)
41. A. Rényi, On measures of information and entropy, in *Proceedings of the Fourth Berkeley Symposium on Mathematical Statistics and Probability, Vol. I* (University of California Press, Berkeley, 1961)
42. V.K. Rohtagi, A.K.Md.E. Saleh, *An Introduction to Probability and Statistics* (Wiley, Hoboken, 2015)
43. S. Ross, *First Course in Probability* (Pearson, London, 2014)
44. C.E. Shannon, A mathematical theory of communication. Bell Syst. Tech. J. **27**, 379–423, 623–656 (1948)
45. C.E. Shannon, W. Weaver, *The Mathematical Theory of Communication* (The University of Illinois Press, Champaign, 1964)
46. P.C. Shields, *The Ergodic Theory of Discrete Sample Paths* (AMS, Providence, 1991)
47. R. Sowers, Stein's lemma–a large deviation approach, Naval research laboratory report 9185, 1989
48. S.M. Stiegler, The epic story of maximum likelihood. Stat. Sci. **22**, 598–620 (2007)
49. L. Szilard, On the decrease of entropy in a thermodynamic system by the intervention of intelligent beings. Z. Phys. **53**, 840–856 (1929). English translation in The Collected Works of Leo Szilard: Scientific Papers, ed. by B.T. Feld, G.W. Szilard (MIT Press, Cambridge, 1972), pp. 103–129
50. W. Thirring, *Quantum Mathematical Physics. Atoms, Molecules, and Large Systems* (Springer, Berlin, 2002)
51. A.W. van der Vaart, *Asymptotic Statistics* (Cambridge University Press, Cambridge, 1998)
52. S. Verdú, Fifty years of Shannon theory. IEEE Trans. Inf. Theory **44**, 2057–2078 (1998)
53. H. Wilming, R. Gallego, J. Eisert, Axiomatic characterization of the quantum relative entropy and free energy. Entropy **19**, 241–244 (2017)

rinted in the United States
y Bookmasters